Homogeneous Photocatalysis

Wiley Series in Photoscience and Photoengineering

Executive Editorial Board

Professor A Braun (*Karlsruhe, Germany*)
Professor M Anpo (*Osaka, Japan*)
Professor J C André (*Nancy, France*)
Professor D Eaton (*Wilmington, USA*)
Professor D F Ollis (*Raleigh, USA*)
Professor E Pelizzetti (*Torino, Italy*)
Professor M Schiavello (*Palermo, Italy*)
Professor H Tributsch (*Berlin, Germany*)

Volume 1
Surface Photochemistry

Edited by

Masakazu Anpo
University of Osaka Prefecture
Osaka, Japan

Volume 2
Homogeneous Photocatalysis

Edited by

M Chanon
Universite de Droit,
D'Economie et des Sciences d'Aix-Marseille,
Marseille, France

Homogeneous Photocatalysis

Edited by
M. CHANON
Universite de Droit, D'Economie et des Sciences d'Aix-Marseille, Marseille, France

WILEY SERIES IN PHOTOSCIENCE AND PHOTOENGINEERING VOLUME 2

JOHN WILEY & SONS
Chichester • New York • Weinheim • Brisbane • Singapore • Toronto

Copyright © 1997 by John Wiley & Sons Ltd,
Baffins Lane, Chichester,
West Sussex PO19 1UD, England

National   01243 779777
International (+44) 1243 779777
e-mail (for orders and customer service enquiries): cs-books@wiley.co.uk
Visit our Home Page on http://www.wiley.co.uk
or http://www.wiley.com

All Rights Reserved. No part of this book may be reproduced, stored in a retrieval system, or transmitted, in any form or by any means, electronic, mechanical, photocopying, recording or otherwise, except under the terms of the Copyright, Designs and Patents Act 1988 or under the terms of a licence issued by the Copyright Licensing Agency, 90 Tottenham Court Road, London, UK WIP 9HE, without the permission in writing of the publisher.

Other Wiley Editorial Offices

John Wiley & Sons, Inc., 605 Third Avenue,
New York, NY 10158-0012, USA

VCH Verlagsgesellschaft mbH,
Pappelallee 3, D-69469 Weinheim, Germany

Jacaranda Wiley Ltd, 33 Park Road, Milton,
Queensland 4064, Australia

John Wiley & Sons (Canada) Ltd, 22 Worcester Road,
Rexdale, Ontario M9W 1L1, Canada

John Wiley & Sons (Asia) Pte Ltd, 2 Clementi Loop #02-01,
Jin Xing Distripark, Singapore 129809

Library of Congress Cataloguing-in-Publication Data

Homogeneous photocatalysis / edited by M. Chanon.
  p. cm.
  Includes bibliographical references and index.
  ISBN 0-471-96753-X
  1. Photocatalysis. I. Chanon, Michel, 1940-
QD716.P45H66 1997
541.39'5 — dc20   96-25955
                                                CIP

British Library Cataloguing in Publication Data

A catalogue record for this book is available from the British Library

ISBN 0 471 96753 X

Typeset in 10/12pt Times by Laser Words, Madras, India
Printed and bound in Great Britain by Biddles Ltd, Guildford, Surrey
This books is printed on acid-free paper responsibly manufactured from sustainable forestation, for which at least two trees are planted for each one used for paper production.

Contents

List of Contributors		vii
Series Foreword		ix
Preface		xi
1	Introduction to Photocatalysis M. Chanon and M. Schiavello	1
2	Fundamentals of Interaction between Light and Matter J. C. Mialocq	15
3	Homogeneous Proton Transfer Photocatalysis L. G. Arnaut and S. J. Formosinho	55
4	Principles and Organic Synthetic Applications of Photoinduced Electron Transfer Photosensitization J. Santamaria and C. Ferroud	97
5	Transition Metal Complexes and Homogeneous Photocatalytic Transformations of Organic Substrates Charles Kutal	135
6	Photocatalytic Aspects of Silver Photography J. Belloni	169
7	Immobilized Photosensitizers and Photocatalysis Michel Julliard	221
8	Water Splitting: from Molecular to Supramolecular Photochemical Systems Edmond Amouyal	263
9	Organized Media and Homogeneous Photocatalysis Isabelle Rico-Lattes and Armand Lattes	309
10	Photosynthesis, a Natural Model for Photocatalysis Paul Mathis	355

List of Contributors

Armand Lattes
Interactions Moléculaires et Réactivité Chimie et Photochimique, Université Toulouse 3 - Bât. 2R1, 118, Route de Narbonne, 31062 — Toulouse, France

L. G. Arnaut
Department of Chemistry, University of Coimbra, Coimbra, Portugal

J. Belloni
Laboratoire de Physico-Chimie des Rayonnements, Université Paris-Sud, Bât. 350, 91405 — Orsay Cedex, France

M. Chanon
*Universite de Droit,
D'Economie et des Sciences d'Aix-Marseille, Marseille, France*

Charles Kutal
Department of Chemistry, University of Georgia, Athens, Georgia, USA

Edmond Amouyal
Laboratoire de Physico-chimie de Rayonnements, Université Paris-Sud, Bât. 350, 91405 — Orsay Cedex, France

S. J. Formosinho
Department of Chemistry, University of Coimbra, Coimbra, Portugal

Isabelle Rico-Lattes
Interactions Moléculaires et Réactivité Chimie et Photochimique, Université Toulouse 3 - Bât. 2R1, 118, Route de Narbonne, 31062 — Toulouse, France

J. C. Mialocq
Commissariat à l'Energie Atomique, Service Chimie Moléculaire, CE Saclay — Bât. 125, 91191 — Gif-sur-Yvette, France

Michel Julliard
Laboratoire d'Activation, Méchanisms, Modelisation Moleculaire, Faculté des Sciences de St. Jérôme, Case 561, 13397 — Marseille Cedex 20, France

Paul Mathis
Section de Bioenérgétique, CEN Saclay, DB-SBPH, Bât. 532, 91191 Gif-sur-Yvette, France

J. Santamaria
Laboratoire de Synthèse des Composés d'Intérêt Biologique URA 476,
Ecole Supérieure de Physique et de Chimie Industrielle de Paris,
10, rue Vauquelin, 75231 Paris Cedex 05, France

M. Schiavello
Università di Palermo, Viale delle Scienze
90128 Palermo, Italy.

Series Foreword

Looking into the history of chemistry, one of the fascinating facts is that discoveries and developments which helped to shape today's civilization and technical standards have frequently been made on a totally empirical basis and it has usually been quite sometime afterwards that fundamental research developed the means for understanding the implications of chemical structures, interactions, reactivities and physical characteristics.

Despite the fact that light plays a key role in our daily life, photochemistry found only limited recognition as a proper domain of chemical sciences and engineering. For most chemists, photochemistry has been a play ground of academic study. But considering technical developments in microelectronics, informatics, (dental) medicine, materials, fine and bulk chemicals, etc., for example, our world would be different if the industries concerned would not have found the relevant results by using light as a reagent in rather complex reaction systems; and academic research has been in most cases bypassed by industrial success.

With the series "Photoscience and Photoengineering", we aim to foster interaction between fundamental research and technical development. Topics of the volumes will be carefully selected depending on the impact of new developments and the knowledge of the related basic principles. Therefore, these books will not be ensembles of loosely related reviews, but self-contained accounts on specific areas.

I sincerely hope that our approach will lead to a better understanding and a closer interaction between scientists and engineers of many disciplines.

Karlsruhe, September 9, 1995
Prof. Dr. André M. Braun
Chairman of the Board of Editors

Preface

There is a kind of magic in being able to induce important changes in a medium just by flashing a short impulse of light, or by adding catalytic amounts of a compound to this medium before irradiating it. It is this phenomenon that the book describes at the molecular level. The reader will see several facets of the molecular origin of these effects but also several practical applications which follow from understanding and mastering such photostimulated molecular gears: from the design of new improved photographic processes to the tapping of solar energy. Nature gives the lead by showing how the most efficient living cells have integrated photocatalytic gears to control the light entering the earth's atmosphere.

1 Introduction to Photocatalysis

M. CHANON
Universite de Droit, D'Economie et des Sciences d'Aix-Marseille, Marseille, France
and
M. SCHIAVELLO
Dipartimento di Ingegneria Chimica dei Processie dei Materiali Universitá di Palermo Viale Delle Scienze Italy

1 Definitions of photocatalysis 1
   1.1 The proton-like approach 2
   1.2 Broader approaches 3
2 Classifications of photocatalytic transformations 5
   2.1 Catalytic processes in photons and in one substance 6
   2.2 Non-catalytic processes in photons 7
3 Specific aspects of heterogeneous photocatalysis 9
4 Photocatalysis a crossroad for fundamental, applied photochemistry and catalysis 11
References 12

1 DEFINITIONS OF PHOTOCATALYSIS

"Photocatalysis" is a composite name of "photo" and "catalysis".

A strict respect of the composition of this word would lead to a typical nonsense. Let us take the definitions of photon and catalysis as given in *Webster's New Universal Unabridged Dictionary* [1]: photon: "a quantum of electromagnetic energy having both particle and wave behavior: it has no charge or mass but possesses momentum: the energy of light, X-rays, gamma rays, etc. is carried by photons"; catalysis: "the action of a catalyst in a chemical reaction"; catalyst: "a *substance* which either speeds up or slows down a chemical reaction, but which itself undergoes no permanent chemical change thereby". Is therefore "photocatalysis" the phenomenom associated with the action of photon speeding up a chemical reaction? Clearly not because:

1. A catalyst has to be a substance ("physical matter of which a thing consists; material") otherwise one cannot weight it to include its concentration in kinetic laws.
2. The photons of which one thinks in photocatalysis are usually not gamma-rays, or X-rays and even less so vibrational photons.

Homogeneous Photocatalysis. Edited by M. Chanon
© 1997 John Wiley & Sons Ltd

3. Suppose that one goes beyond the two preceding inconsistencies and then it is the whole field of photochemistry which would become an illustration of photocatalysis. This possibly explains why you do not find "photocatalysis" in *Webster*.

1.1 THE PROTON-LIKE APPROACH

Nevertheless different approaches have been taken to define, photocatalysis. None is completely satisfactory: the "cleanest" one from a theoretical point of view demands much experimental work on the reaction studied and, at this point, many of the reactions classified as relevant to the field of photocatalysis do not yet actually deserve the label. This definition is derived from that of Bell for a thermal catalyst [2]: "a substance which appears in the velocity expression to a power greater than its coefficient in the stoichiometric equation". Applied to a photochemical reaction this definition becomes: "a substance that appears in the quantum yield expression for reaction from a particular excited state to a power greater than its coefficient in the stoichiometric equation" [3]. Schemes 1.1a and 1.1b illustrate more precisely this definition in terms of quantum yields.

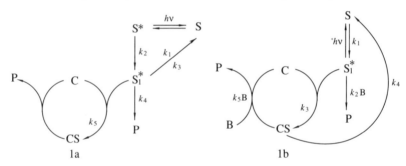

Scheme 1.1

In Scheme 1.1a one starts from a substrate S which is photoexcited to S*; this photochemical state obtained on Franck–Condon excitation transforms into S_1^* whose efficiency of production is given by $\Phi_i = k_2/(k_1 + k_2)$. S_1^* reacts with the catalyst to give CS which transforms into the product(s) P. The efficiency of reaching the product P, from S_1^* in the absence of C is $\Phi_{ii} = k_4/(k_3 + k_4)$. In the presence of a catalyst C reacting irreversibly with S_1^* to form CS, the quantum yield of P production becomes

$$\Phi = \Phi_i \left(\frac{k_4 + k_5[C]}{k_3 + k_4 + k_5[C]} \right)$$

which may be inverted to

$$\frac{1}{\Phi} = \frac{1}{\Phi_i} \left(1 + \frac{k_3}{k_4 + k_5[C]} \right)$$

… INTRODUCTION TO PHOTOCATALYSIS

If $k_4 \gg k_5[C]$ this expression becomes

$$\frac{1}{\Phi} = \frac{1}{\Phi_i}\left(1 + \frac{k_3}{k_4}\right)$$

Scheme 1.1b represents a case in which the product-forming step is bimolecular and the catalyst complex CS decays to starting materials. If f stands for partitioning of CS, i.e. $f = k_5[B]/(k_4 + k_5[B])$ the expression of Φ becomes

$$\Phi = \frac{k_2[B] + f\,k_3[C]}{k_1 + k_2[B] + k_3[C]}$$

if $f\,k_3[C] \gg k_2[B]$ the inverted quantum yield becomes

$$\frac{1}{\Phi} = \frac{1}{f}\left(1 + \frac{k_1 + k_2[B]}{k_3[C]}\right)$$

This definition of 'catalysis of a photoreaction' (Wubbels recommends that the use of "photocatalysis" be discontinued) excludes [3] the following situations:

1. Sensitization by electronic energy transfer;
2. Enhancing the efficiency of reaching a particular excited state (substances containing a heavy atom, or able to make donor–acceptor complex with excited states;
3. Photochemical production of a catalyst in its ground state that catalyzes a ground-state reaction such as isomerization of an olefin;
4. Light absorption by a species which causes a chemical change in another species while returning itself to its original state. This situation applies to photocatalytic semiconductor devices and to photoelectrochemical cells;
5. Many photoreactions of organic substrates mediated by metal ions;
6. "Photostimulated" radical chain reactions, i.e. reactions where the photochemistry serves only to generate chain-propagating paramagnetic or diamagnetic species. The fact that this definition excludes several situations that were widely studied or used under the label "photocatalysis" illustrates the variety of opinion about the word "photocatalysis". Even within its impression of rigorousness, Wubbel's definition has still been submitted to criticism by even, stricter chemists (see Chapter 3, p. 60 and ref. 3b).

1.2 BROADER APPROACHES

The word "catalysis" itself has considerably changed since the first Berzelius definition (1836); an overview of these changes may be gained by scanning refs. [4–13]. These changes arise from a better mechanistic understanding of some aspects of catalysis, but also because catalysis has been studied in an amazing variety of contexts: homogeneous, supported, heterogeneous, biological, with a spirit of practical applications or with one of elaborated kinetic consistency. The IUPAC definition of photocatalysis [14]: "a catalytic reaction involving light absorption by a catalyst or by a substrate" seems broad enough to encompass all the preceding contexts. Its apparent simplicity is deceptive for at least two reasons:

1. In the direction of homogeneous photocatalysis this definition depends directly upon the adopted definition of catalysis and catalysts.
2. In the direction of heterogeneous photocatalysis, to experimentally assess the true catalytic nature of the studied reaction is not easy task. Several chemists have wondered if some so-called photocatalytic heterogeneous reactions are actually catalytic or if the surface/near-subsurface of the non-stoichiometric semiconductors is merely a photoactivated reagent [15].

It is universally written without hesitation "a Lewis acid catalysed Diels–Alder reaction" despite the carbon skeleton construction involved in this reaction which stands in clear contrast with the "lysis" part of "catalysis". We selected the contents of this volume from a definition of photocatalysis which encompasses Wubbels catalysed photoreactions, but includes also the entries (p. 3) 1) (provided that the sensitiser is not consumed), 3), 4), 6) excluded in the Wubbels definition. 'Photocatalysis', in this context, is taken as a word as general as 'catalysis'. It may be described by the statement: 'A reaction is photocatalysed if catalytic quantities of photons or substances (with respect to the transformed substrate) suffice to accelerate or to make possible the photochemical transformation of S into P.' One may discuss at length the inclusion of photoinduced chain reactions: we include it to follow the generalized Prettre definition of thermal catalysis [5]. Scheme 1.2 recalls the difference between a chain (1.2a) and a catalytic (1.2b) reaction. In the chain reaction, the initiator I, in catalytic quantities, plays its role by transforming the substrate S into an intermediate S' from which a closed cycle of elementary steps will develop without further intervention of I (remember a

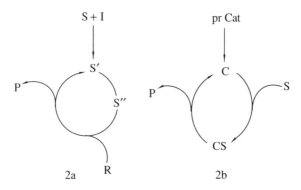

Scheme 1.2. pr Cat stands for procatalyst

INTRODUCTION TO PHOTOCATALYSIS 5

match lighting up a Bunsen burner). In the catalytic reaction, the catalyst C plays its role inside the closed cycle of elementary steps. Experimentally, it is not always easy to recognize if the substance added in small quantities is an initiator or a catalyst: both may be consumed during the reaction, whereas Schemes 1.2a and 1.2b suggest that only I is consumed. These considerations certainly do not close the discussion about what should or should not be included under the label "photocatalysis" [6]. We have retained the most general one because we feel that the connection of approaches from different fields contains more heuristic richness than strict adherence to formal rules. This approach converges with that recently expressed by Henning et al. [7]. Further discussions on this matter in homogeneous photocatalysis may be found in Chapter 3, 5 and 9.

2 CLASSIFICATIONS OF PHOTOCATALYTIC TRANSFORMATIONS

Because of the practical emphasis of this series, we will adopt a classification of photocatalyzed transformations based on the simplest experimental evidence. The price of simplicity is obviously imperfection.

The photocatalytic reactions may be gathered into two large classes. For the first, when the source of light is switched off, the reaction goes on for a while. For the second, the same cause immediately stops the reaction [17]. The first class corresponds to the situation depicted in Scheme 1.3a: the photon plays its role before the succession of cyclic steps making the catalytic or the chain turnover. In the second class (Scheme 1.3b), the photon makes possible or accelerates a step which is part of the catalytic loop.

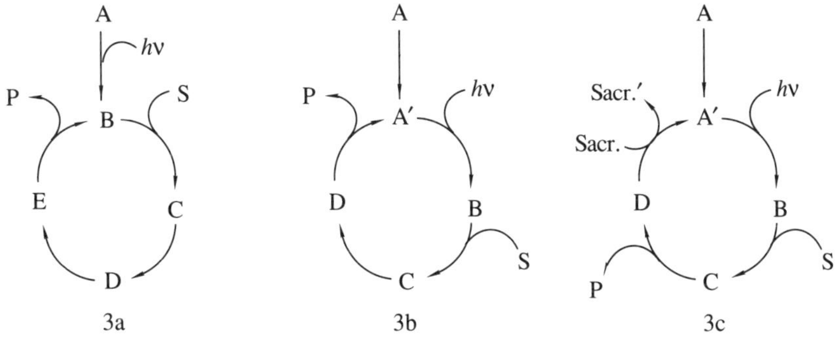

Scheme 1.3

In Schemes 1.3b and 1.3c $h\nu$ has been arbitrarily placed between A' and B, meaning that it makes possible the transformation A' $\longrightarrow$ B; it should be clear that it could have been introduced at any place in the catalytic loop, meaning that it made possible any of the elementary steps described.

A typical example of Scheme 1.3a would be photogeneration of the catalyst which then helps the thermal transformation of substrate + reagent into products. Chapter 5 provides several examples of Scheme 1.3a. If we accept the generalized definition of catalysis [5], the wide-ranging class of photoinitiated chain reactions also falls into the first class. Chapter 4 gives some examples of photoinduced chain reactions; refs [18] and [24] put them in perspective.

Photosensitization of a transformation in a process where the photosensitizer is not consumed would illustrate the operation of Scheme 1.3b. In 1.3c a sacrificial reagent is needed to close the sequence of elementary steps.

2.1 CATALYTIC PROCESSES IN PHOTONS AND IN ONE SUBSTANCE

In these processes, the catalysis part of "photocatalysis" originates both in a catalytic consumption of photons (usually $\Phi > 1$, but see refs [19] and [20] for exceptions) and in the necessary presence of a substance in catalytic quantities with respect to the substrate. Examples of such processes may be found under the labels 'catalysed photoreactions' [13], 'photoinduced catalytic reactions' [21], 'photogenerated catalysis' [22], 'photoinduced chain reaction' [23], 'photoinduced electron transfer chain reaction' [24]. For this last case, although a catalytic quantity of a reducing or oxidizing substance is often needed [24a], it is not compulsory in some examples [25]. In these examples, the role of electronic excitation is to enhance the redox relationship between the substrate and a reagent [26]. The photoinduced electron transfer between this pair of substances initiates a chain which then goes on without further intervention of the photon.

Within the general Scheme 1.3a several roles may be devoted to the photon:

1. It may interact with a procatalyst which is photochemically transformed into a catalyst then playing its role in the dark. Examples of such an action are amply furnished by transition metal complexes used in photocatalysis [21,22,27–33]. Scheme 5.4 provides a clear illustration of such a situation.
2. It may initiate a chain reaction by inducing an homolytic cleavage in the excited state of A. The photochemical chloration of toluene [23b] illustrates such a role.
3. It may cause the decomplexation of a ligand in the excited state of a transition metal complex. The photoinitiated anionic polymerization of ethyl α-cyanoacrylate (Scheme 5.5) is triggered by thiocyanate ions released by $trans$-$Cr(NH_3)_2(NCS)_4^-$ in its excited state.
4. Because excited states are both better reducing and better oxidizing agents than their ground state counterparts (Figure 5.2), the absorption of a photon by A (Scheme 1.3a) may render possible an electron transfer to or from the substrate. This electron transfer may strongly activate the substrate either in an associative or a dissociative mode [34], generating therefore active species able to trigger a chain reaction [24,25].

INTRODUCTION TO PHOTOCATALYSIS 7

5. The acid–base properties of excited states do also differ from those displayed by the ground state [35] (Chapter 3), thus absorption of a photon by a nominal catalyst may yield photochemically the real catalyst which then participates in the subsequent dark reactions (mechanisms V and VI in Chapter 3). It is interesting to note that much less attention has been devoted to photocatalysis involving Lewis acids (BF_3, SbF_5, etc.) and Lewis bases (electron rich olefins, etc.).

2.2 NON-CATALYTIC PROCESSES IN PHOTONS

Kutal proposed to gather all the non-catalytic photocatalytic situations in photons under the label 'catalysed photochemistry' [36]. Three main groups constitute this set (Scheme 1.4.):

1. An excited state of the substrate S reacts with the catalyst (Scheme 1.1a), whereas S in the ground state would not react. Every transformation S $\longrightarrow$ P has to pass through the bottleneck of S_1^*, therefore at least one photon is consumed per catalytic cycle; C may be an acid, a base, or a transition metal complex. Such a type of situation was labelled 'catalyzed photoreaction' by Wubbels [3], Salomon [22], Mirbach [27] and Henning et al. [21].

2. In 'stoichiometric photogenerated catalysis' [22], also labelled 'photoassisted' [21,28] 'photoenhanced' [37] and 'photoactivated' [38], the photon makes possible one of the elementary steps of the catalytic loop (Scheme 1.4a); this elementary step does not involve an added substance which reacts in its excited state (otherwise one would have photosensitization). The photochemistry of a cyclic alkene in the presence of O_2 and μ-oxobis (tetraphenylporphyrinato)iron III (Scheme 5.13) illustrates such a situation.

3. Photosensitization is, by far, the most documented set of reactions relevant to photocatalysis when the photosensitizer is used in catalytic quantities with respect to the substrates S [39]. Some types of photosensitizer cycles are represented in Scheme 1.4.

Scheme 1.4b represents an energy transfer-induced transformation of a substrate into product(s), Scheme 5.10 provides an actual example of such a process. In Scheme 1.4c the photosensitization occurs by electron transfer rather than by energy transfer; several illustrations of Scheme 1.4c are detailed in Chapter 4 and ref. 40 shows the variety of transformations made possible along this direction. In Scheme 1.4c the succession of elementary steps leading from S to P was such that, in one of them, the sensitizer was regenerated. This is no longer true in Scheme 1.4d where a sacrificial reagent must be added to make possible the regeneration of sens. At least 1 mol of sacrificial reagent is consumed by 1 mol of the formed product. The use of sacrificial reagents has been well developed by

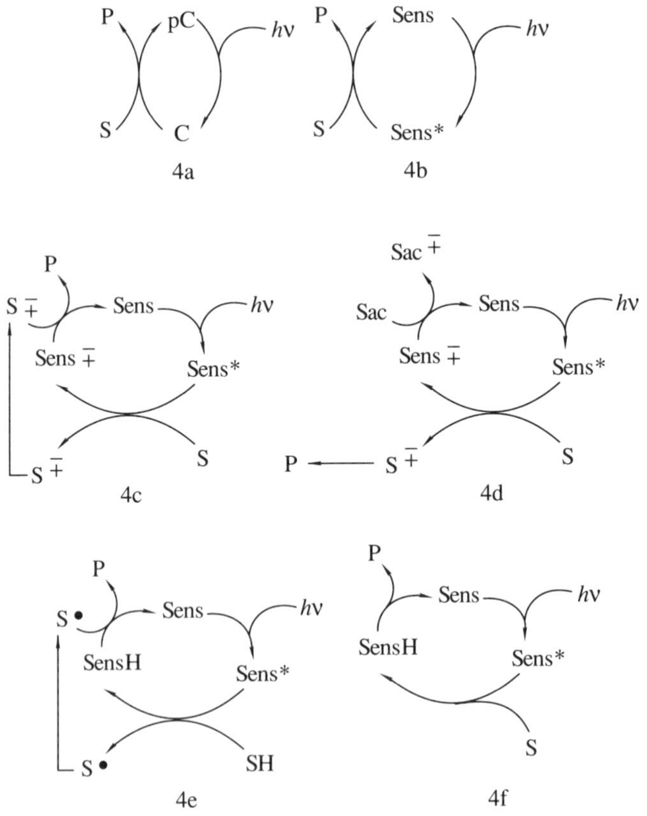

Scheme 1.4. Catalytic cycles of non-catalytic photocatalysis in photons (see also Scheme 1.1). Sens = sentitizer, Sac = Sacrificial reagent, SH = substrate bearing a hydrogen possibly involved in atom transfer

the chemists engaged in the difficult challenge of water activation (Chapter 8). Scheme 1.4e shows that an excited photosensitizer may also play its role by atom (H, X) transfer; in the regeneration step (S° ⟶ P) the radical S takes back its atom from Sens H. Finally, in Scheme 1.4f, the excitation of the sensitizer may lead to an associative activation of the substrate towards the sensitizer, and this substrate–sensitizer complex evolves into product(s) with regeneration of the sensitizer. One of the practical difficulties resulting from the more and more complex mixtures designed in photocatalysis is the separation of the products at the end of the reaction. One way to diminish this burden is to perpetuate the sensitizers; Chapter 7 describes the successes in this field.

The design of more and more complex mixtures also originates in the mimicking of natural photocatalytic gears such as the photosynthetic systems (Chapter 10). This is specially true for systems relying upon redox photosensitization, and the progressive gain in complexity was analyzed in several reviews [40–42] and clearly pervades the content of Chapter 8.

Although we have given mainly examples of homogeneous processes in Schemes 1.4b–1.4f, the most successful results in terms of alkane partial oxidation, waste organic acid and cyanide conversion, chlorinated hydrocarbon decomposition has been obtained with heterogeneous photosensitizers such as titanium dioxide particles. In Chapter 8 of this volume you will see that, given a formidable challenge (photocatalytic production of H_2 from H_2O), the chemists have designed chemical systems which continuously pass over the whole range of homogeneous, semi-homogeneous and heterogeneous media. This is one of the reasons why we have to understand the fundamentals of organized media (Chapter 9). As a reward, we will see in this Chapter 9 that a variety of other transformations can profit from this better understanding.

No classification is without flaws and this one is no exception. First, a sensitizer may sensitize the formation of a catalyst or an initiator, under such conditions we will have a photosensitized process with quantum yields greater than I. Second, there are actual processes of great practical importance for which the mechanistic intricacies defy the simplistic classifications that we have given (Chapter 6). Heterogeneous photocatalysis, with all its practical successes, presents fundamental specificities which are dealt with in Section 3, a specific volume will be devoted to this topic.

3 SPECIFIC ASPECTS OF HETEROGENEOUS PHOTOCATALYSIS

Normally, in practical cases a rate for a process is reported as a parameter to evaluate its activity. This rate is

$$\text{rate} = \frac{dn}{dt} \ (s^{-1}) \qquad \ldots (1.1)$$

where n is expressed in moles.

According to IUPAC, the rates should be referred to volume, mass or area of the catalyst. Therefore for the volume of the solid catalyst:

$$\text{rate} = \frac{1}{V}\frac{dn}{dt} \ (cm^{-3} s^{-1}) \qquad \ldots (1.2)$$

where V is the volume of the solid catalyst and the rate is called *volumetric rate*.

It has been proposed to use the term *specific rate* for the rate, referred to the mass of the catalyst, i.e. to the following expression:

$$\text{rate} = \frac{1}{m}\frac{dn}{dt} \ (g^{-3} s^{-1}) \qquad \ldots (1.3)$$

where m is the mass of the catalyst, expressed in grams. Finally it has been proposed to use the term *areal rate* for the rate referred to the surface area of the catalyst, i.e. the following expression:

$$\text{rate} = \frac{1}{A}\frac{dn}{dt} \quad (\text{cm}^{-2}\,\text{s}^{-1}) \quad \ldots (1.4)$$

where A is the surface area of the catalyst.

For homogeneous processes the quantity "specific rate" can be used, while for heterogeneous processes all three quantities are appropriate. However, it has to be noted that the areal rate could be better referred to as the active sites catalyst, which, while having the same surface area, may have different concentrations of active sites. But also the determination of the active sites is not an easy task. Therefore the areal rate must be considered as a rough approximation to the rate expressed in terms of active sites.

As previously discussed, reporting the values of the rates does not allow the differences between the catalytic and non-catalytic nature of the process to be distinguished. In particular cases, the values of the rates are reported together with other experimental observations as follows:

1. The lack of any apparent reaction in absence of the catalyst;
2. A reasonable duration of the reaction;
3. Other specific observations for photocatalysis, as the role of the radiation;
4. And so on.

All above observations shed some light on the nature of the process. Only the knowledge of the rates of the uncatalyzed and catalyzed processes give a definite reply to the above question. Furthermore, it is better to report "the extent of the reaction", instead of moles designated with the symbol ξ and defined as

$$\xi_{(\text{mol})} = \left(\frac{n_i^o - n_i}{\nu_B}\right) \quad \ldots (1.5)$$

where n_i^o, n_i are quantities of the substance (in moles) at the initial time and at any time, respectively and ν_B is the stoichiometric coefficient (plus for products, minus for reactants) of any product or reactant B.

Finally, it must reported that the rates are often referred to as turnover number, turnover frequencies or turnover rates.

Another parameter used in photocatalysis is the quantum yield (qy), It is defined by the International Union of Pure and Applied Physics (IUPAC) as *quantum yield* (Φ, Y) the number of defined events which occur per photon absorbed by the system. The integral quantum yield is

$$\Phi = \frac{\text{number of events}}{\text{number of photons absorbed}} \quad \ldots (1.6)$$

For a photochemical reaction

$$\Phi = \frac{\text{amount of reactant consumed or product formed}}{\text{amount of photon absorbed}} \quad \ldots (1.7)$$

INTRODUCTION TO PHOTOCATALYSIS

The differential quantum yield is

$$\Phi = \frac{d[x]/dt}{n} \quad \ldots (1.8)$$

where $d[x]/dt$ is the rate of change of a measurable quantities and n the amount of photons (mole or its equivalent, einstein) absorbed per unit of time; Φ can be used for photophysical processes or photochemical reactions. Thus knowledge of qy gives essential information, i.e. how many molecules (moles) are transformed per number (einstein) of absorbed photons.

For homogeneous systems the determination of qy is quite simple, since the experimental determination of both quantities — the molecules (moles) transformed and the number (einsteins) of absorbed photons — is a relatively simple task.

For heterogeneous systems the experimental determination of qy is more complicated, since the photons impinging on such systems are in part absorbed and in part scattered in various ways depending on the particle sizes [43]. So far an experimental method for determining the parameter "absorbed photons" for heterogeneous systems does not exist.

Often values of qy, for heterogeneous systems, are reported using the quantity "impinging photons" (or "total emitted photons"), but these values do not give any true information about the molecules transformed by the photons really absorbed. For heterogeneous systems, for which the particle sizes are in the micron range $50 \div 500$ μm, i.e. for that range in which geometrical optics can be applied, the optical scatter can be considered as due only to reflection phenomenon. When the heterogeneous systems have the features for the particles sizes described above, a method for determining the reflected fraction of the total impinging photons has been proposed [44]. Knowing this fraction, it is easy to determine the really absorbed photons and then evaluate the real qy.

4 PHOTOCATALYSIS A CROSSROAD FOR FUNDAMENTAL, APPLIED PHOTOCHEMISTRY AND CATALYSIS

Research aiming at highly ambitious objectives is seldom confined to its original field. The design of photocatalytic gears to achieve the non-polluting production of an inextinguishable fuel, H_2, has had applications in the direction of activation of inert molecules (Chapter 5), new, mild reactions of synthetic importance (Chapter 4), ecological treatment of wastes (vol. Heterogeneous Photocatalysis) and use of solar energy for heating homes. To do so one had to master several aspects of fundamental photochemistry and catalysis; this is the reason why a special chapter (Chapter 2) deals with the fundamentals of interaction between light and matter. The photophysics of very fast processes [45] play a central role in the modelization of photocatalytic gears: the kinetics constants of elementary steps such as charge transfer reactions in molecular liquids and at interfaces, the kinetics of proton transfer in ground and excited states (Chapter 3) all participate in the overall working or non-working of photocatalytic processes. It is then surprising to see that several principles at work in this main stream also manifest themselves in the totally different field of argentic photography (Chapter 6) with possible cross-fertilization between these seldom confronted fields.

The most interesting contribution remains, however, Chapter 10 where some of the subtleties associated with the photosynthetic machine are exposed. This section gives a hint of the long way still to go for photocatalysis in the lab to become a fully fledged science.

REFERENCES

[1] N. Webster, *Webster's New Universal Unabridged Dictionary*, 2nd edn, revised by J. L. McKechnie, New World Dictionaries (Simon and Schuster, Cleveland, Ohio, 1983).
[2] R. P. Bell, *Acid-Base Catalysis* (Oxford University Press, Oxford, 1941).
[3] (a) G. G. Wubbels, Acc. Chem. Res., **16**, 285 (1983); (b) A. Albini, Acc. Chem. Res., **17**, 235 (1984).
[4] P. G. Ashmore, *Catalysis and Inhibition of Chemical Reactions* (Butterworths, London, 1963), pp. 6–12.
[5] M. Prettre, *Catalyse et Catalyseurs*, 4th edn, Collection Que sais-je. (Presse Universitaire de France, Paris, 1970), pp. 9, 14, 16. For example on p. 9 this author writes. "One may divide in two groups the set of reactions which are accelerated or made possible by the addition of small amount of a substance.... The first group may be gathered under the label "true catalysis". The second group is constituted of chain reaction whose rates are particularly sensitive to the action of a variety of added substances.... We will describe the second group as "generalized" conception of catalysis".
[6] O. V. Gerasimov and V. N. Parmon, Russian Chem. Rev., **61**, 154 (1992).
[7] H. Henning, R. Billing and H. Knoll, in *Photosensitization and Photocatalysis using Inorganic and Organometallic Compounds*, edited by K. Kalyanasundaram, M. Gratzel (Kluwer, Dordrecht, 1993).
[8] E. A. Moelwyn-Hughes, *Physical Chemistry*, (Pergamon Press, Oxford, 1951), p. 1129.
[9] M. Boudart, *Kinetics of Chemical Processes* (Prentice-Hall, New York, 1968), p. 61.
[10] M. Boudart and G. Djega-Mariadassou, "*Kinetics of Heterogeneous Catalytic Reactions* (Princeton University Press, Princeton, New Jersey, 1984).
[11] IUPAC, Definitions, terminology and symbols in colloid and surface chemistry. Part 2. Heterogeneous catalysis, *Pure Appl. Chem.*, **51**, 1213 (1979).
[12] M. L. Bender, *Mechanisms of Homogeneous Catalysis from Protons to Proteins* (Wiley, New York, 1971).
[13] A. A. Frost and R. G. Pearson, *Kinetics and Mechanisms*, 2nd edn, (Wiley, New York, 1961).
[14] IUPAC glossary of terms used in photochemistry, *Pure Appl. Chem.*, **60**, 1055 (1988).
[15] (a) H. van Damme and W. K. Hall, J. Amer. Chem. Soc. **101**, 4373 (1979); (b) L. P. Childs and D. F. Ollis, J. Catal., **66**, 383 (1980).
[16] H. Kirsch, in *Photocatalysis. Fundamentals and Applications*, edited by N. Serpone and E. Pelizetti (Wiley, New York, 1989) p. 1.
[17] K. A. Alexander and D. M. Roundhill, J. Mol. Cat., **19**, 85 (1983).

[18] M. Chanon, Acc. Chem. Res., **20**, 217 (1987).
[19] D. P. Summers, J. C. Luong and M. S. Wrighton, J. Am. Chem. Soc., **103**, 5238 (1981).
[20] S. J. Oishi, Mol. Cat., **40**, 289 (1987).
[21] H. Henning, D. Rehorek and R. D. Archer, Coord. Chem. Rev., **61**, 1 (1985).
[22] R. G. Salomon, Tetrahedron, **39**, 485 (1983).
[23] (a) F. S. Dainton, *Chain Reactions* (Methuen, London, 1956). (b) E. S. Huyser, *Free Radical Chain Reactions* (Wiley, New York, 1970).
[24] (a) J. F. Bunnett, Acc. Chem. Res., **11**, 413 (1978). (b) N. Kornblum, Angew. Chem. Int. Ed., **14**, 734 (1975). (c) M. Chanon and M. L. Tobe, Angew Chem. Int. Ed., **21**, 1 (1982).
[25] M. Julliard and M. Chanon, Chem. Rev., **83**, 425 (1983).
[26] M. Chanon, M. D. Hawley and M. A. Fox, in *Photoinduced Electron Transfer*, edited by M. A. Fox and M. Chanon, (Elsevier, Amsterdam, 1988) Part A, p. 1.
[27] M. J. Mirbach, EPA Newsl., **20**, 16 (1984).
[28] L. Moggi, D. Juris, L. Sandrini and M. F. Manfrin, Rev. Chem. Intermed., 1981 **4**, 171 (1981).
[29] M. S. Wrighton, D. S. Ginley, M. A. Schroeder and D. L. Morse, Pure Appl. Chem., **41**, 671 (1975).
[30] V. Carassiti, EPA Newsl., **21**, 12 (1984).
[31] F. Chanon and M. Chanon, in *Photocatalysis* edited by N. Serpone and E. Pelizzetti, (Wiley, New York 1989) p. 489.
[32] P. C. Ford and A. F. Friedman, in *Photocatalysis* edited by N. Serpone and E. Pelizzetti, (Wiley, New York 1989) p. 541.
[33] A. Heuman and M. Chanon, in *Applied Homogeneous Catalysis by Organometallic Complexes*, edited by J. Cornils, (VCH, Berlin, 1996).
[34] M. Chanon, M. Rajzmann and F. Chanon, Tetrahedron, **46**, 6193 (1990).
[35] P. Wan and D. Shukla, Chem. Rev. **93**, 571 (1993).
[36] C. Kutal, Coord. Chem. Rev., **64**, 191 (1985).
[37] A. W. Adamson, Comments Inorg. Chem., **1**, 33 (1981).
[38] W. Strohmeier and L. J. Weigelt, Organomet. Chem. **133**, 43 (1977).
[39] M. Koizumi, S. Kato, N. Mataga, T. Matsuura and Y. Usui, *Photosensitized Reactions* (Kagakudojin, Kyoto, Japan, 1978).
[40] M. Chanon and L. Eberson, in *Photoinduced Electron Transfer* edited by M. A. Fox and M. Chanon, (Elsevier, Amsterdam, 1988) Part A, p. 409.
[41] M. Chanon, Bull. Soc. Chim. Fr., 209 (1985).
[42] M. Chanon, M. Julliard, J. Santamaria and F. Chanon, New J. Chem. 1992, **16**, 171 (1992).
[43] M. Schiavello, V. Augugliaro and L. Palmisano, An experimental method for the determination of the photon flow reflected and absorbed by aqueous dispersions containing polycrystalline solids in heterogeneous photocatalysis, J. Catal., **127**, 332 (1991).
[44] V. Augugliaro, L. Palmisano and M. Schiavello, Determination of photon absorption rate by aqueous titanium dioxide dispersion contained in a stirred batch photoreactor, AIChE J., **37**, 1096 (1991).
[45] Y. Gauduel and P. J. Rossky, *Ultrafast Reaction Dynamics and Solvent Effects* AIP Conference Proceedings 298 (AIP Press, New York, 1994).

2 Fundamentals of Interaction Between Light and Matter

J. C. MIALOCQ
Commissariat à l'Energie Atomique,/DSM/DRECAM/SCM/URA 331 CNRS Gif-sur-Yvette, France

1	Light and the dualistic theory	16
	1.1 Plane-wave Propagation	16
	1.2 Matter and absorption of light: the Lambert–Beer law	17
2	Quantum theory of molecular systems	19
3	Potential energy surfaces	21
4	Nuclear motions. Vibronic states and vibrational wavefunctions	22
5	Molecular states	26
	5.1 Quantum numbers	26
	5.2 Spin correlation	26
6	Interaction of light with molecular systems	27
	6.1 Molecular dipoles	27
	6.2 Transition moment	27
	6.3 Oscillator strength of the absorption band	29
	6.4 The Franck–Condon principle	30
	6.5 Selection rules	30
7	Nonadiabatic reactions	32
	7.1 Deexcitation processes of electronically excited molecules	32
	7.2 Relationship between absorption intensity and fluorescence lifetime	34
	7.3 Mirror symmetry	35
	7.4 Radiationless processes	38
8	Intermolecular processes	41
	8.1 Energy transfer	43
	8.1.1 Förster theory of coulombic energy transfer	44
	8.1.2 Dexter theory of electron-exchange energy transfer	45
	8.1.3 Energy transfer in disordered materials and restricted geometries	46
	8.2 Electron transfer	46
	8.2.1 Classical theories	46

Homogeneous Photocatalysis. Edited by M. Chanon
© 1997 John Wiley & Sons Ltd

	8.2.2	Quantum effects	50
	8.2.3	The inverted region	50
	8.2.4	The distance dependence of intramolecular electron-transfer rates	51
References			51

1 LIGHT AND THE DUALISTIC THEORY

Over the years, the nature of light switched between corpuscles which interacted with the "ether" (Isaac Newton, 1675) to waves (Christiaan Huygens, 1690 and James Clerk Maxwell, 1867). Modern physics assumes both corpuscles and waves characteristics since Planck (1900), Einstein (1905) and de Broglie (1924) initiated this dualistic theory.

1.1 PLANE-WAVE PROPAGATION

Light is usually understood as electromagnetic radiation in the near-ultraviolet, visible and near-infrared spectral range, characterized by its frequency v expressed in hertz (Hz) as illustrated in Figure 2.1. This is the region of interest (about 10^{15} Hz) for photochemists, but the electromagnetic spectrum extends over a vast range of energies and hence frequencies, from γ-rays to radiofrequencies [1]. Light is also regarded as a beam of photons or quanta of energy E moving along trajectories determined by fundamental optics. Here E is related to the radiation frequency v by the Planck equation

$$E = hv \qquad \ldots(2.1)$$

where h is Planck's constant (6.63×10^{-34} J s). Photons are bosons and have no mass in contrast to hadrons (protons and neutrons). They possess spin quantum

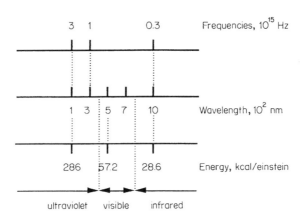

Figure 2.1. Spectrum of the electromagnetic radiations from the ultraviolet to the infrared

INTERACTION BETWEEN LIGHT AND MATTER

numbers $s = 1$ and two possible values of the spin angular momentum $m_s = +1$ and -1 ($m_s = 0$ is forbidden by the relativistic quantum theory) which correspond to left and right circularly polarized light [1h]. Photons may be viewed as chiral reagents, and the most convincing evidence is that racemic mixtures may be resolved if one enantiomer absorbs more efficiently polarized light [1c].

Let us consider a monochromatic, linearly polarized electromagnetic plane wave, propagating in a homogeneous and nonabsorbing medium. The electric field vector (which is the primary concern of the photochemist) can be represented by the equation

$$\mathbf{E} = \mathbf{E}_0 \exp[i(2\pi z/\lambda - \omega t)] \quad \ldots (2.2)$$

where $\mathbf{E}_0$ is a constant vector in the xy plane, z is the coordinate along the direction of propagation. The wavelength λ and the angular frequency ω in radians per second are related through the phase velocity c/n, where c is the velocity of light in vacuum ($2.997\,924\,58 \times 10^8$ m s^{-1}) and n the refractive index of the medium

$$\lambda = (2\pi/\omega)(c/n) \quad \ldots (2.3)$$

The wavelength λ and the frequency ν are related by the equation

$$\lambda = c/\nu \quad \ldots (2.4)$$

In the SI units (Système International d'Unités), λ is given in meters, but the wave number $\bar{\nu} = 1/\lambda$ (defined as the number of waves per centimeter) in reciprocal centimeters or Kaysers is also used, particularly in infrared and Raman spectroscopy. In photochemistry, the value of λ is in the 100–1500 nm spectral range. In the following example, using eqs. 2.1 and 2.4, one can calculate for 532 nm (the wavelength at the second harmonic of the familiar Nd–Yag laser), the photon energy

$$E_{h\nu}^{532\,\text{nm}} = 6.63 \times 10^{-34} \times 2.998 \times 10^8 / 532 \times 10^{-9} = 3.74 \times 10^{-19} \text{ J}$$

1.2 MATTER AND ABSORPTION OF LIGHT: THE LAMBERT–BEER LAW

The term "matter" includes a large variety of objects, from microscopic (atoms and molecules) to macroscopic media (gases, liquids and solids). The interaction between light and matter is the subject of theoretical and experimental researches. Since 1960, the discovery of lasers as a new source of light opened a new chapter due to the specific properties of laser radiation, spatial and time coherence, directivity, divergence (10^{-3} rad), monochromaticity, short pulse duration (as short as a few femtoseconds, 1 fs = 10^{-15} s), high intensity.

The light intensity I of a monochromatic beam can be defined as the number of photons crossing a surface A in unit time. Let us consider the intensity I_0 of the light incident on the front surface of an optical cell containing an absorbing solution, the transmitted intensity I_T and the absorbed intensity I_A. If the intensity of the reflected light can be regarded as negligible, the intensity of the absorbed light is

$$I_A = I_0 - I_T \qquad \ldots (2.5)$$

The amount of light absorbed is proportional to the light intensity, the number M of absorbing molecules per cubic centimeter and the width dl of the solution

$$-dI = \alpha I M \, dl \qquad \ldots (2.6)$$

where α is the cross-section in square centimeters per molecule at the wavelength λ. The cross-section is the area accessible to being struck by one photon. Integration of eq. (2.6) leads to eq. (7)

$$I_T = I_0 \exp(-\alpha M l) \qquad \ldots (2.7)$$

where l is the cell length in centimeters or eq. (2.8) which is the Lambert–Beer law, by introducing the molar concentration C of the solute expressed in mol dm^{-3}, $C = 1000M/N$ where N is Avogadro's number (6.02×10^{23}), and the wavelength-dependent decadic molar extinction coefficient ε in dm^3 mol^{-1} cm^{-1}; I_T/I_0 is called the transmittance and $\varepsilon C l$ is the absorbance or optical density of the sample. Maximal values of ε are $1\text{–}5 \times 10^5$ dm^3 mol^{-1} cm^{-1}.

$$I_T = I_0 10^{-\varepsilon C l} \qquad \ldots (2.8)$$

The relationship between ε and α is

$$\frac{\varepsilon}{\alpha} = \frac{N}{2.303 \times 10^3} \qquad \ldots (2.9)$$

Assuming for a molecule an experimental molar extinction coefficient ε of 10^5 dm^3 mol^{-1} cm^{-1} or a cross-section $\alpha = 2.3 \times 10^3 \times 10^5 / 6.02 \times 10^{23} = 3.8 \times 10^{-15}$ cm^2 molecule^{-1}, the cross-section per molecule is thus 38 A^2, corresponding roughly to a length of 6 A, a few conjugated double bonds.

Departure from the Lambert–Beer equation is observed when the number of photons absorbed is too large with respect to the number of molecules in the illuminated volume or when two photons are absorbed by molecule. The interaction of light with molecules is generally between one molecule A and one photon $h\nu$ leading to an electronically excited molecule A* [2].

$$A + h\nu \longrightarrow A^* \qquad \ldots (2.10)$$

The excitation energy of the molecule is obtained by inspection of its *absorption* or *emission spectrum*

$$\Delta E = E_2 - E_1 = h\nu \qquad \ldots (2.11)$$

where E_2 and E_1 are the energies of the molecule in the final and initial states. The maximums of an absorption spectrum or an emission spectrum are expressed by

INTERACTION BETWEEN LIGHT AND MATTER 19

their wavelengths or their wavenumbers. Equation (2.11) is sometimes referred as Bohr's frequency condition.

Extending the concept of mole to photons, an *einstein* is defined as 1 mol of photons (6.02×10^{23}). The energy of 1 einstein of 532 nm light is thus (see above)

$$E^{532\,nm} = 3.74 \times 10^{-19} \text{ J} \times 6.02 \times 10^{23} \quad \text{or}$$

$$22.5 \times 10^4 \text{ J, i.e. } 225 \text{ kJ } (53.8 \text{ kcal})$$

This energy and the energy required to break chemical bonds are of the same order of magnitude. Single N–N and C–N bonds have strengths of 46 and 57 kcal/mole respectively, an excited molecule may undergo bond breaking [2,3] and the efficiency of the process will depend on the various deactivating processes in competition. The near UV region which extends from 390 nm to 200 nm is the region where the energies encompass π and σ bond energies. Below 190 nm, spectroscopy and photochemistry have to be performed in the absence of air because molecular oxygen absorbs.

2 QUANTUM THEORY OF MOLECULAR SYSTEMS

Classical mechanics considers position and momentum. Quantum mechanics considers wavefunctions Ψ solutions of a wave equation. Let us consider a molecule consisting of N nuclei and n electrons. The Hamiltonian of the molecule can be written [4] as follows:

$$H = \sum_{i=1}^{n} \frac{\mathbf{p}_i^2}{2m} + \sum_{s=1}^{N} \frac{\mathbf{P}_s^2}{2M_s} + V(\mathbf{r}_i, \mathbf{R}_s) \qquad \ldots (2.12)$$

where m is the mass of the electron, $\mathbf{p}_i$ the linear momentum of the ith electron, M_s and $\mathbf{P}_s$ respectively the mass and the linear momentum of the sth nucleus and $V(\mathbf{r}_i, \mathbf{R}_s)$, the sum of the mutual potential energy of the electrons, the mutual potential energy of the nuclei, the potential energy of the electrons with respect to the nuclei.

The Schrödinger eq. (2.13) belonging to the Hamiltonian (2.12) is

$$H\Psi(\mathbf{r}_i, \mathbf{R}_s) = E\Psi(\mathbf{r}_i, \mathbf{R}_s) \qquad \ldots (2.13)$$

where $\Psi(\mathbf{r}_i, \mathbf{R}_s)$ are wavefunctions which depend on the electronic ($\mathbf{r}_i$) and nuclear ($\mathbf{R}_s$) coordinates. The solutions are the following

$$\Psi(\mathbf{r}_i, \mathbf{R}_s) = \chi(\mathbf{R}_s)\psi(\mathbf{r}_i, \mathbf{R}_s) \qquad \ldots (2.14)$$

In the *adiabatic approximation*, also called the *Born–Oppenheimer approximation*, abbreviated BO, the function which represents the motion of the electrons varies slowly with the nuclear coordinates since the electrons have much smaller masses than the nuclei [4,5]. The BO approximation is a particular case of the general *Ehrenfest* principle which states that if a system is perturbed slowly, it remains in a definite stationary state. This approximation leads to the separation between a fast subsystem (electronic motions) and a slow one (nuclear motions). Equation (2.13) is reduced to eq. (2.15)

$$-\frac{\psi}{\phi}\sum_{S}\frac{\hbar^2}{2M_S}\nabla_S^2\chi + \left[-\sum_{i}\frac{\hbar^2}{2m}\nabla_i^2 + V\right]\psi = E\psi \qquad \ldots(2.15)$$

where the terms in square brackets represent the electronic Hamiltonian operator H_{el} if the nuclei are kept fixed in space. The solution of eq. (2.15) reduces to the solutions of two eqs. (2.17) and (2.18) whose eigenfunctions represent respectively the motion of the electrons when the nuclei are kept fixed in space and the motion of the nuclei.

$$H_{el} = -\sum_{i}\frac{\hbar^2}{2m}\nabla_i^2 + V(\mathbf{r},\mathbf{R}) \qquad \ldots(2.16)$$

$$-\sum_{i}\frac{\hbar^2}{2m}\nabla_i^2\psi_k(\mathbf{r},\mathbf{R}) + V(\mathbf{r},\mathbf{R})\psi_k(\mathbf{r},\mathbf{R}) = \varepsilon_k(\mathbf{R})\psi_k(\mathbf{r},\mathbf{R}) \qquad \ldots(2.17)$$

$$-\sum_{S}\frac{\hbar^2}{2M_S}\nabla_S^2\chi_v^k(\mathbf{R}) + \varepsilon_k(\mathbf{R})\chi_v^k(\mathbf{R}) = E_{kv}\chi_v^k(\mathbf{R}) \qquad \ldots(2.18)$$

where $\psi_k(\mathbf{r},\mathbf{R})$, an eigenfunction of the electronic Hamiltonian, describes the motion of the electrons and is associated with an energy $\varepsilon_k(\mathbf{R})$ which is an eigenvalue of eq. (2.17) depending on the electronic quantum number k. For each k, the energy values define the adiabatic potential energy surface of the kth *electronic state*. The energy spectrum is discrete for lower k (electronically bound states) and continuous for higher k (ionized states plus a free electron or autoionizing states) [1].

The Hamiltonian H_V of eq. (2.18) is

$$H_V = -\sum_{S}\frac{\hbar^2}{2M_S}\nabla_S^2 + \varepsilon_k(\mathbf{R}) \qquad \ldots(2.19)$$

and $\chi_v^k(\mathbf{R})$ are nuclear eigenfunctions, where v is a nuclear quantum vibrational number. Inspection of eqs. (2.18) and (2.19) shows that the solutions describe the motion of the nuclei in an effective potential field of electrons, i.e. $\varepsilon_k(\mathbf{R})$ is acting as the potential energy for nuclear motion.

The exact Hamiltonian H is invariant under all spatial transformations, translations and rotations. The Hamiltonian H_V is invariant under all operations which send identical nuclei into one another. In the BO approximation, the relevant symmetries are those related to the nuclear coordinates. Diatomics are of symmetry $C_{\infty v}$ or $D_{\infty h}$, triatomics are of symmetry C_s or higher and for higher polyatomics the point group is C_1.

As the nuclei move around, electrons adjust their motion to the instantaneous position of the nuclei. However, the BO approximation fails when the nuclei move very rapidly.

3 POTENTIAL ENERGY SURFACES

A molecule which contains N atoms has $3N$ degrees of freedom. To describe the motion of the atoms, $3N$-6 coordinates are needed since three represent the translation of the center of gravity and three the rotation of the molecule as a whole about its center of gravity. In the case of linear molecules, $3N$-5 coordinates are needed since it is not necessary to specify rotation about the molecular axis. The set of all possible vectors **R** describes the space of the nuclear configuration of the molecule. For triatomic molecules such as H_2O or CO_2, this space is thus three dimensional. The total dimensionality of the space containing also energy is $3N$-5. The electronic state is described by a ($3N$-5)-dimensional hypersurface. The fundamental or ground state has a minimum which corresponds to the stable molecular geometry. Other minima may be found when conformers or isomers exist. The concept of potential energy surfaces provides a useful visualization of molecular energies, transitions between states and photochemical reactions. Two-dimensional energy curves are more easily visualized as in diatomic molecules. But already for triatomic molecules, it is impossible to plot the potential energy as a function of all the bond angles and bond lengths. In the case of polyatomic molecules, a two-dimensional plot of the potential energy versus the "reaction coordinate" is generally presented by choosing the more appropriate coordinate as that which undergoes the most important variation during the reaction. The obvious advantage is the possibility of having the potential energy curves of several electronic states on the same diagram.

A point moving along a potential energy curve (Figure 2.2) represents a specific nuclear configuration. Kinetic energy and potential energy of the particle interchange and the total energy E, sum of the kinetic energy T and potential energy V is conserved. For a motionless particle ($T = 0$), the total energy is equal to the potential energy. The particle is attracted toward the minimum by a driving force $F(r)$ which is the slope of the potential energy curve as given by eq. (2.20):

$$F(r) = -\frac{dV(r)}{dr} \qquad \ldots (2.20)$$

This driving force is associated with the attractive force between electrons and nuclei. Because $E - V > 0$, the representative point cannot be below the potential energy curve. This idealized description holds for a frictionless curve, but frictional forces can remove kinetic energy. If the particle experiences some friction, the friction is an energy sink. The speed and acceleration of the particle are diminished with respect to the case of a frictionless surface. Molecular collisions or molecular vibrations constitute the friction experienced by a molecule [1c].

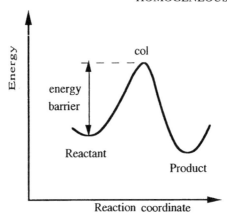

Figure 2.2. Potential energy curve as a function of the reaction coordinate

Let us now examine the potential energy curve for a reaction in the ground state (Figure 2.2). The reaction coordinate may be an internuclear distance (dissociation) or a bond angle (isomerization). The curve has a minimum for the reactant(s), an energy barrier and a saddlepoint or a col [1d] and another minimum for the product(s).

When two reaction coordinates are involved, the potential energy curves as a function of both coordinates may be plotted defining a two-dimensional framework. The reaction proceeds from the reactant valley to the product valley. The kinetic behaviour of chemical reactions is governed by the Arrhenius rate law:

$$k(T) = A \exp(-E/RT) \qquad \ldots (2.21)$$

where $k(T)$ is the temperature-dependent rate constant, A the preexponential factor or frequency factor, E the activation energy per mole, R the product of the Avogadro number by the Boltzmann constant. In the Eyring theory, the rate constant of the reaction is derived from the height of the potential energy barrier, difference between the energy of the *activated complex* or *activated state* and that of the reactants. Evans and Polanyi called *transition state* the col on the potential energy surface.

4 NUCLEAR MOTIONS. VIBRONIC STATES AND VIBRATIONAL WAVEFUNCTIONS

When a fixed and rigid nuclear geometry is assumed, a zeroth-order description of electronic structure and electronic energy is given. The effect of nuclear motion has to be considered by replacing the vibrationless molecule with a vibrating molecule. The states of a vibrating molecule are called *vibronic states* [1]. Two types of nuclear motions are distinguished: (1) small-amplitude motions around

INTERACTION BETWEEN LIGHT AND MATTER

the minimum of the potential hypersurface; (2) large-amplitude motions leading to highly distorted geometries. Small-amplitude nuclear motions are described as a superposition of normal modes of vibration, i.e. oscillations along combinations of bond length and bond angle changes. The set of 3N-6 normal coordinates, Q_i, express the nuclear vibrations. The earliest descriptions of chemical bonding used pictures where atoms are represented by weights and intramolecular forces by springs. Each of the normal modes of vibration can be treated as a one-dimensional harmonic oscillator with a potential energy

$$V = \tfrac{1}{2}kr^2 = 2\pi^2 v^2 \mu r^2. \qquad \ldots (2.22)$$

where k is the force constant and μ the reduced mass or harmonic mean of the masses of the two vibrating atoms [1h,1i,1j,3]. A more accurate description of the nuclear motion uses a quantum mechanical viewpoint where the nuclear wavefunction of the initially prepared state is dictated by the time-dependent Schrödinger equation. This wavefunction is also called a wave packet [1f]. Stationary states are found by solving the Schödinger equation:

$$\left(-\frac{\hbar^2}{2\mu}\frac{d^2}{dr^2} + \frac{1}{2}kr^2\right)\chi = E\chi \qquad \ldots (2.23)$$

giving a set of vibrational wavefunctions χ_v which describe the geometry and the motion of the nuclei [1g,1i,1k]. The equally spaced vibrational energy levels E_v are quantized with the vibrational quantum number v which may take integral values 0, 1, 2, 3 ...:

$$\chi_v(r) = (\alpha/\pi)^{1/4}(2^v v!)^{-1/2}\exp(-\alpha r^2/2)H_v(\alpha^{1/2}r) \qquad \ldots (2.24)$$

$$\alpha = 2\pi v\mu/\hbar \qquad \ldots (2.25)$$

$$E_v = hv\left(v + \tfrac{1}{2}\right) \qquad \ldots (2.26)$$

v, the fundamental vibrational frequency of the classical oscillator depends on the curvature of the potential energy hypersurface at the minimum along the normal coordinate. The lowest energy level for $v = 0$ is not zero but $\tfrac{1}{2}hv$, the zero point energy of the oscillator which exists at the temperature of absolute zero and represents quantum fluctuations of the nuclei, a manifestation of the Heisenberg uncertainty principle [1h,1j]. For a diatomic molecule AB, where A and B are attached by a spring, the relative motion of A and B, stretching and contracting, is thus a periodic function of time. The shape of the potential energy curve depends on the strength of the bond and the weights of the atoms as shown in Figure 2.3.

The amplitude of the classical vibrational motion is given by the intersections of the potential energy curve with the vibronic energy level. At these "turning points" (extreme stretching or compression) [1c], the kinetic energy is zero and the total vibrational energy is equal to the potential energy. The potential energy is

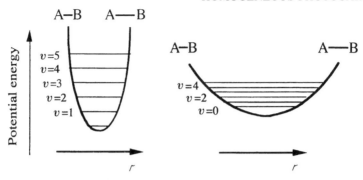

Figure 2.3. Classical potential energy curves as a function of the internuclear distance for light atoms and/or strong bonds (left) and for heavy atoms and/or weak bonds (right) [1c]

minimum and the kinetic energy is maximum close to the equilibrium internuclear distance.

In eq. (2.24), the functions H_v are Hermite polynomials of the vth degree [1g,1i,1k]. The mathematical forms of the vibrational ground state ($v = 0$) and ($v = 1$) wavefunctions are given [1i,1k] by eqs. (2.27) and (2.28)

$$\chi_0 = (\alpha/\pi)^{1/4} \exp(-\alpha r^2/2) \qquad \ldots (2.27)$$

$$\chi_1 = (\alpha/\pi)^{1/4} 2^{1/2} \exp(-\alpha r^2/2) \qquad \ldots (2.28)$$

The wavefunctions χ_v for $v = 0, 1, 2, 3, 4$ are represented [1c,1h,1k] in Figure 2.4. The value of χ_v is positive above the line of the vibrational energy level, negative below [1c]. The number of times the representative curve crosses

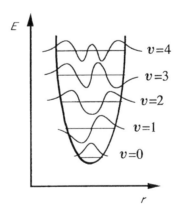

Figure 2.4. Harmonic-oscillator wavefunctions χ_v superimposed on the respective vibrational levels ($v = 0, 1, 2, 3, 4$)

the line is equal to the vibrational quantum number. The χ_1 wavefunction has one node [1k]. For $v = 0, 2, 4, \ldots$, the wavefunction for $-r$ is the same as that for $+r$, but for $v = 1, 3, 5, \ldots$ the wavefunctions have opposite signs [1k]. For $v = 0$, the quantum mechanical description differs from the classical viewpoint. Instead of two turning points, a broad probability maximum is observed at the equilibrium distance. The probability function χ_v^2 represents the probability of finding the nuclei at a given distance r during vibration in the vibrational level v [1c].

The vibrational overlap integral $\langle \chi \rangle$ is the probability of transition between two different vibrational states as given by eq. (2.29):

$$\int \chi_i \chi_j \, d\tau = \langle \chi_i | \chi_j \rangle \equiv \langle \chi \rangle \qquad \ldots (2.29)$$

The harmonic oscillator model gives a good representation of bond vibrations for small displacements from the equilibrium distance. However, for sufficiently large distances, the "restoring force" decreases and the potential energy will depart from the $V = \frac{1}{2}kr^2$ function and reach, as the bond breaks, a limited value called the dissociation energy as shown in Figure 2.5. For short internuclear distances, the rise of coulombic repulsions will increase the potential energy more rapidly than the harmonic oscillator function does. An alternative description is the nonharmonic oscillator [3]. The energy levels found for the Schrödinger equation are given by

$$E = h\nu \left(v + \tfrac{1}{2}\right) - h\nu x \left(v + \tfrac{1}{2}\right)^2 \qquad \ldots (2.30)$$

where x is the anharmonicity constant which varies from 0.01 to 0.05 for single bonds [1h,1i,1j].

For electronically excited states, molecular vibrations can be described by the same scenario as that used for the electronic ground state. Typical potential

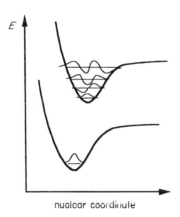

Figure 2.5. Potential energy curves of two electronic states of different equilibrium geometry

energy curves for the ground and the electronically excited state are given in Figure 2.5.

The adiabatic BO approximation works well when the electronic wavefunctions change slowly with the nuclear coordinates. This is not the case when the electronic wavefunctions change abruptly as in the region of a crossing of potential energy surfaces.

The representation of vibrations of polyatomic molecules is more difficult since the vibrations involve not only bond stretching but also group scissoring, twisting, wagging, etc. A polydimensional potential energy surface is needed.

5 MOLECULAR STATES

5.1 QUANTUM NUMBERS

If we consider the hydrogen atom wavefunction within spherical polar coordinates (r, θ, ϕ), the radial part of the wavefunction depends on two quantum numbers, the principal quantum number n which may take integer values, $n = 1, 2, 3, \ldots$, etc. and the electron's orbital angular momentum quantum number l, which can take the values $l = 0, 1, 2, \ldots, n - 1$. The other part depends on two quantum numbers, l and the magnetic quantum number m_l which may take the $(2l + 1)$ values $l, l-1, l-2, \ldots, -l$ [1h]. In quantum mechanics, the angular momentum of the electron moving around the proton is represented by the vector **l** and is quantized both in magnitude and direction and the allowed values are given by

$$\mathbf{l} = [l(l+1)]^{1/2}\hbar \qquad \ldots (2.31)$$

A fourth quantum number was proposed by Dirac, the spin quantum number, s which has a single value $s = \frac{1}{2}$ in the case of the electron (particles with half-integral values of the spin quantum number are fermions). However, the electron should not be viewed as spinning around its axis [1h]. The **s** vector has for magnitude

$$\mathbf{s} = [s(s+1)]^{1/2}\hbar \qquad \ldots (2.32)$$

and it can take two orientations with respect to a z axis, corresponding to two values of the z component of s, $s_z = m_s\hbar$, where m_s may take the values $+\frac{1}{2}$ and $-\frac{1}{2}$. The two orientations are called "spin up", ↑ (spin wavefunction α), and "spin down", ↓ (spin wavefunction β).

According to the Pauli exclusion principle, two electrons cannot possess the same quantum numbers n, l, m_l and m_s. In an orbital, the electrons are spin paired.

5.2 SPIN CORRELATION

In a ground state molecule, all the electrons have their spin paired, i.e. the spins of two electrons occupying the same level are antiparallel. This property is fundamental to explain chemical bonds and reactivity. The resulting spin is thus $S = 0$

INTERACTION BETWEEN LIGHT AND MATTER

and the ground state is a singlet state (the spin multiplicity defined as $2S+1$ is 1) which is labelled S_0. However, if one of the electrons is raised to a higher energy level, a parallel arrangement of the spins is made possible and the resulting spin is $S=1$, giving a triplet state (the spin multiplicity is 3).

The energy of a triplet state is lower than that of the corresponding singlet state (T_1 is lower than S_1) because the repulsion between two electrons occupying different orbitals is minimized when the spins are opposed.

The total electronic wavefunction Ψ, including the spatial wavefunction Φ and the spin wavefunction S, must be antisymmetric with respect to the exchange of any two electrons [7].

$$\Psi = \Phi S = \prod_i \varphi_i s_i \qquad \ldots (2.33)$$

The φ_i and the s_i are respectively the one-electron orbital wavefunctions and the spin wavefunctions. Let us consider a two-electron system with molecular orbitals φ_a and φ_b. The arguments 1 and 2 refer to electrons (1) and (2). We find four antisymmetrical product wavefunctions for the electronic configuration $\varphi_a \varphi_b$:

$$\Psi_S = \frac{1}{2} \left[\varphi_a(1)\varphi_b(2) + \varphi_a(2)\varphi_b(1) \right] \left[\alpha(1)\beta(2) - \alpha(2)\beta(1) \right] \quad \ldots (2.34)$$

$$\Psi_{T,+1} = \frac{1}{\sqrt{2}} \left[\varphi_a(1)\varphi_b(2) - \varphi_a(2)\varphi_b(1) \right] \left[\alpha(1)\alpha(2) \right] \qquad \ldots (2.35)$$

$$\Psi_{T,0} = \frac{1}{2} \left[\varphi_a(1)\varphi_b(2) - \varphi_a(2)\varphi_b(1) \right] \left[\alpha(1)\beta(2) + \alpha(2)\beta(1) \right] \quad \ldots (2.36)$$

$$\Psi_{T,-1} = \frac{1}{\sqrt{2}} \left[\varphi_a(1)\varphi_b(2) - \varphi_a(2)\varphi_b(1) \right] \left[\beta(1)\beta(2) \right] \qquad \ldots (2.37)$$

where Ψ_S is the singlet wavefunction and $\Psi_{T,+1}$, $\Psi_{T,0}$, $\Psi_{T,-1}$ are the triplet wavefunctions.

6 INTERACTION OF LIGHT WITH MOLECULAR SYSTEMS

6.1 MOLECULAR DIPOLES

A pair of $+q$ and $-q$ charges separated by a distance r is called an electric dipole. The magnitude μ of the dipole moment vector, given by

$$\mu = q \times r \qquad \ldots (2.38)$$

is expressed in C m or in debyes (D). For two charges $+e$ and $-e$ separated by 1 Å, the dipole moment is $\mu = 1.602 \times 10^{-19} \times 10^{-10}$ C m $= 4.8$ D.

6.2 TRANSITION MOMENT

As discussed in Section 1.1, light can be considered as an electromagnetic wave, i.e. an electric field vector $\mathbf{E}$ and a magnetic field vector $\mathbf{H}$. It may exert electric and magnetic forces on charged particles and magnetic dipoles. The total force exerted on an electron by an electromagnetic field is the sum of an electrical force eE and of a magnetic force eHv/c, where v is the velocity of the electron. The magnetic force is negligible [1c]. The electric field vector $\mathbf{E}(t)$ varies according to eq. (2.39)

$$\mathbf{E}(t) = \mathbf{E}_0 \sin 2\pi \nu t \qquad \ldots (2.39)$$

and generates a time-dependent force field $\mathbf{F} = e\mathbf{E}$. In molecules, electrons may be set into motion by the oscillating electric field which generates a transitory dipole moment in the molecule. The interaction of light with molecules may be viewed as an exchange of energy between a collection of oscillating dipoles (electrons) and an oscillating electric field [1c]. This interaction depends on resonance conditions. The law of energy conservation must be satisfied (eq. 2.11), i.e. the energy gap between two electronic states of the molecule is related to the frequency of oscillation of the electric field of the light wave. Electronic transitions under light excitation in the UV-visible give rise to oscillations of the electrons in chemical bonds in the range 10^{15}–10^{16} s^{-1}.

If initially for $E = 0$, the centers of positive and negative charges of a molecule coincide, they are displaced under the effect of the electric field E. The molecule becomes polarized and the induced dipole moment or transition dipole moment μ_i is a vector given by

$$\mu_i = \alpha E \qquad \ldots (2.40)$$

where α is the polarisability.

An example is the interaction of a light wave with a hydrogen atom. The lowest energy state is the 1s spherically symmetrical. The interaction of the electric field of the light leads to an excited atom which possesses an electron set into vibration in the 2p orbital along the electric vector direction [1c]. The strength of the interaction between E and an electron is related to the magnitude of the maximal charge separation, i.e. to the transition dipole moment μ_i. In the case of the diatomic H_2 molecule, the interaction of light converts an electron from a cylindrically symmetric σ orbital into a π orbital or a σ^* orbital [1c]. The $\sigma \longrightarrow \pi$ transition corresponds to an electronic oscillation perpendicular to the bond axis, and the $\sigma \longrightarrow \sigma^*$ transition to an electronic oscillation parallel to the bond axis [1c].

Let us now consider the interaction of the electric field vector of the wave with an oscillating electric dipole moment of a molecule [1e,1f,1h,6]. The frequency of oscillation is dictated by the molecular electronic structure. A simple two-state system can be used to describe the absorption transition between two electronic states of wavefunctions Ψ_m and Ψ_n, and energies E_m and E_n; m and n are different quantum numbers, $m < n$. The light interaction provides a time-dependent perturbation of the molecular Hamiltonian $\hat{H}^0$. The magnitude of the perturbation is the scalar product of the electric field and dipole moment vectors [1e,1h,6]

$$\hat{H}' = -\mu \cdot E \qquad \ldots (2.41)$$

INTERACTION BETWEEN LIGHT AND MATTER

The total Hamiltonian is thus

$$\hat{H} = \hat{H}^0 + \hat{H}' \qquad \ldots (2.42)$$

The electronic transition moment vector M_{mn} is related [1f,1h] to the electronic wavefunctions of the initial state Ψ_m and the final state Ψ_n by eqs. (2.43) and (2.44)

$$M_{mn} = \langle \Psi_n | \hat{\mu} | \Psi_m \rangle \qquad \ldots (2.43)$$

$$\hat{\mu} = -|e| \sum_i r_i + |e| \sum_\alpha Z_\alpha R_\alpha \qquad \ldots (2.44)$$

where $\hat{\mu}$ is the dipole moment operator, e represents the electron charge, r_i the position vector of the ith electron, R_α the position vector and Z_α the charge of the αth nucleus. The total absorption probability is proportional to $|M_{mn}|^2$, and the direction of the transition moment M_{mn} gives the polarization direction of the transition.

The probability of transition from the ground state Ψ_1 to the final state Ψ_2 is given by the coefficient B_{12} (Einstein transition probability of absorption):

$$B_{12} = (8\pi^3/3h^2c)g_2|M_{12}|^2 \qquad \ldots (2.45)$$

where g_2 is the degeneracy of the final state ($= 1$ for singlet–singlet transitions).

6.3 OSCILLATOR STRENGTH OF THE ABSORPTION BAND

If an electron excited by a light wave possessed perfect oscillating properties, its probability of excitation, the "oscillator strength" of the electron, would be unity. The relationship between the Einstein transition probability and the oscillator strength (integrated experimental molar extinction coefficient) [1h] is given by

$$f = \left[\frac{8\pi^2 mc}{3e^2 h} \right] \bar{\nu}_{12} |M_{12}|^2 \qquad \ldots (2.46)$$

where $\bar{\nu}_{12}$ is the average wavenumber of the absorption band. The term in brackets is 4.7×10^{-7} D^{-2} cm giving, for a transition at 16 000 cm^{-1} with an oscillator strength of unity, a transition dipole moment $|M_{12}|$ of about 11.5 D.

For some dye molecules (symmetrical polymethines and related compounds used as laser dyes and saturable absorbers), the one-dimensional free electron gas model of Kuhn [7–9] gives a satisfactory agreement between calculated and experimental values of the wavelength of the absorption maximum

$$\lambda_{\max} = \frac{8mc}{h} \frac{L^2}{K+1} \qquad \ldots (2.47)$$

where L is the length of the conjugated chain and K the number of π electrons. The oscillator strength f can also be calculated using this model and the transition moments X and Y along and normal to the long molecular axis which yield the orientation of the transition moment in the molecule

$$f = 2\frac{8m\pi^2}{3h^2}\Delta E(X^2 + Y^2) \qquad \ldots (2.48)$$

where ΔE is the energy of the transition. The oscillator strength can be calculated using the equation

$$f = 4.32 \times 10^{-9} \int_{\bar{\nu}_1}^{\bar{\nu}_2} \varepsilon(\bar{\nu})\,d\bar{\nu} \qquad \ldots (2.49)$$

The units of the constant term are $\mathrm{mol\,dm^{-3}\,cm^2}$.

6.4 THE FRANCK–CONDON PRINCIPLE

The absorption transition is a vertical process which preserves the geometry of the molecule. The separation of the electronic and nuclear wavefunctions presented in Section 2 (BO approximation) and based on a fast subsystem (electronic motions) and a slow one (nuclear motions) justifies the electronic "vertical" absorption transition. This is the Franck–Condon principle which postulates [2,3,5] that "when light is absorbed the nuclei can alter only their potential energy" [2], "the absorption of light merely substitutes a new law of nuclear interaction, say $V_2(r)$ for the old one $V_1(r)$" [3] and "immediately after the transition the nuclei will have the same separation as before" [5].

Physically, under light excitation with the frequencies needed for electronic excitation, electrons follow the changes of the oscillating electric field, whereas the inertia of the nuclei precludes any resonant excitation [1f].

6.5 SELECTION RULES

The efficiency of the absorption transition depends on the square of the transition moment as shown by eqs. (2.45) or (2.46), i.e. on the charge redistribution. The latter depends on the electronic orbitals involved. Although the calculations of the transition moments are complicated, they can be estimated using selection rules. A transition forbidden by a selection rule has no intensity but a real case is often more complicated. Forbidden transitions are weakly allowed and present some intensity in their absorption spectra.

The transition moment vector M_{mn} given by eq. (2.43) is related to the electronic spatial wavefunction Φ, the electronic spin wavefunction S and the nuclear wavefunction χ. Equation (2.43) should be replaced by

$$M_{mn} = \langle \Phi_n|\mu|\Phi_m\rangle\langle S_n|S_m\rangle\langle \chi_n|\chi_m\rangle \qquad \ldots (2.50)$$

The transition moment vector is thus the product of three terms, the electronic transition moment and two other overlap integrals concerning the electron spin wavefunctions and the nuclear wavefunctions. For the first term, the symmetry properties of the wavefunctions have to be considered.

The first term can be deduced using the symmetry properties of molecular orbitals and the group theory; $n \longrightarrow \pi^*$ transitions are symmetry-forbidden transitions. The symmetry of the n orbitals precludes the overlap with π orbitals. Conversely, the $\pi \longrightarrow \pi^*$ transition is symmetry-allowed. For $n \longrightarrow \pi^*$ transitions (from a n orbital of a keto group C=O to a π orbital, as in formaldehyde for example), the molar extinction coefficient ranges from 1 to 10^3 dm^3 mol^{-1} cm^{-1} and for $\pi \longrightarrow \pi^*$ transitions, between 10^3 and 10^6 dm^3 mol^{-1} cm^{-1}.

In the second term, we have seen in Section 5 that we have to consider two electronic spin wavefunctions α and β for each electron of the outermost pair of electrons involved in the transition. In the ground state, organic molecules are in the singlet state (spin wavefunction S_m). Upon excitation, the spin orientation is preserved (spin wavefunction S_n). These spin wavefunctions are given in eq. (2.34). The relations $\langle\alpha|\alpha\rangle = \langle\beta|\beta\rangle = 1$ and $\langle\alpha|\beta\rangle = \langle\beta|\alpha\rangle = 0$ lead to $\langle S_n|S_m\rangle = 1$, showing that the singlet–singlet transition is spin allowed [1f]. Similarly, it is easy to show that the singlet–triplet transition is strictly forbidden since $\langle S_n|S_m\rangle = 0$ (orthogonality of the spin wavefunctions). Triplet–triplet transitions are spin-allowed and can be observed after photoexcitation of organic molecules. Spin–orbit coupling enables the breakdown of selection rules and spin-forbidden transitions to become weakly allowed. Spin–orbit coupling and mixing of the singlet and triplet excited states increase with the nuclear charge. This effect is known as the internal heavy atom effect in organic molecules containing Br or I atoms, and the external heavy atom effect when the heavy atoms belong to neighboring molecules incorporated into the solvent (potassium iodide, KI or xenon rare gas, Xe).

The third term is the nuclear overlap integral also called Franck–Condon overlap integral, the square of which is the Franck–Condon factor $\langle\chi_n|\chi_m\rangle^2$. A good illustration of the Franck–Condon factors and the spectral intensities of the electronic transitions from the lowest vibrational level ($v = 0$) of the singlet electronic ground state ($m = 0$) to vibrational levels v of the first electronically excited singlet state ($n = 1$) is given in Figure 2.6. In the ground electronic state at room temperature, the $v = 0$ vibrational level is indeed favored by the Boltzmann distribution. The overlap of the $\chi_{m=0,v=0}$ and $\chi_{n=1,v}$ nuclear wavefunctions depends on the relative position of the potential energy surfaces S_0 and S_1. The 'vertical' absorption transition is consistent with the inertia of the nuclei.

For molecules with similar nuclear geometries and potential energy surfaces identical in shape and equilibrium geometry, the overlap integral is large for the $\chi_{m=0,v=0}$ and $\chi_{n=1,v=0}$ vibrational wavefunctions as shown in Figure 2.6A. The $0 \longrightarrow 0$ transition is Franck–Condon allowed and appears with a high intensity in the absorption spectrum. In Figure 2.6B, the displacement of the potential energy curve of the electronically excited singlet state S_1 to larger equilibrium internuclear distances compared to the ground state is due to the weakening of

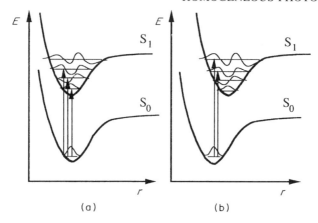

Figure 2.6. Quantum mechanical interpretation of the Franck–Condon principle. Electronic transition between states of similar equilibrium geometry (A), between states of different equilibrium geometry (B)

the bond strength. The largest overlap occurs for the vibrational levels $v \geq 2$ of the S_1 state. However, the greater the difference between the vibrational quantum numbers of the nuclear wavefunctions, the smaller is the Franck–Condon overlap and the more difficult the transition.

7 NONADIABATIC REACTIONS

Many reactions can be interpreted in terms of the motion of the system along a potential-energy surface. In a *nonadiabatic* reaction, a transition occurs between two potential-energy surfaces. Photoreactions leading to ground-state molecules are nonadiabatic since photoexcitation gives an excited molecule characterized by a different potential-energy surface, i.e., a transition has occurred. Two classes of nonadiabatic transitions can be distinguished, radiative (fluorescence, phosphorescence) and radiationless (internal conversion, intersystem crossing,...).

7.1 DEEXCITATION PROCESSES OF ELECTRONICALLY EXCITED MOLECULES

Electronically excited molecules tend to return to the ground state via radiative processes which involve the emission of light and nonradiative (or radiationless) processes. The two types of radiative transitions are fluorescence from an excited singlet state of the same multiplicity as the lower state (for example, the $S_1 \longrightarrow S_0 + h\nu'$ in an organic molecule) and phosphorescence from an excited triplet state of a different multiplicity ($T_1 \longrightarrow S_0 + h\nu''$). Fluorescence is spin-allowed so that it occurs on a fast time scale (10^{-12}–10^{-6} s). Phosphorescence is spin-forbidden and occurs on a slower time scale (10^{-6} to 1 s). Spin–orbit interactions may invalidate the spin-forbidden character of phosphorescence.

INTERACTION BETWEEN LIGHT AND MATTER

Fluorescence and phosphorescence are vertical and follow the Franck–Condon principle since electron motion is more rapid than nuclear motion.

Following the light excitation of a molecule A (reaction 2.10), the photophysical processes are written:

$$A^*(S_1) \longrightarrow A(S_0) + h\nu' \quad \text{(fluorescence)} \quad \ldots (2.51)$$

$$A^*(S_1) \longrightarrow A(S_0) \quad \ldots (2.52)$$

$$A^*(S_1) \longrightarrow A^*(T_1) \quad \ldots (2.53)$$

$$A^*(S_1) + Q \longrightarrow A(S_0) + Q \quad \text{(quenching)} \quad \ldots (2.54)$$

The nonradiative deactivation processes are called *internal conversion* (2.52) between states of the same multiplicity and (2.53) with change of the multiplicity (*intersystem crossing*). Reaction (2.54) is a quenching process which involves the transfer of energy from the initially excited molecule to another molecule. If we consider the excited molecule A^* to be deactivated by the first-order processes (eqs. 2.51–2.53) and by the pseudo-first order process (2.54), the rate of decay of A^* is given by

$$-\frac{d}{dt}[A^*] = k_0[A^*] + k_{ic}[A^*] + k_{isc}[A^*] + k_Q[Q][A^*] \quad \ldots (2.55)$$

where k_0 is the radiative fluorescence rate constant ($k_0 = \tau_0^{-1}$, τ_0 is the radiative lifetime), k_{ic} the rate constant of internal conversion, k_{isc} the rate constant of intersystem crossing, k_Q the bimolecular (or second order) rate constant and $[Q]$ the concentration of the quencher. The A^* population decays exponentially to zero with a rate constant k_F according to

$$[A^*] = [A^*]_0 \exp(-k_F t) \quad k_F = \frac{1}{\tau_F} = \sum_i k_i + k_Q[Q] \quad \ldots (2.56)$$

where τ_F is the experimentally measured lifetime in a time-correlated single photon counting experiment, for example [10].

The quantum yield of each first-order process is given by

$$\phi_i = \frac{k_i}{\sum_i k_i + k_Q[Q]} \quad \ldots (2.57)$$

and $\sum_i k_i$ is the measured decay rate constant in the absence of quencher. The fluorescence quantum yield ϕ_F defined [11] as the ratio of the number of emitted photons to the number of absorbed quanta given by eqs. (2.5) and (2.8):

$$\phi_F = \frac{\int I_F(\bar{\nu}) \, d\bar{\nu}}{I_E(1 - 10^{-\varepsilon C d})} \quad \ldots (2.58)$$

where I_E is the intensity of exciting radiation in quanta per second and $I_F(\bar{v})$ the fluorescence intensity in quanta per second and per wavenumber interval, thus reduces to

$$\phi_F = \frac{k_0}{k_F} = \frac{\tau_F}{\tau_0} \qquad \ldots (2.59)$$

7.2 RELATIONSHIP BETWEEN ABSORPTION INTENSITY AND FLUORESCENCE LIFETIME

The relationship between the transition probabilities for induced absorption, induced (or stimulated) emission and spontaneous emission was derived by Einstein in 1917 using the radiation density $\rho(v)$ (erg cm^{-3} per unit frequency) within the medium given by Planck's blackbody radiation law (eq. 2.60).

$$\rho(v) = (8\pi h v^3 n^3 / c^3)[\exp(hv/kT) - 1]^{-1} \qquad \ldots (2.60)$$

The spontaneous emission probability A_{21} is related to the probability $B_{21}\rho(\bar{v}_{21})$ of induced emission ($B_{21} = B_{12}$), i.e. to the third power of the frequency of the transition. In solutions, a correction for the refractive index was first added by Perrin [12].

$$A_{21} = 8\pi h n^3 \bar{v}_{21}^3 B_{12} \qquad \ldots (2.61)$$

The spontaneous emission probability A_{21} which is the reciprocal of the radiative lifetime τ_0 can be written according to the Strickler–Berg eq. (2.62):

$$A_{21} = 1/\tau_0 = 8 \times 2303 \pi c N^{-1} n^2 \frac{\int I(\bar{v}) \, d\bar{v}}{\int \bar{v}^{-3} I(\bar{v}) \, d\bar{v}} \frac{g_1}{g_2} \int \varepsilon(\bar{v}_{12}) \, d(\ln \bar{v}_{12}) \qquad \ldots (2.62)$$

where the ratio of the integrals over the fluorescence spectrum is the reciprocal of the mean value of $\bar{v}_{21}^{-3}$ in the fluorescence spectrum, g_1 and g_2 are the degeneracies of the ground state 1 and upper state 2. Equation (2.62) is considered valid for broad molecular bands when the transition is strongly allowed [12]. If the fluorescence band is sharp and the absorption and fluorescence transitions occur at the same wavenumber, eq. (2.62) reduces to eq. (2.63), valid for most atomic transitions [12]

$$A_{21} = 1/\tau_0 = 8 \times 2303 \pi c N^{-1} n^2 \bar{v}_{21}^2 \frac{g_1}{g_2} \int \varepsilon(\bar{v}_{12}) \, d\bar{v}_{12} \qquad \ldots (2.63)$$

The coefficient A_{21} and the radiative lifetime τ_0 can be determined experimentally from the measurements of the fluorescence quantum yield ϕ_F and lifetime τ_F (eq. 2.59).

In fact, the fluorescence quantum yield (eq. 2.59) is given by a more general equation which considers the formation efficiency η of the fluorescing state:

$$\phi_F = \eta \times \tau_F \times 1/\tau_0 \qquad \ldots (2.64)$$

INTERACTION BETWEEN LIGHT AND MATTER

Table 2.1. Recommended fluorescence quantum yields and lifetimes [13,14]

Fluorescence quantum yields (Φ_F)			
Region (nm)	Compound	Solvent	Φ_F
300–400	Naphthalene	Cyclohexane	0.23 ± 0.02
360–480	Anthracene	Ethanol	0.27 ± 0.03
400–600	Quinine bisulfate	1 N H_2SO_4	0.546
600–650	Rhodamine 101	Ethanol	1.0 ± 0.02
Fluorescence lifetimes (τ_F)			
Fluorescence wavelength (nm)	Compound	Solvent	τ_F (ns)
320	Naphthalene	Cyclohexane	100
>400	Anthracene	Ethanol	5.1
460	Coumarin 450	Ethanol	4.3 ± 0.2
600	Rhodamine B	Ethanol	2.85

Following the recommendations of IUPAC [13,14], some reference materials for the measurements of fluorescence quantum yields and fluorescence lifetimes are given in Table 2.1.

7.3 MIRROR SYMMETRY

Both the absorption spectrum and the fluorescence spectrum of an organic compound may exhibit features which can be assigned to Franck–Condon vibronic transitions.

When the nuclear configuration of the excited state is not too different from that of the ground state, the 0 ⟶ 0 vibronic transition is the most intense (Section 6.5). However, photoexcitation to higher vibrational levels $v = 1, 2, 3,$ ... of the electronically excited state is possible. In solution, vibrational relaxation is very rapid through collisions with solvent molecules, leading to the $v = 0$ vibrationally relaxed energy level of the fluorescent excited state (Figure 2.7A).

Similarly, the emission intensity is greatest for the 0 ⟶ 0 fluorescence transition, but vibrationally excited $v = 1, 2, 3, \ldots$ energy levels of the electronic ground state can be reached. The absorption spectrum and the fluorescence spectrum appear symmetrical and a mirror can be set at the 0 ⟶ 0 energy. The *Stokes shift* (or Stokes loss) is the difference between the wavenumbers of the absorption and fluorescence maxima. Naphthalene and anthracene have a small Stokes shift [15]. As an example, the corrected and normalized excitation and fluorescence spectra of rhodamine 6G in ethanol presented in Figure 2.8 [16] display a 800 cm^{-1} Stokes shift. A noticeable difference between the 0 ⟶ 0 absorption and fluorescence transitions of molecules may be due to the different solute–solvent interactions in the two electronic states.

When the nuclear configuration of the excited state is different from that of the ground state, the absorption maximum is at a higher energy than the 0 ⟶ 0

36   HOMOGENEOUS PHOTOCATALYSIS

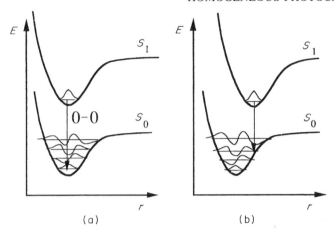

Figure 2.7. Quantum mechanical interpretation of the Franck-Condon principle. Fluorescence transition between states of similar equilibrium geometry (a), between states of different equilibrium geometry (b)

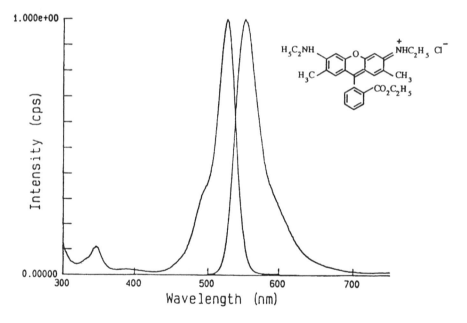

Figure 2.8. Corrected and normalized excitation and fluorescence spectra of rhodamine 6G in ethanol solution. The wavelengths of the maxima are: excitation (or absorption) 530 nm, fluorescence 553 nm [16]

INTERACTION BETWEEN LIGHT AND MATTER 37

transition and the fluorescence maximum is at a lower energy (Figure 2.7B). Similarly to the absorption transition (see Section 6.5), the larger the difference of the vibrational quantum numbers, the smaller is the Franck–Condon overlap and the less efficient is radiative transition.

In polar solvents, the evolution of the fluorescence spectrum of solvated molecules following an instantaneous change in their charge or dipole moment can be monitored using picosecond and femtosecond laser techniques [17–29]. Molecular dynamic simulations and femtosecond fluorescence up-conversion studies of dipolar solvation in polar solvents have recently revealed an ultrafast <150 fs inertial component which contributes >50% of the solvent relaxation [18,21].

For example, the DCM (4-dicyanomethylene-2-methyl-6-p-dimethyl-aminostyryl-4H-pyran) laser dye presents in a polar solvent such as methanol, a large Stokes shift of its fluorescence spectrum which is due to solvent relaxation [22–29]. The peak of intensity shifts at early times from 560 nm to about 600 nm within 2 ps and reaches the steady-state position at 630 nm after ≈ 20 ps [28,29]. A three-dimensional picture of the time-dependent fluorescence spectra of DCM is shown in Figure 2.9. These spectra recorded using the "upconversion" or "sum-frequency generation" technique [19] are corrected for the spectral response of the detection system. In methanol, the mean frequency shifts to the red biexponentially (175 fs and 3.2 ps) as shown in Figure 2.10 [28]. In chloroform, the dynamic Stokes shift is negligible [28]. It is also interesting to notice a spectral narrowing with a 10 ps time constant in methanol (7 ps in chloroform) [28] due to vibrational energy dissipation from the excited solute to the surrounding solvent [30–32].

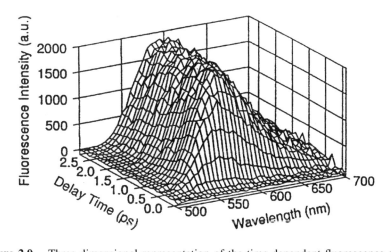

Figure 2.9. Three-dimensional representation of the time-dependent fluorescence spectrum of DCM in methanol with a time-step of 100 fs [29]

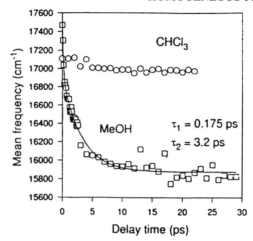

Figure 2.10. The temporal evolution of the mean frequency of the DCM fluorescence spectrum in methanol and chloroform [28]

7.4 RADIATIONLESS PROCESSES

Transitions which do not give emission are radiationless (nonradiative) processes: vibrational relaxation, energy transfer, inelastic collisions. Whereas radiative processes are vertical, radiationless processes are "horizontal", i.e. they occur between isoenergetic quantum states. We have seen (Section 7.1) that internal conversion (i.e., reaction 2.52) is a radiationless transition between electronic states of the same spin multiplicity and that, conversely, intersystem crossing (i.s.c., reaction 2.53) involves electronic states of different multiplicity. Because the transition is isoenergetic, the $S_1(v=0) \longrightarrow S_0(v=n)$ process leads to a vibrationally excited electronic ground state. The excess vibrational energy of the "hot" molecule is removed by rapid collisions (10^{13} s^{-1}) with the surrounding solvent molecules, a process called *vibrational relaxation*. Radiationless processes are thus characterized by the degradation of a single large quantum in many small quanta. The inverse also occurs in the case of thermal excitation or anti-Stokes fluorescence. Fluorescence spectroscopy is a power tool for studying radiative transitions; absorption spectroscopy and photoacoustic spectroscopy are the techniques of choice for radiationless transitions.

The theory of radiationless processes is based on the adiabatic approximation for the separation of "fast" and "slow" modes. A further separation may be based on the orbital states of the electrons "fast", vibrons (vibrational and rotational modes of molecules and local modes) "intermediate" and phonons (normal modes) "slow" [33]. The radiationless transition is a transfer of energy between a fast subsystem and a slow subsystem.

Some types of radiationless transitions are presented in Figure 2.11, a non-activated electronic deexcitation (Figure 2.11a), an electronic deexcitation

INTERACTION BETWEEN LIGHT AND MATTER 39

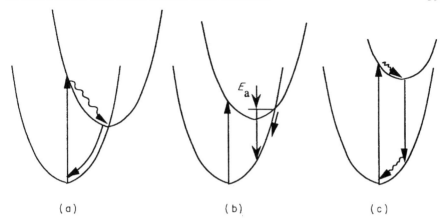

Figure 2.11. Radiationless transitions

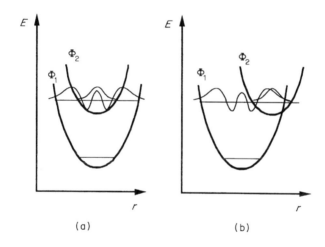

Figure 2.12. Schematic representation of radiationless transitions. Overlap of vibrational wavefunctions: (a) with surface noncrossing, (b) with surface crossing

activated with activation energy E_a (Figure 2.11b) and a radiationless relaxation after a radiative absorption transition or a radiative fluorescence transition (Figure 2.11c) [33].

A quantum mechanical visualization of radiationless transitions in terms of the net positive overlap of the nuclear wavefunctions χ_2 and χ_1 given by the Franck–Condon integral $\langle \chi_2 | \chi_1 \rangle$ is shown in Figure 2.12. The "vertical jump" from the initial excited surface of electronic spatial wavefunction Φ_2 to a lower surface of wavefunction Φ_1 refers to electronic motion. The "horizontal jump" is due to some overlap of χ_2 and χ_1 [1c]. It strongly affects the nuclear motion of the molecule. The radiationless transition between noncrossing (or "matching" [1c]) surfaces involves for the state Φ_1 a vibrational wavefunction which oscillates between negative and positive values, whereas the vibrational wavefunction of the state Φ_2 is always positive (Figure 2.12a). Due to the resulting poor overlap of χ_2 and χ_1, the kinetics of the radiationless transition will be very slow. In the case of surface crossing (Figure 2.12b), a fast radiationless transition at geometries near the zero order intersection point can occur without a strong modification of the momentum of the nuclei. Organic molecules with a rigid structure and a long conjugated chain of alternated single and double bonds are often highly fluorescing molecules. This results from the molecular rigidity which precludes the motion of the nuclei: internal conversion $S_1 \longrightarrow S_0$ and intersystem crossing $S_1 \longrightarrow T_1$ are thus very slow.

The probability of the transition between two states ("state-to-state transition" probability) is given by Fermi's golden rule:

$$k_{nr} = (2\pi/h)\langle\Psi_2|H'|\Psi_1\rangle^2\langle\chi_2|\chi_1\rangle^2 \qquad \ldots(2.65)$$

where H' represents the perturbation which couples the initial and final states. The $\langle\Psi_2|H'|\Psi_1\rangle$ term is much smaller for intersystem crossing compared to internal conversion due to the different spin multiplicity (see eq. 2.33). The consideration of the symmetry properties of electronic wavefunctions leads to El-Sayed's rules given in Table 2.2 [1h,34]: "The rate of the intersystem crossing process between two states of different origin (e.g. (n, π^*) and (π, π^*)) is found to be 1000 times faster than that between two states of the same electronic origin (e.g. $(\pi, \pi^*) \longrightarrow (\pi, \pi^*)$ or $(n, \pi^*) \longrightarrow (n, \pi^*)$)."

For matching surfaces, the Franck-Condon factor $\langle\chi_2|\chi_1\rangle^2$ depends on the energy gap between the states, as shown by eq. (2.66):

$$\langle\chi_2|\chi_1\rangle^2 = \alpha \exp(-\Delta E) \qquad \ldots(2.66)$$

The higher the energy gap, the higher the vibrational quantum number in the electronic ground state and the smaller the Franck-Condon overlap. As a result, the Franck-Condon factor also depends on the vibrational energy spacings: the higher the vibrational energy spacing, the smaller the vibrational quantum number and the better the Franck-Condon overlap. An example is the effect of high-energy vibrations, C–H and O–H stretching in molecules, which favor deactivation via energy transfer.

Radiationless processes among excited states $S_n \longrightarrow S_m$ are much more efficient than the $S_1 \longrightarrow S_0$ transition because the high excited states are more

Table 2.2. El-Sayed's rules for intersystem crossing transitions

Forbidden transitions	Allowed transitions
$S_1(\pi, \pi^*) \longrightarrow T_1(\pi, \pi^*)$	$S_1(\pi, \pi^*) \longrightarrow T_1(n, \pi^*)$
$S_1(n, \pi^*) \longrightarrow T_1(n, \pi^*)$	$S_1(n, \pi^*) \longrightarrow T_1(\pi, \pi^*)$

INTERACTION BETWEEN LIGHT AND MATTER 41

closely spaced with frequent crossing situations and they have more distorted geometries.

Certain vibrations are acceptors of the energy difference between the electronic states involved. The stretching or the twisting of a bond may mix the Φ_2 and Φ_1 wavefunctions and induce a radiationless transition at some critical value of the coordinate. The stretching vibration is analogous to a "loose bolt", the twisting motion to a "free rotor". The fluorescence efficiency of stilbene and styrene derivatives can be explained using the free rotor scenario [35].

8 INTERMOLECULAR PROCESSES

The representation of potential energy surfaces in the space of nuclear configuration is the same for one molecule or several. Two parts of a single molecule may be viewed as two separate molecules, i.e. intramolecular energy or charge transfer from one chromophore to the other can be treated as if they were intermolecular. A significant difference between inter- and intramolecular processes is the small degree of overlap between the encounter partners. Another difference is that intermolecular processes involve relatively slow rotational and translational motions. We shall focus our attention to bimolecular processes.

Returning to the quenching process (eqs. 2.54–2.57), the fluorescence quantum yield given by eq. (2.57) is reduced from the value ϕ_F^0 measured in the absence of quencher ([Q] = 0) to the value ϕ_F measured in the presence of quencher. The ratio of the two values is given by the Stern–Volmer equation,

$$\frac{\phi_F^0}{\phi_F} = \frac{k_0 / \sum_i k_i}{k_0 / \left(\sum_i k_i + k_Q[Q] \right)} = 1 + \frac{k_Q[Q]}{\sum_i k_i} \qquad \ldots (2.67)$$

which can be treated using a Stern–Volmer plot of the ratio of the measured fluorescence intensities or lifetimes as a function of the concentration of the quencher. The bimolecular rate constant k_Q can be calculated from the slope of the straight line and the measured fluorescence lifetime in the absence of quencher. In solution, k_Q is limited to an upper value which is diffusion controlled ($k_{\text{diff}} = 10^9$–10^{10} M^{-1} s^{-1} in nonviscous solvents). The Stern–Volmer relation has been recently generalized to describe systems in which the decay process is not a simple exponential and systems in which the quenching must be described with a time-dependent rate constant [36]. However, the Stern–Volmer plot is no longer valid if a static quenching intervenes, i.e. when a complex is formed between the ground state A and the quencher.

A simplified representation of the potential energy surfaces involved in a collisional energy transfer [1c] is shown in Figure 2.13. The energy of the encounter pair A*Q or AQ* is generally lower than that of the separated pairs A* + Q and

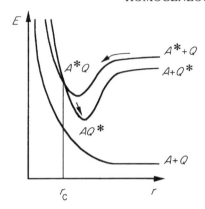

Figure 2.13. Schematic representation of collisional energy transfer. A*Q and AQ* are exciplexes

$A + Q^*$. The energy transfer occurs at the crossing point followed by a relaxation leading to the minimum (AQ*) of the lower surface. (AQ*) may eventually separate into the reaction products $A + Q^*$.

Examples of bimolecular quenching reactions are the quenching via external heavy atom effect and molecular oxygen. In the presence of a quencher containing a heavy atom, the formation of an exciplex may occur and lead to an intersystem crossing to the triplet state via internal heavy atom effect.

$$^1A^* + Q \longrightarrow {}^1(AQ)^* \longrightarrow {}^3(AQ)^* \longrightarrow {}^3A + Q \quad \ldots (2.68)$$

The quenching of excited singlet states of aromatic hydrocarbons by molecular oxygen [37] which is a triplet ground state ($^3\Sigma_g$) may intervene and lead to intersystem crossing to the triplet state of A. It is called "oxygen-induced intersystem crossing" (reactions 2.69 and 2.71) [38–42]. Although both the quenching reactions of singlet and triplet states by molecular oxygen with singlet oxygen formation are spin-allowed processes, triplet state quenching more often results in $^1O_2(^1\Delta_g)$ formation (reaction 72) [38,40–42] because the energy difference $\Delta E(S_1 \longrightarrow T_1)$ (reaction 2.71) rarely exceeds the energy of singlet oxygen (7880 cm^{-1}). Moreover, when the energy difference is large enough, the lifetimes of the S_1 states are too short [38]. In the case of 9,10-dicyanoanthracene [40], perylene and pyrene [42], $^1O_2(^1\Delta_g)$ is directly generated from the first excited singlet state. The electron transfer reaction (2.73) also competes in the triplet quenching [38].

$$^1A^* + {}^3O_2 \longrightarrow {}^3A^* + O_2 \quad \ldots (2.69)$$

$$^3A^* + {}^3O_2 \longrightarrow {}^1A + O_2 \quad \ldots (2.70)$$

$$^1A^* + {}^3O_2 \longrightarrow {}^3A^* + {}^1O_2^* \quad \ldots (2.71)$$

INTERACTION BETWEEN LIGHT AND MATTER

$$^3A^* + {}^3O_2 \longrightarrow {}^1A + {}^1O_2^* \qquad \ldots (2.72)$$

$$^3A^* + {}^3O_2 \longrightarrow A^{+\cdot} + O_2^{-\cdot} \qquad \ldots (2.73)$$

The most important bimolecular quenching processes in photochemistry are energy transfer and electron transfer from a donor (D) to an acceptor (A) molecule. The energy level of A* must be lower than that of D*. Singlet–singlet and triplet–triplet energy transfer processes are both spin-allowed processes. Energy transfer is important for sensitized reactions. Reaction 2.74, called "photosensitization", widens the applications of organic photochemistry. For example, triplet–triplet energy transfer is used to generate triplet excited states of molecules which possess a low quantum yield of intersystem crossing (i.s.c.) to the triplet state $S_1 \longrightarrow T_1$. The donor will be a molecule chosen for its high i.s.c. efficiency.

$$D^* + A \longrightarrow D + A^* \qquad \ldots (2.74)$$

Electron transfer reactions (2.75) and (2.76) involve either oxidation or reduction of the excited state.

$$D^* + A \longrightarrow D^{+\cdot} + A^{-\cdot} \qquad \ldots (2.75)$$

$$D + A^* \longrightarrow D^{+\cdot} + A^{-\cdot} \qquad \ldots (2.76)$$

8.1 ENERGY TRANSFER

Different types of energy transfer are distinguished: radiative energy transfer and nonradiative energy transfer. In a radiative energy transfer, emission of light by the donor is reabsorbed by the acceptor. A simple requirement is that the emission spectrum of D* and the absorption spectrum of A* overlap. Singlet–triplet and triplet–triplet radiative energy transfer are spin forbidden as the $S_0 \longrightarrow T_1$ absorption transition, but singlet–singlet and triplet–singlet are spin allowed. The efficiency of a radiative energy transfer depends on the fluorescence or phosphorescence quantum yield of the donor, the concentration of the acceptor and the oscillator strength of its absorption transition.

In a nonradiative energy transfer, two mechanisms may be considered which involve different types of interaction between the donor and the acceptor: *coulombic energy transfer* (also called "induced dipole mechanism", which extends over long distances up to 100 Å) and *electron-exchange energy transfer* (also called "collisional mechanism" via orbital overlap, i.e. at short distances, 6–20 Å). These mechanisms also have specific selection rules. In a coulombic energy transfer, the "virtual" transitions of the donor and the acceptor must be spin allowed as in reaction (2.77). In an exchange energy transfer, the total spin must be conserved as in reaction (2.78).

$$D^*(S_1) + A(S_0) \longrightarrow D(S_0) + A^*(S_1) \qquad \ldots (2.77)$$

$$D^*(T_1) + A(S_0) \longrightarrow D(S_0) + A^*(T_1) \qquad \ldots (2.78)$$

In a coulombic energy transfer, the dipole oscillation of the donor D* induces a dipole oscillation of the acceptor, but the electrons of D* remain on D after the energy transfer. In an electron-exchange energy transfer, the excited electron of D* is transferred into the lowest unoccupied molecular orbital (LUMO) of A, whereas an electron from the highest occupied molecular orbital (HOMO) of A is transferred into the corresponding orbital of D. This simultaneous exchange of two electrons requires overlap of the orbitals.

Energy transfer involves two Franck-Condon transitions. The energy transfer is vertical in both partners. The rate of energy transfer falls off if the positions of the nuclei have to change during the transfer [1].

As in eq. (2.65), the probability of energy transfer can be written in the form of Fermi's golden rule:

$$k(D^*A \longrightarrow DA^*) = \frac{2\pi}{h} \rho \langle \Psi_1 | H' | \Psi_2 \rangle^2 \qquad \ldots (2.79)$$

where ρ is the density of initial and final states capable of interaction. The Hamiltonian H' can be broken in two Hamiltonians H_c for the coulombic energy transfer and H_e for the electron-exchange energy transfer [1c,1h].

8.1.1 Förster Theory of Coulombic Energy Transfer

In the coulombic mechanism, the dipole–dipole interaction energy is proportional to the magnitude of the dipoles of the donor D and the acceptor A (μ_D and μ_A) and inversely proportional to the third power of the distance between both molecules [43]:

$$E = \frac{\mu_D \mu_A}{R^3} \qquad \ldots (2.80)$$

where μ_D and μ_A have been identified by Förster with the oscillator strengths for the radiative transitions $D^* \longrightarrow D$ and $A \longrightarrow A^*$, and the rate constant of energy transfer is proportional to the square of the interaction energy [43–45].

$$k_{ET} = \frac{9000 \ln 10 \kappa^2}{128 \pi^6 n^4 N \tau_D^0 R^6} \int_0^\infty f_D(\nu) \varepsilon_A(\nu) \frac{d\nu}{\nu^4} \qquad \ldots (2.81)$$

where $f_D(\nu)$ is the fluorescence spectral distribution of the donor, τ_D^0 its radiative lifetime and κ an orientation factor given by eq. (2.82):

$$\kappa = \cos \alpha_{DA} - 3 \cos \alpha_D \cos \alpha_A \qquad \ldots (2.82)$$

where α_{DA} is the angle between the transition moments vectors of D and A, α_D and α_A are the angles between these vectors and the direction $D \longrightarrow A$. For a random distribution, the average value of κ^2 is $\frac{2}{3}$. The transfer process is resonant, i.e. the two transitions occur simultaneously.

INTERACTION BETWEEN LIGHT AND MATTER

The efficiency of energy transfer is measured by defining a critical transfer distance R_0 for which excitation transfer and spontaneous deactivation of the donor have equal rates:

$$k_{ET}[D^*][A] = k_D[D^*] \quad \text{or} \quad k_{ET}[A] = k_D = 1/\tau_D \qquad \ldots (2.83)$$

where τ_D is the experimental lifetime of the donor given by eq. (2.59) (without transfer), i.e. $\tau_D = \phi_D \times \tau_D^0$. By varying the concentration of the acceptor, one finds the value k_{ET} for which eq. (2.83) holds and thus the value of R_0, which increases with the fluorescence quantum yield of the donor and the overlap of the spectra:

$$R_0^6 = \frac{9000 \ln 10 \kappa^2 \phi_D}{128\pi^6 n^4 N} \int_0^\infty f_D(\nu)\varepsilon_A(\nu)\frac{d\nu}{\nu^4} \approx \frac{9000 \ln 10 \kappa^2 \phi_D}{128\pi^6 n^4 N \bar{\nu}^4} \int_0^\infty f_D(\nu)\varepsilon_A(\nu)\,d\nu \qquad \ldots (2.84)$$

For any separation distance of the donor and the acceptor, the rate constant for energy transfer is thus given by

$$k_{ET} = \frac{1}{\tau_D}\left(\frac{R_0}{R}\right)^6 \qquad \ldots (2.85)$$

The time-dependent fluorescence decay of the donor becomes [45]

$$I(t) = I(0)\exp\{-[t/\tau_D + 2\gamma(t/\tau_D)^{1/2}]\} \qquad \ldots (2.86)$$

where $\gamma = C_A/C_A^0$ is the ratio of the acceptor concentration C_A to the acceptor critical concentration $C_A^0 = 1500/\pi^{3/2} N R_0^3$.

8.1.2 Dexter Theory of Electron-exchange Energy Transfer

In the case of collisions, the electron orbitals of the reacting partners may overlap and electron exchange may occur. The upper limit of the rate constant of energy transfer is thus "diffusion-controlled". A simplified form of the equation is

$$k_{\text{diff}} = \frac{8RT}{3000\eta} \qquad \ldots (2.87)$$

where η is the viscosity of the solution, T the temperature in K and R the gas constant.

The rate constant of energy transfer falls off exponentially with the separation distance of the donor and the acceptor (short-range process) and is also related to the spectral overlap, as represented by the theory of Dexter

$$k_{ET} = K^2 \frac{2\pi}{\hbar} \exp(-2R_{D^*A}/L) \int_0^\infty f_D(\nu)\varepsilon_A(\nu)\,d\nu \qquad \ldots (2.88)$$

where K is a constant related to the interaction of the orbitals involved and L the "effective average Bohr radius" [46]. The absorption spectrum of the acceptor and the fluorescence spectrum of the donor are normalized to unity, i.e. the rate constant does not depend on the oscillator strength of the acceptor. The expression for the donor emission intensity as a function of time was given by Inokuti and Hirayama [47]:

$$I(t) = \exp\left[-\frac{t}{\tau_D} - \frac{C}{C_0}\frac{1}{\gamma^3}g\left(\exp(\gamma)\frac{t}{\tau_D}\right)\right] \quad \ldots (2.89)$$

where $\gamma = 2R_{D^*A}/L$, τ_D is the deactivation rate of the donor in the absence of the acceptor and $g(z)$ a convergent Taylor series.

8.1.3 Energy Transfer in Disordered Materials and Restricted Geometries

Energy transfer has been extensively applied to random materials and restricted geometries. Klafter and co-workers [48, 49] have proposed a general Förster equation for the survival probability of the excited donor in an "exact" fractal medium

$$\ln[\Phi(t)] = -\frac{t}{\tau_D} - A\Gamma\left(1 - \frac{\bar{d}}{s}\right)\left(\frac{t}{\tau_D}\right)^{\bar{d}/s} \quad \ldots (2.90)$$

where $\Gamma(x)$ is the gamma Euler function, $\bar{d}$ the fractal dimensionality of the medium (the Hausdorff fractal dimension) and s the exponent of the mutual distance between the donor and the acceptor (s equals 6 for dipolar interactions). For example, for cationic porphyrins adsorbed on the negatively charged surface of vesicles, the fluorescence decay function derived for multistep excitation migration on fractal and fast trapping at the first trap encounter can reproduce the observed decays [50].

8.2 ELECTRON TRANSFER

The importance of electron-transfer reactions in photochemistry is well recognized, as shown by excellent reviews [51–53]. A classification of electron-transfer processes based on the charges present on the electron donor and the electron acceptor as reactants can be proposed [1f]:

1. Charge recombination of a negatively charged donor and a positively charged acceptor;
2. Charge separation, the donor and the acceptor being initially neutral;
3. Charge translocation (charge shift) involving a negatively charged donor and a neutral acceptor or a neutral donor and a positively charged acceptor.

8.2.1 Classical Theories

In his Nobel lecture (1983), Henry Taube [54] classified chemical reactions into two categories: substitution and oxidation-reduction, the latter involving electron transfer. Two mechanisms are operative for electron-transfer reactions of metal complexes [55,56]: the *inner-sphere mechanism* in which the transition state involves interpenetration of the coordination spheres and a bridging ligand and the *outer-sphere mechanism* in which the two coordination spheres remain separated. The theory of adiabatic electron transfer for exchange between ions in an ionizing solvent outlined by Hush [56] is applicable to processes in which the coordination shells around the ions are not disrupted on electron transfer. Most of the research of Hush's group has been done on systems reacting by inner-sphere mechanisms. In the outer-sphere mechanism, the interaction of the electronic orbitals of the two centers is weak. Marcus introduced a reorganization of the coordination shells prior to and following the electron transfer [55,57–60]. In solution, a bimolecular collision between the excited donor and the acceptor may form with the diffusion rate constant k_{12} a close-contact encounter complex which may then dissociate with the dissociation rate constant k_{21}. If electron transfer occurs with a first-order rate constant k_{23} and if the reverse electron transfer rate is negligible relative to the rate of separation of products, the observed bimolecular rate constant k_q for electron transfer (fluorescence quenching) [57,58,61,62] is given by

$$D^* + A \underset{k_{21}}{\overset{k_{12}}{\rightleftarrows}} D^* \ldots A \underset{k_{32}}{\overset{k_{23}}{\rightleftarrows}} D^+ \ldots A^- \longrightarrow D^+ + A^- \quad \ldots (2.91)$$

$$k_q = \frac{k_{12}k_{23}}{k_{21} + k_{23}} \quad \text{or} \quad \frac{1}{k_q} = \frac{1}{k_{12}} + \frac{1}{K_{D^*A}k_{23}} \quad \ldots (2.92)$$

where $K_{D^*A} = k_{12}/k_{21}$ is the equilibrium constant for the encounter complex formation. If the reverse electron-transfer rate k_{32} is not negligible, a more complete treatment can be found in the Rehm and Weller treatment [61,62].

When the reaction is partially diffusion controlled, the bimolecular rate constant k_q is the same as $K_{D^*A}k_{23}$. In the Marcus theory which predicts a parabolic relationship between the driving force and the rate [58], the rate constant of the electron transfer reaction is given by

$$K_{D^*A}k_{23} = \kappa A\sigma^2 \exp(-\Delta G^*_{23}/RT) \quad \ldots (2.93)$$

The transmission coefficient κ is the function $\kappa(r)$ averaged over the distance r of the reactants. The efficiency of electron transfer which requires overlap of the orbitals of the donor and the acceptor decreases with increasing distance as $\exp(-\beta r)$; $A\sigma^2$ has dimensions of collision frequency; ΔG^*_{23} is the free energy of activation [57–60] expressed by

$$\Delta G^* = w^r + \frac{\lambda}{4}\left(1 + \frac{\Delta G^{\circ\prime}}{\lambda}\right)^2 \quad \ldots (2.94)$$

$$\Delta G^{\circ\prime} = \Delta G^\circ + w^p - w^r \quad \ldots (2.95)$$

where ΔG° is the standard free energy in the "prevailing" medium [57], w^p and w^r are work terms for bringing the reactants together and for separating the products and λ includes the contribution λ_i of changes in the bond lengths and the reorganizational parameter λ_0 which reflects changes in the polarization of the solvent molecules during electron transfer [55,59]. In nonpolar solvents, λ_0 vanishes.

$$\lambda_0 = (\Delta e^2) \left(\frac{1}{2r_D} + \frac{1}{2r_A} - \frac{1}{r_{AD}} \right) \left(\frac{1}{\varepsilon_{op}} - \frac{1}{\varepsilon_s} \right) \quad \ldots (2.96)$$

where Δe is the charge transferred from one reactant to the other, r_D and r_A are the radii of the reactants, r_{AD} is the center-to-center distance, ε_{op} and ε_s are the optical (square of refractive index) and static dielectric constants of the solvent.

The excitation energy of the donor or the acceptor and the redox potentials $E(D^+/D)$ and $E(A/A^-)$ of the ground state couples determine the driving force of the reaction. The redox potentials for the excited-state couples (reactions 2.75 or 2.76) are given by

$$E(D^+/D^*) = E(D^+/D) - E_{0,0} \quad \ldots (2.97)$$

$$E(A^*/A^-) = E(A/A^-) + E_{0,0} \quad \ldots (2.98)$$

where $E_{0,0}$ is the zero-zero excitation energy, i.e. the energy difference of the ground and the excited molecule which can be obtained from absorption and fluorescence spectra [61,62].

In the case of oxidative quenching (Figure 2.14) for example, the free energy change ΔG_{23} of the electron-transfer reaction leading to the solvent-separated

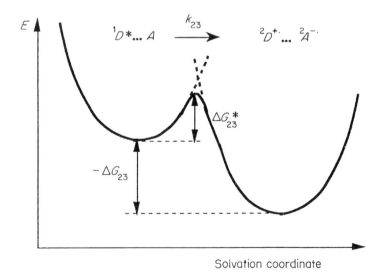

Figure 2.14. Potential energy curves for the outer-sphere electron transfer process

INTERACTION BETWEEN LIGHT AND MATTER

ion pair (SSIP) can be calculated using the Rehm–Weller eq. [52,62]

$$\Delta G_{23}(\text{kcal/mol}) = 23.06 \left[E(\text{D}^{+\cdot}/\text{D}) - E(\text{A}/\text{A}^{-\cdot}) - \frac{e^2}{\varepsilon_S d_{\text{SSIP}}} \right] - E_{00} \quad \ldots (2.99)$$

where the term $e^2/\varepsilon_S d_{\text{SSIP}}$ represents the energy gained by bringing the two radical ions to the distance d_{SSIP} in the solvent of dielectric constant ε_S. If the solvent-separated ion pair dissociates into free ions in a polar solvent, this term can be neglected at distances $d > 7$ Å.

Rehm and Weller have carried out quenching experiments in acetonitrile with a number of excited acceptor–donor systems varying in ΔG_{23} between -25 and $+6$ kcal/mol [61,62]. They have been able to fit their k_q values using eqs. (2.100) and (2.101)

$$k_q = \frac{2 \times 10^{10}}{1 + 0.25 \left[\exp\left(\frac{\Delta G_{23}^*}{RT}\right) + \exp\left(\frac{\Delta G_{23}}{RT}\right) \right]} \quad \ldots (2.100)$$

$$\Delta G_{23}^* = \left(\left(\frac{\Delta G_{23}}{2}\right)^2 + \left(\Delta G_{23}^*(0)\right)^2 \right)^{1/2} + \frac{\Delta G_{23}}{2} \quad \ldots (2.101)$$

where the activation energy when $\Delta G_{23} = 0$ is $\Delta G_{23}^*(0) = 2.4$ kcal/mol. Their results given in Figure 2.15 are in good agreement with the Marcus theory in the so-called Marcus "normal" region when $\Delta G_{23} > -15$ kcal/mol, i.e. the rate constant increases as the reaction becomes more exothermic. However, for more negative values of ΔG_{23}, in the Marcus "inverted" region, where the rate constant

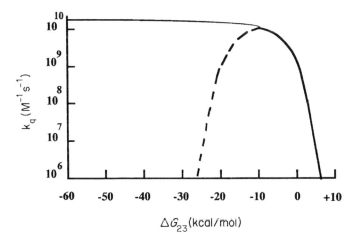

Figure 2.15. Electron transfer rate constant as a function of the free energy (according to refs [61,62]

should begin to decrease, the rate constant remains constant and equal to the diffusion-controlled rate constant value of 2×10^{10} $M^{-1} s^{-1}$ in acetonitrile.

8.2.2 Quantum Effects

The quantum theory of nonadiabatic electron transfer reactions was developed in 1959 by Levich and Dogonadze using the theory of radiationless transitions [57,63]. The rate constant for electron transfer is given by

$$k = \frac{2\pi |V^2|}{\hbar}(F.C.) \qquad \ldots (2.102)$$

where V is the electronic matrix element describing the electronic coupling of the reactants' electronic state with the products' electronic state and $(F.C.)$ the Franck–Condon factor for the electron transfer reaction [57–60,63]. $(F.C.)$ is the sum of products of overlap integrals of the vibrational and solvational wavefunctions of the reactants with those of the products [58]. The quantum effect on the rate constants was found relatively unimportant in the "normal" region, but the possible importance of quantum effects was considered to understand the failure of a series of highly exothermic reactions to exhibit no decrease of the rate constant in the "inverted" region [64–66]. Recently, Jortner and Bixon [67] have described long-range photoinduced electron transfer in isolated solvent-free supermolecules and derived an energy gap law for the microscopic electron transfer rate

$$k(E) = (2\pi V^2/\hbar)\, AFD(E) \qquad \ldots (2.103)$$

where the averaged Franck–Condon density (AFD) is calculated for a simple harmonic model system.

8.2.3 The Inverted Region

As emphasized by Marcus [57], eqs. 2.93 and 2.94 show a decrease of rate with increasing driving force in the inverted region. The quantum corrections in the inverted region reduce the inverted effect but they do not eliminate it. Apart a few number of reports claiming some evidence of the inverted effect, the search for this effect has given a constant diffusion-controlled value [57,59]. Several explanations have been proposed which have been recently reviewed [59,68]. Among them, the dielectric saturation effect in polar solvents invoked by Kakitani and Mataga has been critically examined by Tachiya [69] who showed that the Marcus theory is still valid. Carter and Hynes [70] indicated that the solute charge dependence of solvent force constants have considerable impact on electron-transfer rate–free energy gap relations. For the NP (neutral pair) $\longrightarrow$ IP (ion pair) reaction, their derivation of the activation free energy relation is as follows:

$$\Delta G^* = \beta\{[(1+\beta)\Delta G_r]^{1/2} - [\beta \Delta G_r - \Delta G]^{1/2}\}^2 \qquad \ldots (2.104)$$

where ΔG_r is the IP reorganization energy, ΔG the reaction free energy and the force constant parameter $\beta = k_{NP}/(k_{IP} - k_{NP})$ is calculated from the solvent force constants. For small β values, ΔG^* should grow more slowly with increasing free energy thus suppressing the "anomalous region", but according to a treatment of Tachiya the curve of the rate constant as a function of the free energy change is still symmetrical [71].

The inverted region for highly exoergic electron transfer reactions has been experimentally revealed by Miller in the case of electron transfer from the biphenyl$^{-\bullet}$ radical anion to a number of acceptors in rigid MTHF glass [72] and by McLendon and Miller in protein-protein complexes [73]. A recent conclusion is that the bell-shape dependence of the rate constant on ΔG is observed for charge shift reactions and charge recombinations but not for charge separation reactions [74-76].

The influence of the static and dynamic properties of the solvent on the energetics and dynamics of the charge transfer process has been recently discussed [75-77].

8.2.4 The Distance Dependence of Intramolecular Electron-transfer Rates

In metalloproteins reactions, for instance those of cytochrom c, eq. (2.93) can be used [58], but the transmission coefficient $\kappa(r)$ varies at large distances. The first-order rate constant k for transfer of an electron between molecules held at fixed distances varies with the electron exchange coupling $V(r)$ and $V^2(r)$ decreases exponentially with the separation distance [58,72,78]. Equation (2.93) can be written as

$$k = \kappa(r)\nu \exp(-\Delta G^*/RT) \qquad \ldots (2.105)$$

where ν has dimensions of a frequency (s^{-1}) and $\kappa(r)\nu$ is given by [58,78]

$$\kappa(r)\nu = 10^{13} \exp[-\beta(r - r_0)] \qquad \ldots (2.106)$$

where r_0 is the value of r at which $\kappa(r)\nu$ is equal to 10^{13} s^{-1}. The value of β varies between 10 and 20 nm^{-1} for many reactions between solvated electrons and acceptors [58]: β is about 9 nm^{-1} for systems of the type [M(NH$_3$)$_5$]$_2$ (bridge) [78], zinc/ruthenium-modified myoglobins [79], 20 nm^{-1} for ruthenium-modified cytochrom derivatives [80] and 8.8 nm^{-1} for intercalated dyes and electron acceptors bound to the outer surface of the DNA coil [81].

REFERENCES

[1] See textbooks: (a) W. Heitler, *The Quantum Theory of Radiation*, 2nd edn (Oxford University Press, 1944) pp. 82-112; (b) B. Di Bartolo, D. Pacheco and V. Goldberg, *Spectroscopy of the Excited State*, (Plenum Press, New York, 1976) pp. 1-46; (c) N. J. Turro, *Modern Molecular Photochemistry* (University

Science Books, Mill Valley, California, 1991); (d) L. Salem, *Electrons in Chemical Reactions: First Principles*, (Wiley-Interscience, New York, 1982); (e) G. J. Ferraudi, *Elements of Inorganic Photochemistry*, (Wiley-Interscience, New York, 1988). Ch. III; (f) J. Michl and V. Bonacic-Koutecky, *Electronic Aspects of Organic Photochemistry*, (Wiley-Interscience, New York, 1990); (g) I. N. Levine, *Molecular Spectroscopy*, (Wiley-Interscience, New York, 1975) Ch. 4; (h) A. Gilbert and J. Baggott, *Essentials of Molecular Photochemistry* (Blackwell, Oxford, 1991); (i) K. Nakamoto, *Infrared and Raman Spectra of Inorganic and Coordination Compounds*, [1i] K. Nakamoto, Infrared and Raman Spectra of Inorganic and Coordination Compounds (Wiley, New-York, 1978) 3rd ed., Section I. 2; (j) N. B. Colthup, L. H. Daly and S. E. Wiberley, *Introduction to Infrared and Raman Spectroscopy* (Academic Press, New York, 1964) Ch. 1; (k) R. S. Berry, S. A. Rice and J. Ross, *Physical Chemistry* (Wiley, New York, 1980) pp. 135–142. (l) V. Balzani and V. Carassiti, *Photochemistry of Coordination Compounds* (Academic Press, London, 1970), pp. 6–33.

[2] J. Franck, Trans. Far. Soc. **21**, 536–542 (1925).
[3] E. Condon, Physical Review, **28**, 1182–1201 (1926).
[4] M. Born, R. Oppenheimer, Ann. Phys., **84**, 457–484 (1927).
[5] E. U. Condon, Physical Review, **32**, 858–872 (1928).
[6] Y. J. Yan, S. Mukamel, J. Chem. Phys., **85**, 5908 (1986).
[7] F. P. Schäfer, in *Dye Lasers*, Topics in Applied Physics, Vol. 1, 3rd edn, edited by F. P. Schäfer (Springer-Verlag, Berlin, 1990) Ch. 1.
[8] H. Kuhn, J. Chem. Phys. **16**, 840–841 (1948).
[9] H. Kuhn, J. Chem. Phys. **17**, 1198–1212 (1949).
[10] D. V. O'Connor, D. Phillips, *Time-correlated Single Photon Counting* (Academic Press, London, 1984).
[11] H. V. Drushel, A. L. Sommers and R. C. Cox, Analytical Chemistry, **35**, 2166 (1963).
[12] S. J. Strickler and R. A. Berg, J. Chem. Phys. **37**, 814–822 (1962).
[13] D. F. Eaton, Pure & Appl. Chem., **60**, 1107–1114 (1988).
[14] D. F. Eaton, J. Photochem. Photobiol., B: Biology, **2**, 523–531 (1988).
[15] I. B. Berlman, *Handbook of Fluorescence Spectra of Aromatic Molecules* (Academic Press, New York, 1965).
[16] J. C. Mialocq and X. Armand, In *Laser à Colorant. Fluorescence des Colorants Rhodamines. II* (Commissariat à l'Energie Atomique, CEA-R-5538, 1990).
[17] E. W. Castner, M. Maroncelli and G. R. Fleming, J. Chem. Phys. **86**, 1090–1097 (1987).
[18] S. J. Rosenthal, X. Xie, M. Du and G. R. Fleming, J. Chem. Phys. **95**, 4715 (1991).
[19] P. F. Barbara and W. Jarzeba, in *Advances in Photochemistry*, edited by D. H. Volman, G. S. Hammond and K. Gollnick, (Wiley-Interscience, New York, 1990) Vol. 15, pp. 1–68.
[20] P. Hébert, G. Baldacchino, T. Gustavsson and J. C. Mialocq, Chem. Phys. Lett. **213**, 345–350 (1993).
[21] *Ultrafast Reaction Dynamics and Solvent Effects*, edited by Y. Gauduel and P. J. Rossky (AIP Conference Proceedings 298, 1994) Part IV.
[22] M. Meyer and J. C. Mialocq, Opt. Commun. **64**, 264–268 (1987).
[23] M. Meyer, J. C. Mialocq and B. Perly, J. Phys. Chem. **94**, 98–104 (1990).
[24] D. C. Easter and A. P. Baronavski, Chem. Phys. Lett. **201**, 153–158 (1993).
[25] S. Pommeret, T. Gustavsson, R. Naskrecki, G. Baldacchino, P. D'Oliveira, P. Meynadier and J. C. Mialocq, in Opto 94 (ESI Publications, Paris, 1994) pp. 249–254.
[26] H. Zhang, A. M. Jonkman, P. van der Meulen, M. Glasbeek, Chem. Phys. Lett., **224**, 551–556 (1994).

[27] M. Martin, P. Plaza and Y. H. Meyer, Chem. Phys. **192**, 367-377 (1995).
[28] T. Gustavsson, G. Baldacchino, J. C. Mialocq and S. Pommeret, Chem. Phys. Lett. **236**, 587-594 (1995).
[29] S. Pommeret, T. Gustavsson, R. Naskrecki, G. Baldacchino and J. C. Mialocq, Journal of Molecular Liquids, **64**, 101-112 (1995).
[30] A. Seilmeier, P. O. J. Scherer and W. Kaiser, Chem. Phys. Lett. **105**, 140-146 (1984).
[31] A. Mokhtari, A. Chebira and J. Chesnoy, J. Opt. Soc. Am. B **7**, 1551-1557 (1990).
[32] U. Sukowski, A. Seilmeier, T. Elsaesser and S. F. Fischer, J. Chem. Phys. **93**, 4094-4101 (1990).
[33] See textbooks: (a) F. Williams, D. E. Berry and J. E. Bernard, in *Radiationless Processes*, edited by B. DiBartolo, (Plenum Press, New York, 1980) pp. 1-37; (b) E. S. Medvedev and V. I. Osherov, *Radiationless Transitions in Polyatomic Molecules* (Springer-Verlag, Berlin, 1995).
[34] M. A. El-Sayed, Accounts of Chem. Res., **1**, 8-16 (1968).
[35] Ph. Hébert, G. Baldacchino, Th. Gustavsson and J. C. Mialocq, J. Photochem. Photobiol. A: Chem., **84**, 45-55 (1994) and references cited therein.
[36] N. J. B. Green, S. M. Pimblott and M. Tachiya, J. Phys. Chem., **97**, 196-202 (1993).
[37] M. Kasha and D. E. Brabham, in $^1O_2$ *Singlet Oxygen*, Organic Chemistry, a series of monographs, Vol. 40, edited by H. H. Wasserman and R. W. Murray, (Academic Press, New York, 1979) Ch. 1.
[38] R. V. Bensasson, E. J. Land and T. G. Truscott, *Excited States and Free Radicals in Biology and Medicine. Contributions from Flash Photolysis and Pulse Radiolysis* (Oxford Science Publications, Oxford, 1993) Ch. 4.
[39] R. Potashnik, C. R. Goldschmidt and M. Ottolenghi, Chem. Phys. Lett. **9**, 424-425 (1971).
[40] D. C. Dobrowolski, P. R. Ogilby and C. S. Foote, J. Phys. Chem. **87**, 2261-2263 (1983).
[41] A. J. McLean and T. G. Truscott, J. Chem. Soc. Faraday Trans., **86**, 2671-2672 (1990).
[42] A. J. McLean, D. J. McGarvey, T. G. Truscott, C. R. Lambert and E. J. Land, J. Chem. Soc. Faraday Trans, **86**, 3075-3080 (1990).
[43] T. Förster, Discussions Faraday Society, **27**, 7-17 (1959).
[44] T. Förster, Z. Naturforsch., **4a**, 321-327 (1949).
[45] G. Porter and C. J. Tredwell, Chem. Phys. Lett., **56**, 278-282 (1978).
[46] D. L. Dexter, J. Chem. Phys., **21**, 836 (1953).
[47] M. Inokuti and F. Hirayama, J. Chem. Phys., **43**, 1978-1989 (1965).
[48] J. Klafter and A. Blumen, J. Chem. Phys. **80**, 875-877 (1984).
[49] P. Levitz, J. M. Drake and J. Klafter, Chem. Phys. Lett., **148**, 557-561 (1988).
[50] A. Takami and N. Mataga, J. Phys. Chem. **91**, 618-622 (1987).
[51] M. Julliard and M. Chanon, Chem. Rev. **83**, 425-506 (1983).
[52] G. J. Kavarnos and N. J. Turro, Chem. Rev. **86**, 401-449 (1986).
[53] M. A. Fox and M. Chanon, editors, *Photoinduced Electron Transfer*, (a) Part A. Conceptual basis; (b) Part B. Experimental techniques and medium effects; (c) Part C. Photoinduced electron transfer reactions: organic substrates; (d) Part D. Photoinduced electron transfer reactions: inorganic substrates and applications. (Elsevier, Amsterdam, 1988).
[54] H. Taube, in *Nobel lectures. Chemistry 1981-1990*, edited by T. Frängsmyr, (World Scientific) Singapore, 1993 pp. 120-140.
[55] R. A. Marcus, J. Chem. Phys. **24**, 966-978 (1956).
[56] N. S. Hush, Trans. Faraday Soc. **57**, 557-580 (1961).

[57] R. A. Marcus, Faraday Discuss. Chem. Soc., **74**, 7-15 (1982).
[58] R. A. Marcus and N. Sutin, Biochimica et Biophysica Acta **811**, 265-322 (1985).
[59] R. A. Marcus and P. Siddarth, in *Photoprocesses in Transition Metal Complexes, Biosystems and other Molecules. Experiment and Theory*, edited by E. Kochanski, (Kluwer, Dordrecht, 1992), pp. 49-88.
[60] R. A. Marcus, Angew. Chem. Int. Ed. Engl., **32**, 1111-1121 (1993).
[61] D. Rehm and A. Weller, Ber. Bunsenges. Physik. Chem., **73**, 834-839 (1969).
[62] D. Rehm and A. Weller, Israel Journal of Chemistry, **8**, 259-271 (1970).
[63] M. Bixon and J. Jortner, Faraday Discuss. Chem. Soc., **74**, 17-29 (1982).
[64] P. Siders and R. A. Marcus, J. Am. Chem. Soc., **103**, 741-747 (1981).
[65] P. Siders and R. A. Marcus, J. Am. Chem. Soc., **103**, 748-752 (1981).
[66] R. A. Marcus, J. Phys. Chem., **93**, 3078-3086 (1989).
[67] J. Jortner and M. Bixon, J. Photochem. Photobiol. A: Chem., **82**, 5-10 (1994).
[68] G. B. Schuster, in *Advances in Electron Transfer Chemistry*, (JAI Press, 1991) Vol. 1, pp. 163-197.
[69] M. Tachiya, Chem. Phys. Lett., **159**, 505-510 (1989).
[70] E. A. Carter and J. T. Hynes, J. Phys. Chem., **93**, 2184-2187 (1989).
[71] M. Tachiya, J. Phys. Chem., **93**, 7050-7052 (1989).
[72] J. R. Miller, New J. of Chemistry, **11**, 83-89 (1987) and references cited therein.
[73] G. McLendon and J. R. Miller, J. Am. Chem. Soc., **107**, 7811-7816 (1985).
[74] M. Tachiya and S. Murata, J. Phys. Chem., **96**, 8441-8444 (1992).
[75] T. Kakitani, N. Matsuda, T. Denda, N. Mataga and Y. Enomoto, ref. 21, pp. 395-409.
[76] M. Tachiya and M. Hilczer, ref. 21, pp. 447-459.
[77] H. Sumi, ref. 21, pp. 485-495.
[78] S. S. Isied, A. Vassilian, J. F. Wishart, C. Creutz, H. A. Schwarz and N. Sutin, J. Am. Chem. Soc., **110**, 635-637 (1988).
[79] A. W. Axup, M. Albin, S. L. Mayo, R. J. Crutchley and H. B. Gray, J. Am. Chem. Soc. **110**, 435-439 (1988).
[80] B. A. Jacobs, M. R. Mauk, W. D. Funk, R. T. A. MacGillivray, A. G. Mauk and H. B. Gray, J. Am. Chem. Soc. **113**, 4390-4394 (1991).
[81] A. M. Brun and A. Harriman, J. Am. Chem. Soc., **114**, 3656-3660 (1992).

3 Homogeneous Proton Transfer Photocatalysis

L. G. ARNAUT and S. J. FORMOSINHO
Chemistry Department, University of Coimbra, Portugal

1 Introduction	55
2 Excited-state specific acid-base catalysis	61
3 Excited-state general acid-base catalysis	65
4 The Förster cycle	71
5 The Intersecting-state model	75
6 pH and pOH jump experiments	84
7 Case-studies of proton-transfer photocatalysis	88
7.1 Acid catalysis in the photohydration of styrenes and phenylacetylenes	88
7.2 Base catalysis in the photoketonization of dibenzosuberenol	91
7.3 Static and dynamic catalysis in the phototautomerization of lumichrome	93
8 Conclusion	93
References	94

1 INTRODUCTION

The acid and base catalysis of chemical reactions occurring in the ground state is a subject of unquestionable importance that has been thoroughly studied. Many reports on such reactions are available in the literature. In the history of this field there are two landmarks that deserve a special mention: the quantitative relation between the acid–base strength of a species and its catalytic constant first proposed and verified by Brönsted and Pederson in 1924 [1], and the first modern book on this field written by Bell in 1941 [2]. Although the fundamental concepts on acid–base catalysis are now available in standard textbooks, it is nevertheless important to review some of them here before addressing the analogous photochemical reactions.

Acid catalysis takes place when the velocity of a chemical reaction is accelerated by the presence of an acid. This effect is quantitatively measured by the phenomenological rate expression of that chemical reaction, where the order of

Homogeneous Photocatalysis. Edited by M. Chanon
© 1997 John Wiley & Sons Ltd

reaction with respect to the acid is larger than its power in the stoichiometric equation. The acid has a more pronounced effect in lowering the energy barrier to the reaction than it has on the position of the equilibrium between reactants and products. When this acid is the hydrogen ion, H^+, this acceleration is called specific acid catalysis. Similar concepts apply to base catalysis. When the base is the hydroxyl ion, OH^-, the rate acceleration is named specific base catalysis. The hydrogen and hydroxyl ions are not the only acids and bases able to accelerate reaction rates. Other acids or bases may participate in the proton transfer rate-determining step of an acid–base catalysis. When the reaction rate depends on the concentration of acid–base buffers, the reaction is said to be subject to general acid catalysis. In general, for a reaction of the type indicated in Mechanism I:

$$R \xrightarrow{HA} P$$

Mechanism I

the reaction rate in aqueous solution can be written

$$v = -\frac{d[R]}{dt} = \frac{d[P]}{dt} = (k_0 + k_{H_3O^+}[H_3O^+] + k_{OH^-}[OH^-]$$
$$+ k_{HA}[HA] + k_{A^-}[A^-])[R] \quad \ldots (3.1)$$

where k_0 is the reaction rate constant of the uncatalysed reaction, which in fact is the rate constant of the catalysis by water molecules. The other rate constants represent the specific catalysis by the hydronium and hydroxyl ions, and the general catalysis by the AH acid and by the A^- base.

The rate eq. (3.1) does not correspond to an elementary reaction. Acid–base catalysis requires a reaction mechanism involving two or more elementary steps, one of them being a proton transfer step. Catalysis can only be observed if the transfer of a proton is the rate-limiting elementary step or occurs before this step in the reaction mechanism. The sequence of essential steps in specific acid catalysis is shown in Mechanism II, where X represents a reagent attacking the intermediate cation; X may be a water molecule or another nucleophile. A similar mechanism can be proposed for specific base catalysis. The detection of general

$$R \xrightarrow{k_0} P$$

$$R + H_3O^+ \underset{k_{-p}}{\overset{k_p}{\rightleftharpoons}} RH^+ + H_2O$$

$$RH^+ \xrightarrow{k_1} P + H^+$$

$$RH^+ + X \xrightarrow{k_2} P + H^+$$

Mechanism II

acid–base catalysis depends on how much the reaction rate depends on the acidic or basic strength of the catalyst, and will be discussed in detail later in the light of the relation proposed by Brönsted.

The reaction rate equation for Mechanism II is rather complex, but usually it can be simplified with reasonable estimates of the relative rates of the five reactions involved. If $k_2[X] \gg k_{-p}$ and $k_1 \gg k_{-p}$ then the protonation of the substrate is the rate-determining step and the rate equation becomes

$$v = (k_0 + k_p[H_3O^+])[R] \qquad \ldots (3.2)$$

If $k_2[X] \ll k_{-p}$ and $k_1 \ll k_{-p}$ a relatively fast pre-equilibrium is attained and the rate-determining step is the formation of the products from the intermediate RH^+. If this intermediate predominantly decomposes through a first-order reaction, $k_1 \gg k_2[X]$, the rate equation is given by

$$v = \left(k_0 + \frac{k_1}{K_{RH^+}}[H_3O^+]\right)[R] \qquad \ldots (3.3)$$

If the attack of the nucleophile X onto the intermediate is required to yield the products, $k_2[X] \gg k_1$, then the rate equation is

$$v = \left(k_0 + \frac{k_2}{K_{RH^+}}[H_3O^+][X]\right)[R] \qquad \ldots (3.4)$$

The rate constants of the elementary reactions shown in Mechanism II can be calculated using the equation given by the transition state theory:

$$k = \frac{k_B T}{h} \chi_{el} c_0^{1-m_e} \exp\left(-\frac{\Delta G^{\ddagger}}{RT}\right) \qquad \ldots (3.5)$$

if the free energy of activation, $\Delta G^{\ddagger}$, and the electronic factor χ_{el} can be estimated from theoretical models. The other symbols in eq. (3.5) have their usual meanings: k_B is the Boltzmann constant, h is the Planck constant, T is the temperature, c_0 is the unit concentration (= 1 M), m_e is the molecularity of the elementary reaction, and R is the molar gas constant. The electronic factor χ_{el} is normally equal to unity for ground state reactions, but it can be lower than unity when the electronic states of reactants and products do not correlate.

The calculation of rate constants using eq. (3.5) emphasizes the importance of accurate estimations of free energies of activation. Several theoretical models have been proposed to estimate energies of activation of chemical reactions. Here, we will discuss in detail the application of one of them, the intersecting-state model (ISM), which we believe that is the best compromise presently available between computational simplicity and the capacity to predict absolute values and explain trends of the experimental energies of activation of proton-transfer reactions in solution. Before dealing with the theoretical estimation of proton-transfer rates, this account will focus on the reaction mechanisms leading to phenomenological reaction rates of the type described by eq. (3.1), restricted to the systems where the acid, the base or the substrate R are in an electronically excited state. The study of such reaction mechanisms will be guided by a detailed discussion of the thermodynamic and kinetic acidities of photoexcited organic molecules.

Such an emphasis on reaction mechanisms is justified by the conceptual complexity of photocatalysis. Unlike the consensus presently existing on what is meant by 'catalysis' and 'catalyst' in thermal reactions, the definitions of 'photocatalysis' and "photocatalyst" remain controversial. Such concepts will be discussed in detail elsewhere in this volume. However, we cannot avoid an incursion in this subject in order to delimitate the contents of our account.

The use of light to produce electronically excited species with sufficiently long lifetimes to react opens new routes to the molecules. The most interesting electronic transitions leading to excited states are promoted by ultraviolet and visible light. The energy available in this region of the spectrum is very high for chemical purposes. For example, the lowest energy absorption band of benzene has an absorption maximum around $\lambda_{max} = 255$ nm. A mole of photons with this wavelength carries an energy of 469.1 kJ ($= 112.1$ kcal), which is more than the bond strength of a carbon–hydrogen bond or a carbon–carbon single bond. Following light absorption from the ground state to an excited singlet state, most molecules decay very rapidly by internal conversion to their lowest excited singlet state (S_1). The energy dissipated by most of the aromatic molecules of interest for proton transfer photocatalysis in their internal conversion to S_1 may be up to one-half of that absorbed by an aromatic ring. The intrinsic lifetime of S_1 states is very short, less than a few nanoseconds for most "stable" aromatic molecules, because they may fluoresce or decay nonradiatively to the ground state, or undergo intersystems crossing to the triplet manifold. Here, fast internal conversion will bring the molecules to their lowest excited triplet state, T_1. The intrinsic lifetime of triplet states may reach a few microseconds, because their conversion to the ground state is spin-forbidden. Once in the S_1 (or T_1) state, the molecules may transfer their energy to the singlet (or triplet) state of another molecule, provided that the excited state energy of this second molecule is lower than the energy of S_1 (or T_1), and that the molecules are sufficiently close together in solution. This type of energy transfer, shown in Mechanism III, is called photosensitization and the molecule that initially absorbs the radiation is called the photosensitizer, S.

$$S + h\nu \longrightarrow {}^*S$$

$${}^*S + R \longrightarrow S + {}^*R$$

$${}^*R \longrightarrow P$$

Mechanism III

Photosensitization may also occur by electron transfer from/to the photosensitizer to/from a substrate in the ground state as shown in Mechanism IV.

$$S + h\nu \longrightarrow {}^*S$$
$$^*S + R \longrightarrow S^+ + R^-$$
$$R^- \longrightarrow P$$

Mechanism IV

The photosensitized substrate may have an increased reactivity and the role of the photosensitizer has been compared to that of a catalyst [3].

Hennig et al. [4] classified photocatalytic reactions into photoinduced catalytic and photoassisted reactions. In the first type, a catalytically active species is generated photochemically and the catalytic conversion of the substrate R in the product P may occur in subsequent dark reactions, such that the quantum yield of the substrate consumption (Φ_S) is larger than the quantum yield of catalyst generation (Φ_C), $\Phi_S > \Phi_C$. In the second type, a sensitizer (or the substrate) absorbs a photon and energy transfer to the substrate (or reaction with a ground state catalyst) leads to the conversion of R into P, but $\Phi_S < \Phi_C$. Photoinduced catalytic reactions include chain reactions, which are not regarded as catalytic in thermal chemistry and that we do not include in our treatment of proton transfer reactions. Photoassisted reactions embrace the photosensitized reactions of Mechanisms III and IV. The unsatisfying feature of this definition is that a photosensitizer contributes more to increase the exothermicity of the subsequent reaction than it does to reduce its energy barrier. Photosensitized reactions have been subject to extensive studies and, in principle, it is always possible to initiate any photoreaction by a suitable photosensitizer. Thus, conceptually it is not appropriate to extend the definition of photocatalysis to include photosensitized reactions, although a good knowledge of photosensitizers and their action can be very useful to design alternative mechanisms to slow thermal reactions. We should retain, from the definitions of Hennig, two possible mechanisms for proton transfer photocatalysis: light is absorbed by a nominal catalyst (C), which yields photochemically the real catalyst that participates in the subsequent dark reactions (Mechanism V, applied to base catalysis), or light is absorbed by a substrate that reacts in an excited state with a ground state catalyst (Mechanism VI, showing acid catalysis).

$$C + h\nu \longrightarrow {}^*C$$
$$^*C + H_2O \longrightarrow {}^*CH^+ + OH^-$$
$$OH^- + R \longrightarrow P$$

Mechanism V

$$R + h\nu \longrightarrow {}^*R$$
$$^*R + H_3O^+ \longrightarrow P$$

Mechanism VI

Wubbels [5] proposed that a catalyst for a photochemical reaction should be defined as a substance that appears in the quantum yield expression for reaction from a particular excited state to a power greater than its coefficient in the stoichiometric equation. This definition clearly distinguishes between a catalyst and a sensitizer, because the sensitizer increases the quantum yield by increasing the efficiency of reaching the reactive excited state, but not the efficiency of reaction from that reactive state to the products. There is a caveat in referring the catalytic effect to quantum yield increases rather than to the acceleration of reaction rates. Quantum yields are relative quantities that measure the competition between alternative reaction channels. The quantum yield of a given product may increase because the formation of the competing products is inhibited, rather than because its rate of formation is accelerated. Quantum yields may also be dependent on the efficiencies of the crossing between different electronic states in the course of the reaction, as has been shown in ketone photochemistry [6] and *cis-trans* photoisomerizations. Thus, it is better to adhere to a definition based in reaction rates rather than quantum yields, even if the increase in the quantum yield of the product by the presence of another species is an important diagnostic for the assignment of that species as a catalyst of a photoreaction. Furthermore, in proton transfer photocatalysis, the quantum yield increase may be due to the interaction of the acid or base with a primary photoproduct, rather than with the electronically excited reactant. For example, Wubbels [7,8] showed that photo-Smiles rearrangements can be subject to general base catalysis. The catalytic step involves a proton transfer from a nitrogen acid to nitrogen base. A series of nitrogen bases were used, and the logarithms of the proton transfer rate constant plotted against the pK_a of the conjugated acids of the base catalysts, give a curve with asymptotic limits of 1 and 0 for low and high pK_a values, respectively. The effect of the bases in this system follows the definition given by Wubbels for a catalyst of a photochemical reaction [5] but this author also showed that the proton transfer step involves a ground state intermediate of the photochemical reaction. Thus, this system should, in fact, be regarded as thermal catalysis of a photogenerated intermediate.

This general overview of photochemical processes shows that a good understanding of excited state kinetics is very important for rationalize and predict proton transfer photocatalysis. One of the most crucial differences between ground and excited state proton transfers is that the former has to compete with the rate of the uncatalysed reaction, while the latter has to be competitive with the other decay channels of the excited state. This places a very restrictive kinetic limitation on the observation of proton transfer photocatalysis. The kinetics of intermolecular excited-state proton transfer reactions were recently reviewed [9], with special emphasis on quantitative theoretical models. Their intramolecular analogs were also reviewed [10] and, in some cases, the presence of a mediator (catalyst) was discussed. Some of the reactions included in these reviews will be revisited here, but emphasizing the role of the catalyst.

2 EXCITED-STATE SPECIFIC ACID–BASE CATALYSIS

The reciprocal of the lifetime of a singlet state subject to protonation can be expressed as

$$\frac{1}{\tau_s} = k_{ic} + k_f + k_{isc} + k_Q[Q] + k_p^*[H_3O^+] + k_p^o[H_3O^+] + k_{HA}^*[HA] \quad \ldots (3.6)$$

where k_{ic}, k_f and k_{isc} are the first-order rate constants for internal conversion, fluorescence and intersystems crossing respectively, k_Q is the second-order rate constant for the excited state quenching by the quencher Q, k_{AH}^* is the rate constant for acid catalysis by the AH acid, and k_p^* and k_p^o are proton transfer rate constants defined in Mechanism VII. Examples of bimolecular quenching processes are energy and electron transfer reactions. Even when the bimolecular reactions are negligible, singlet lifetimes are usually short because k_{ic}, k_f and/or k_{isc} are relatively high. This is probably the most important limitation of proton transfer photocatalysis. In principle, photocatalysis of species in the triplet state should be easier to observe.

A molecule in an electronically excited state may be regarded as an electronic isomer of that molecule in the ground state. The energy available for reaction and the multiple reaction channels open to the molecule also contributes to change its reactivity, and make photochemistry a broad and complex field. Light can be used to provide alternative reaction channels leading to new products, to drive unfavorable thermal reactions or to accelerated slow thermal reactions. In the first case the photochemical products will be different from the thermal ones and the role played by light is to change the nature of the reactants, which is distinguishable from a catalytic effect. In the second case, it is the extra energy available photochemically that promotes the fast and efficient formation of a

$$R \xrightarrow{k_0} P$$

$$R + H_3O^+ \underset{k_{-p}}{\overset{k_p}{\rightleftharpoons}} RH^+ + H_2O \xrightarrow{k_2} P + H_3O^+$$

$$R + h\nu \xrightarrow{k_2} {}^*R$$

$$^*R \xrightarrow{k_0^*} P$$

$$^*R + H_3O^+ \underset{k_{-p}^*}{\overset{k_p^*}{\rightleftharpoons}} {}^*RH^+ + H_2O \xrightarrow{k_2^*} P + H_3O^+$$

$$^*R + H_3O^+ \xrightarrow{k_p^o} RH^+ + H_2O \xrightarrow{k_2} P + H_3O^+$$

Mechanism VII

product difficult to obtain by the corresponding thermal reaction; in this case, molecular or ionic reactants and products of thermal and photochemical reactions are the same, and light could be seen as a catalyst; however, the absorption of light makes the reaction thermodynamically more favorable, so it should be included in the reaction mechanism as a "reagent", rather than as a catalyst. This corresponds to the first four steps of the general mechanism illustrated in Mechanism VII, which does not include the physical decay channels of the excited states. We will not be concerned here with reactions occurring predominantly by these four steps.

The fifth step of Mechanism VII introduces the possibility of achieving an excited state acid–base equilibrium. That is, the proton transfer from an acid in solution to *R and the reverse proton transfer from *RH$^+$ to the conjugated base of that acid, take place within the lifetimes of *R and *RH$^+$. This not a trivial observation, given the intrinsically short lifetimes of most singlet states. Thus, acid–base equilibrium in singlet states is only possible if the proton transfer is very fast. Such acid–base equilibrium and the corresponding proton transfers are called adiabatic, because they do not involve any change in the electronic states of the species present in the system. Two requirements for the adiabaticity of a photochemical reaction have been proposed [11]: (1) the reaction must involve minor structural changes; (2) the backbone molecular structure must be rigid. A third requirement, related to the others, is that the reaction must be fast. The most direct evidence for the adiabaticity of proton transfers comes from the observation, in a given pH window, of fluorescence from *R and *RH$^+$ in equilibrium. If *RH$^+$ is a weaker acid than RH$^+$, the ground-state acidity constant (K_a) will be larger than the excited-state one (K_a^*), and eqs. (3.3) and (3.4) show that the excited state reaction will be faster than that of the ground state. In fact, the photoinduced redistribution of the electrons in the molecular orbitals may bring about dramatic changes in the acid–base properties of the molecules and lead to $\Delta pK_a (= pK_a^* - pK_a)$ values very different from zero. Very positive ΔpK_a values offers the possibility of observing excited state acid catalysis according to this mechanism. *Mutatis mutandis*, the same can be said for very negative ΔpK_a values and excited state basic catalysis.

The sixth step of Mechanism VII associates the quenching of the reactants by H_3O^+ with the formation of the ground state products. Such reactions, involving a change of electronic state in the course of the reaction, are called diabatic (or nonadiabatic). A simple qualitative account of diabatic and adiabatic processes was offered by Förster [12]. The rates of diabatic processes depend on the height of reaction barriers just like adiabatic ones, but they may also involve electronic preexponential factors much lower than unity, $\chi_{el} \ll 1$, especially when the initial and final electronic states have different spin multiplicities. The observation of diabatic reactions, thus the occurrence of large k_p^o values, may be explained by two reasons: the enhanced reactivity of *R in the presence of an acid is due to an intrinsically higher reactivity of *R and/or to the large exothermicity of this reaction. Only the first of these effects is a truly photocatalytic effect. However, in practice, the distinction between a mechanism leading directly to RH^+ and another one leading first to $^*RH^+$ followed by rapid decay to RH^+, depends on the time resolution of the experiment. In many reactions studied by techniques with a nanosecond time resolution, species of the type RH^+ were proposed to be formed directly from $^*R$; however, it is possible that studies with pico- or femtosecond resolution would detect the presence of $^*RH^+$ as an intermediate. This situation is illustrated in Figure 3.1. In order not to be a hostage of a definition of proton transfer photocatalysis that requires a very detailed knowledge of the reaction mechanism and that is very demanding on the techniques employed to study it, we will consider here that step 6 in Mechanism VII also corresponds to a photocatalytic reaction, independently of the observation of an intermediate $^*RH^+$.

A clearer understanding of the photocatalytic processes following steps 5 or 6 of Mechanism VII, can be obtained from a few simple examples of reactions that fit into this classification. Step 5 is followed by oxygen atom centered acids and bases, which are known to undergo excited state reversible proton transfer reactions, and step 6 seems to be followed by carbon atom centered acids and bases, whose proton transfers are associated with fluorescence quenching of the

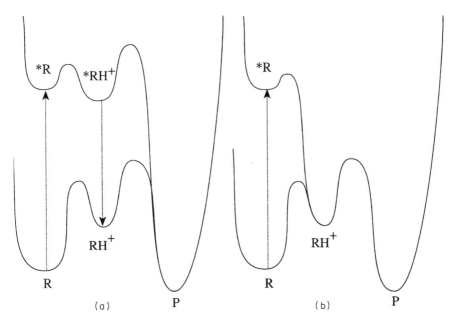

Figure 3.1. Free energy profile corresponding to mechanisms of proton transfer photocatalysis. (a) Step 5 of Mechanism VII; (b) step 6 of Mechanism VII. With the reduction of the barrier for conversion of $^*RH^+$ into P, it becomes increasingly difficult to detect the presence of this intermediate. Eventually, the minimum corresponding to $^*RH^+$ coalesces with that of RH^+

acid (or base) without the corresponding enhancement of the fluorescence of the conjugated base (or conjugated acid).

Proton transfer from photoexcited 2-naphthol or cyano substituted 2-naphthols to water molecules in aqueous tetrahydrofuran [13], is one of the latest and most elegant examples of adiabatic proton transfer photocatalysis. In these systems, the naphthols correspond to *RH and H_2O to the base in the mechanism that can be established for base catalysis by analogy to Mechanism VII. A very interesting feature of these systems is that the solvent is not water, thus the rate constant and reaction order of the proton transfer to H_2O can be followed by increasing the concentration of water in tetrahydrofuran. Tolbert has found that the proton transfer follows a fourth order in the transfer from 2-naphthol and a second order in the transfer from cyano-substituted 2-naphthols.

The proton transfer to a neutral carbon atom yields a carbocation and the deprotonation of a carbon acid gives a carbanion. Both these reactions are of synthetic utility. The excited state properties of carbon acids and bases have been explored in order to use of the dramatic changes in acidity induced by light as new tools in synthetic organic chemistry. Recently Wan and Shukla [14] reviewed this field emphasizing the utility of acid–base behavior of excited states of organic molecules. The π-system of alkenes, alkynes and aromatic systems can accept a proton, thus acting as carbon bases. Protonation of aromatic systems usually requires moderate to strong acids. Calculations show that the pK_a^* values of polycyclic aromatic hydrocarbons are 7–30 powers of 10 higher than their pK_a values. However, there are not many unambiguous examples of proton transfers to electronically excited carbon bases. This is related to the fact that proton transfers to or from carbon atoms are much slower than the analogous reactions at oxygen or nitrogen atoms. No cases are known of simultaneous observation of fluorescence from the acidic and basic forms of aromatic hydrocarbons, so evidence for proton transfer to electronically excited carbon bases has to be from isotope exchange upon photolysis. Such exchanges have been observed by Shizuka [15] in methoxynaphthalenes and by Wan and co-workers [16] in dimethoxybenzenes, but they are accompanied by fluorescence quenching. Shizuka proposed a reaction mechanism involving the initial formation of a singlet state carbocation followed by very fast decay to its ground state, but was unable to detected such carbocation in the nanosecond timescale. Presumably the internal conversion of the carbocation is much faster than any competing processes. Alternatively, the protonation of these π-systems and the conversion from the singlet state to the ground state occur in only one elementary step, that is, the reaction is diabatic and follows step 6 of Mechanism VII. Although Wan reported pK_a^* values for the photoprotonation of dimethoxybenzenes, namely placing them in the -1 to $+0.5$ range as opposed to the ground state pK_a of -6, it is not appropriate to define acidity (equilibrium) constants for these reactions, because they are not reversible. This does not invalidate the fact that such photoprotonations are much more facile than the analogous thermal reactions. Indeed, the interest on these reactions resides in the fact that they are much faster in the excited state than in the ground state and can be efficiently carried out under mild conditions. For example, isotope exchange quantum yields of 0.05 at room temperature were obtained for 1-methoxynaphthalene in $D_2O:CH_3CN$ (1 : 4) solution with pD = 1.3, and for 1,3-dimethoxybenzene in $D_2O:CH_3CN$ (4 : 1) solution with pD = 1.6.

The deprotonation of aromatic hydrocarbons in the ground state is very slow and thermodynamically unfavorable. For example, fluorene is a member of the group of the more reactive hydrocarbons; yet its deprotonation rate to the methoxide ion in methanolic sodium methoxide at 298 K is only 3.1×10^{-5} $M^{-1} s^{-1}$, and its pK_a in cyclohexylamine is 23.04 [17]. Although irradiation of fluorene in $D_2O:CH_3CN$ (1 : 1) solution at 295 K does not lead to measurable isotope exchange and its lifetime has only a minor dependence on the presence of water, the same procedure with suberene leads to exchange of the benzylic proton by a deuteron with a quantum yield of 0.03, concomitant with a fluorescence quenching rate constant of 1.7×10^8 $M^{-1} s^{-1}$, by water. Here, again, it would be interesting to know whether or not the deprotonation is adiabatic.

3 EXCITED-STATE GENERAL ACID-BASE CATALYSIS

General acid catalysis of a chemical reaction can be observed if a pH range can be found where k_{AH} [AH] can compete with $k_{H_3O^+}$ [H_3O^+], and general base catalysis can be observed if a pH range can be found where k_{A^-} [A^-] can compete with k_{OH^-} [OH^-], as defined in eq. (3.1). In order to take general acid-base catalysis explicitly into consideration, we can consider an acid AH and a base B^- in equilibrium with the conjugated base A^- and conjugated acid BH (Mechanism VIII). The effect of substituents in the equilibrium and rates of Mechanism VIII can be investigated by two different strategies: either B^- is kept constant and AH is varied within a series of structurally related acid catalysts, or AH is maintained while B^- is varied within a family of basic catalysts. The first situation corresponds to the study, in the forward direction, of general acid catalysis and the latter to the study of general base catalysis.

$$B^- + HA \underset{k_r}{\overset{k_f}{\rightleftharpoons}} BH + A^-$$

Mechanism VIII

It is useful to recall some of the fundamental thermodynamic and kinetic relations that apply to Mechanism VIII. The equilibrium constant of the system is, by definition,

$$K_{eq} = \frac{k_f}{k_r} = \frac{[BH][A^-]}{[B^-][AH]} \qquad \ldots (3.7)$$

Also by definition

$$pK_{AH} = -\log_{10}\left\{\frac{[H_3O^+][A^-]}{[AH]}\right\} \qquad \ldots (3.8)$$

and

$$\Delta G^0 = -2.303RT(pK_{BH} - pK_{AH}) \qquad \ldots (3.9)$$

In order to compare proton transfer rates of acids with a different number of equivalent acidic protons and bases with a different number of equivalent basic sites, it is useful to make statistical corrections. Defining p_A (p_B) as the number of equivalent acidic protons of AH (BH), and q_A (q_B) the number of equivalent basic sites of A^- (B^-), the statistically corrected rates become

$$k'_f = \frac{k_f}{p_A q_B} \qquad k'_r = \frac{k_r}{p_B q_A} \qquad \ldots (3.10)$$

This correction has to be reflected in the equilibrium constant and, consequently, in the free-energy of the reaction, giving

$$\Delta G^0 = -2.303RT(pK'_{BH} - pK'_{AH}) \qquad \ldots (3.11)$$

where

$$K'_{AH} = \frac{q_A}{p_A}K_{AH} \qquad K'_{BH} = \frac{q_B}{p_B}K_{BH} \qquad \ldots (3.12)$$

In an acid catalyzed reaction such as that shown in Mechanism VIII, a linear relationship between the logarithm of the statistically corrected protonation rate of the substrate B and the pK of the acid catalyst is frequently found. This was first formulated by Brönsted and Pederson and is known as the Brönsted relation for acid catalysis:

$$k'_f = G_f (K'_{AH})^\alpha \qquad \ldots (3.13)$$

where G_f is a constant dependent on the temperature, pressure, medium and substrate, and α is a constant for acids of the same type, often called the Brönsted coefficient for acid catalysis. The reverse reaction in Mechanism VIII can be regarded as a base catalyzed reaction, and should follow a similar relationship:

$$k'_f = G_f \left(\frac{1}{K'_{AH}}\right)^\beta \qquad \ldots (3.14)$$

Brönsted and Pederson also pointed out that the relations (3.13) and (3.14) could only be followed by in limited range of K_{AH} values, because the rates can only increase until the diffusion-controlled limit is reached. Once this limit is reached, k_f becomes independent of K_{AH} and $\alpha = 0$. Now, k_f will be constant and, given the equilibrium condition expressed by eq. (3.7), k_r will change in inverse proportion to K_{eq}, which also means that it changes with $(K_{AH})^{-1}$ and $\beta = 1$. Thus, the sum of the Brönsted coefficient for acid catalysis in the forward direction with that of base catalysis in the reverse direction should be equal to unity, $\alpha + \beta = 1$. The current meaning of these coefficients was first proposed by Leffler [18]. They are associated with the localization of the transition state in a series of reactions between structurally related species. For acid catalysis α should approach the lower limit of zero for very exothermic reactions with rates controlled by diffusion, and should approach the upper limit of unity for very endothermic, slow, reactions (Figure 3.2). Although this reasoning restricts the values of both α and β to the interval [0,1], Bordwell and co-workers [19] have shown that in the deprotonation of a series of substituted nitroalkanes (AH in Mechanism VIII) by hydroxide ion (B^- in that mechanism) in methanol : water solution, $\alpha = 1.61$ and $\beta = -0.61$ for the reverse reaction. This was recently interpreted in terms of the overlapping of electronic effects with genuine free-energy effects [20,21].

It should be emphasized that the study of a forward reaction rate by general acid catalysis or by general base catalysis should yield the same localization for the transition state. That is, if the Brönsted coefficients are a measure of the extent of proton transfer in the transition state, the value of α for general acid catalysis of the reaction proceeding in the forward direction (AH varied, α_A), should be identical to the value of β obtained by general base catalysis of the

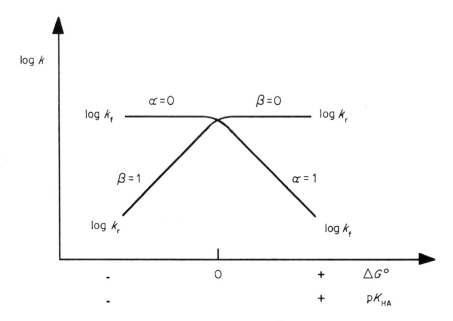

Figure 3.2. Eigen plots for the acid–base catalysis of strong (oxygen or nitrogen) acids or bases (Mechanism VIII). The forward direction ($\log k_f$) corresponds to acid catalysis by a series of acid catalysis AH as a function of thier pK_{AH}. The reverse direction ($\log k_r$) refers to base catalysis by a series of bases A^-

reaction proceeding in that same direction (B$^-$ varied, β_B), that is, $\alpha_A = \beta_B$. In fact, Bordwell and Boyle [22] found that the deprotonation of 1-arylnitroethanes by a series of amine bases have β_B values close to 0.55, thus completely unrelated to the anomalous α_A values obtained in the general acid catalysis. Given these anomalies, Brönsted coefficients should not be regarded as a measure of the extent of proton transfer in the transition state. They remain useful to classify and predict rates in families of structurally related catalysts.

As first pointed out by Bell [2] and later discussed in more detail by Yates and co-worker [23], the detection of general acid catalysis depends on the magnitude of the Brönsted α coefficient as well as on the strength of the acid catalyst. When α is close to 0 most of the catalysis is due to the solvent and the reaction should be regarded as uncatalysed, since the reaction rate is only slightly accelerated, even in the presence of strong acids. If α approaches 1, the catalytic effect of the buffer solution is dominated by that of the hydrogen ion it contains, and it is difficult to detect general acid catalysis by the acid buffer. Given the relation between forward and reverse reactions, when $\alpha = 1$ in the forward direction, the reverse reaction will have $\beta = 0$, so it will be diffusion controlled and uncatalysed. On the other hand, for base catalysis in the forward direction, when $\beta = 1$ (only specific base catalysis observed), the reverse reaction will have $\alpha = 0$ and will be diffusion controlled. Some cases are known where proton transfers from a series of carboxylic acids to a carbon base give $\alpha = 0$ but rates are much lower than the diffusion-controlled limit [24,25]. Jencks and Murray [25] argued that such cases demand that in order to account for the amount of bond formation to the base catalyst in the transition state, and consequently to account for k_f (k_r for the acid catalyst in the reverse direction), the observed value of β (α) should be corrected for a solvation (desolvation) step of the catalyst.

The solvation (desolvation) step can be analyzed in more detail dissecting the proton transfer shown in Mechanism VIII, either taking place in the ground or in the excited state, into three steps shown in Mechanism IX [26]. For simplicity of the notation, henceforward we will consider that the necessary statistical corrections have been made in all the rates and equilibria shown. In the first step of Mechanism IX, the solvated substrate and the acid catalyst diffuse together, partially desolvate and form an encounter complex. The following step is the actual proton transfer, within the complex. Finally, in the last step, the complex is broken down and the products diffuse apart.

$$B^- + HA \underset{k_{-a}}{\overset{k_{-a}}{\rightleftarrows}} B^- \cdots HA \underset{k_{-p}}{\overset{k_{-p}}{\rightleftarrows}} BH \cdots A^- \underset{k_{-s}}{\overset{k_s}{\rightleftarrows}} BH + A^-$$

Mechanism IX

According to this mechanism the forward rate constant for the proton transfer from HA to B is given by

$$k_f = \frac{k_a k_p k_s}{k_p k_s + k_{-a} k_s + k_{-p} k_{-a}} \qquad \ldots (3.15)$$

and the rate of the reverse proton transfer, from BH to A$^-$, is

$$k_r = \frac{k_{-a} k_{-p} k_{-s}}{k_p k_s + k_{-a} k_s + k_{-p} k_{-a}} \qquad \ldots (3.16)$$

When the proton transfer rate in the encounter complex, k_p, is much faster than the rates of complex formation and breakdown and $k_p > k_{-p}$, a limiting situation is reached where the observed forward rate constant is identical to the rate of complex formation:

$$k_f = k_a \qquad k_r = \frac{k_{-s}}{k_s} \frac{k_{-p}}{k_p} k_{-a} \qquad \ldots (3.17)$$

In this case, the forward rate constant is independent of the acidity of AH (or of the basicity of B$^-$), and $\alpha_A = 0$ ($\beta_B = 0$). Thus, no catalysis is observed. The dependence of the reverse reaction rate on the equilibrium constants reveals the features dictated by the equilibrium conditions already discussed. When the couple AH/A$^-$ is changed along the reaction series and B$^-$/BH is the substrate (acid catalysis in the forward direction, but base catalysis in the reverse one), the Brönsted coefficient affecting k_r has to be $\beta_A = 1$. However, if the reaction series involves changes in the B$^-$/BH couple and AH/A$^-$ is the substrate (base catalysis in the forward direction and acid catalysis in the reverse one), then $\alpha_B = 1$, as illustrated in Figure 3.3. According to considerations given by Bell and Yates, in these cases where $\beta_A = 1$ or $\alpha_B = 1$ occur, only specific catalysis by OH$^-$ or H$_3$O$^+$ respectively, can be observed.

Another limit for eqs. 3.15 and 3.16 is obtained if k_p much higher than the rates of complex formation and breakdown, but $k_p < k_{-p}$. In this case, the forward rate constant is given by a product of equilibrium constants and the rate constant for the breakdown of the successor complex:

$$k_f = \frac{k_a}{k_{-a}} \frac{k_p}{k_{-p}} k_s \qquad k_r = k_{-s} \qquad \ldots (3.18)$$

Here the forward rate depends on the equilibrium constant of the proton transfer step, that is, it is directly proportional to (pK_{AH} − pK_{BH}). As opposed to the limit given by eq. (3.17), the forward reaction is now endothermic. The consequences are analogous to those discussed above: acid catalysis will give $\alpha_A = 1$ and base catalysis will give $\beta_B = 1$. The reverse reaction will not be catalyzed.

In summary, when the energy barrier for proton transfer in the encounter complex is small, the proton transfer rates of the reactions in the exothermic direction approach the diffusion-controlled limit and are not sensitive to catalysis, while the rates corresponding to proton transfer in the endothermic reaction vary

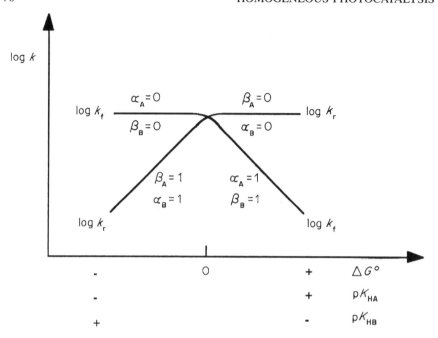

Figure 3.3. Eigen plots for the acid–base catalysis of strong acids or bases (Mechanism IX). For acid catalysis HA is the catalyst and is changed, while B^- is the substrate and is kept constant, in the forward direction the coefficient α_A is obtained and in the reverse direction β_A is obtained. For base catalysis B^- is changed while AH is kept constant, the coefficient β_B corresponds to the reaction in the forward direction and α_B to the reverse direction

in the same proportion as the pH (specific acid catalysis) or the pOH (specific base catalysis) of the solution.

Naphthols and pyrene derivatives in the S_1 state in aqueous solution have small, positive, pK_a^* values, and they meet the conditions leading to eq. (3.18). Gutman and co-workers [27] first pointed out that in the case of the forward proton transfer from these electronically excited oxygen acids to water molecules, the logarithm of the rate is linearly dependent on the pK_a^* of the acid, with a slope approaching unity. Later, Gutman [28] showed that structurally related molecules in the ground state could be included in that correlation. The rate of the reverse adiabatic proton transfer from H_3O^+ to the oxygen anions is diffusion controlled. Thus, the slope of the plot of $\log k_f$ vs. pK_{AH} for the forward, endothermic, reaction is $\alpha = 1$ and it formally a "specific acid catalyzed reaction", while the reverse, exothermic, reaction is uncatalyzed. This behavior is typical of oxygen acids [9] and it has been used in pH jump experiments. Heterocyclic compounds in S_1 are more basic than in their ground state and have been used in pOH jump experiments. This type of experiments will be discussed in more detail on p. 84.

A third limit to eqs. (3.15) and (3.16) is obtained if the interconversion between the precursor and successor complex is slow, $k_p < k_{-a}$ and $k_{-p} < k_s$. Then, the proton transfer becomes rate-limiting and a simple equation is obtained.

$$k_f = \frac{k_a}{k_{-a}} k_p \qquad k_r = \frac{k_{-s}}{k_s} k_{-p} \qquad \ldots (3.19)$$

The occurrence of this limit offers the possibility of observing general acid–base catalysis because intermediate values of α and β can be expected. Slow proton transfers often involve carbon acids or bases and they are the best candidates for general acid–base photocatalysis.

It has been frequently argued in the literature that the slowness of the proton transfers to/from carbon atoms is due to the need to reshape their solvation shell in the course of the reaction. However, high barriers for proton transfer to/from carbon atoms were also found in the gas phase and attributed to the low polarity of the C–H bonds [29]. In the light of these results, it seems more reasonable to view the first and last steps of Mechanism IX as diffusion of the species together/apart, than as solvation/desolvation processes respectively.

One of the best examples of general acid photocatalysis is the photohydration of aromatic alkenes and alkynes, investigated by Yates and co-workers [23]. The acid catalysis introduces curved Brönsted plots with α coefficients centered in the range 0.14–0.18. The analogous ground state reactions have α coefficients in the range 0.5–0.86. Several theoretical models were tested for their capacity to interpret these results [30]. A quantitative account of the variation of the rates with the free energy of the reactions was offered [31], presuming that these photohydrations are diabatic processes.

4 THE FÖRSTER CYCLE

In 1950 Förster [32] proposed a method to calculate the pK_a^* of excited species that became very popular. This cycle combines thermodynamic and spectroscopic data to relate the proton transfer equilibrium in the ground state to that in the excited state. The application of the cycle consists essentially in measuring the frequencies of the electronic transitions between the lowest vibrational levels of the ground and excited states, both for the acidic, ν_{AH}, and basic, ν_{A^-}, forms of the species. As shown in Figure 3.4, the enthalpy changes of the acid ionization equilibrium in the ground, ΔH, and excited state, ΔH^*, are related to the frequencies ν_{AH} and ν_{A^-}.

$$N_A h \nu_{AH} + \Delta H^* = N_A h \nu_{A^-} + \Delta H \qquad \ldots (3.20)$$

where N_A is the Avogadro number. If the solutions are dilute enough for ΔH to approximate the standard value ΔH^0, and if ΔS^0 does not change appreciably

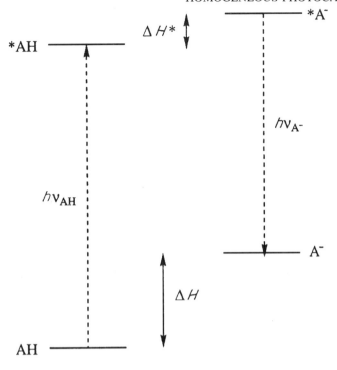

Figure 3.4. Förster cycle for the acid–base equilibrium $^*AH + H_2O \rightleftharpoons {}^*A^- + H_3O^+$

upon excitation, that is, given

$$\Delta H^* - \Delta H \approx \Delta G^{*0} - \Delta G^0 \qquad \ldots (3.21)$$

then

$$pK_a^* = pK_a + \frac{N_A h(\nu_{A^-} - \nu_{AH})}{2.303RT} \qquad \ldots (3.22)$$

In principle, either the absorption or the emission bands of the conjugated acid–base pair can be used to calculate its ΔpK_a values. However, even when the absorption and emission bands involve the same electronic configurations, the ΔpK_a values obtained from the absorption or emission bands differ because the solvent shell, bond lengths and bond angles may relax following the electronic transition. The recommended procedure to calculate ΔpK values using the Förster cycle is to take as ν_{AH} the intersection point of the mutually normalized absorption and emission spectra of the acidic form in solution, and identically for and ν_{A^-} using the spectra of the basic form. A good discussion of the Förster cycle was presented by Grabowski and Grabowska [33].

An authoritative account of photoinduced acidity constant changes, most of them calculated by the Förster cycle was offered by Ireland and Wyatt [34] in 1976 and still is the most exhaustive source of acidity constants of electronically excited molecules. There are updates [9,35] for some of the acidity constants reported by Ireland and Wyatt, but the general picture given by these authors remains essentially correct: in the first singlet excited state of aromatic molecules the functional groups R−OH, R−NH$_2$, R−NH$_3{}^+$ become stronger acids (weaker bases), while the groups R$_2$C=OH$^+$, R−CO$_2$H, R−CO$_2$H$_2{}^+$, R−SO$_3$H$_2{}^+$, R−PO$_3$H$_3{}^+$, R−AsO$_3$H$_3{}^+$ and R−NO$_2$H become stronger bases (weaker acids). The acidity constants of the triplet states are usually intermediate between those of the singlet and ground states. These changes in the acidity constants result from changes in the equilibrium position of the acids and their conjugated bases, both in the same electronic state. Simultaneous luminescence of the acidic and basic forms of some excited state species have been observed and shown to be in equilibrium. Given eq. (3.22), any theory that can account for the relative displacements of the electronic transitions of the acidic vs. basic forms in equilibrium, will also be able to explain the changes in their equilibrium upon electronic excitation.

The reliability of the Förster cycle can be assessed comparing the p$K_a{}^*$ values calculated through this cycle with the values obtained from the direct measurement of excited state protonation and deprotonation rates [9]. That comparison shows that for naphthylamines the p$K_a{}^*$ values obtained through the Förster cycle are ca. 5 pK units lower than the presently accepted values, while for 1-naphthol the Förster cycle gives a value 1.5 pK units too high. This discrepancy have been assigned to an inversion of the electronic levels during the lifetime of the excited state. A similar comparison for phenanthrylamines shows that the pK^* obtained through the Förster cycle is 1−5 pK units too low, and this was correlated to the large Stokes shifts exhibited by these molecules, meaning that the relaxation of the excited state may release up to 30 kJ/mol ($\approx$7 kcal/mol).

Daudel [36] argued that when the wavefunction corresponding to the transfer of one electron from an electron donor substituent to an electron acceptor substrate makes a small contribution to the ground-state wavefunction of the whole molecule, that charge-transfer wavefunction will become more important in the first singlet excited state. Thus, it can be anticipated that electron donor substituents, like OH and NH$_2$, become stronger donors in the excited state, while acceptors, like COOH and CO, become stronger acceptors. In the first case, the oxygen and nitrogen atoms will lose electron density in the excited state, the protons attached to them will be less firmly bound, and they will become stronger acids (or weaker conjugated bases). The decrease of the pK_a of aromatic alcohols from the S$_0$ to the S$_1$ states ranges from 7 to 9 pK_a units and that of aromatic amines is about half of this. The opposite reasoning applies to the deprotonation of carboxylic acids, for which increases in pK_a may reach 6 units upon electronic excitation, and to the protonation of carbonyl groups, where pK_a increases in the range 6−11 have been reported [34].

Some of the most dramatic ΔpK_a changes that have been reported involve the protonation of carbon bases (pK_a increases of 16 and 23 units were calculated for

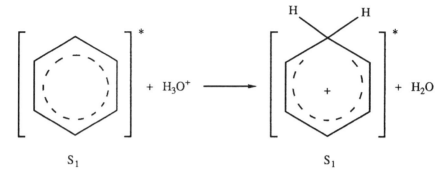

Figure 3.5. Protonation of an aromatic hydrocarbon in the S_1 state, facilitated by the loss of electron delocalization

naphthalene and phenanthrene respectively) [34] or the deprotonation of carbon acids (pK_a decreases of nearly 30 and 40 units was claimed for fluorene and suberene respectively) [37].

The increase in basicity observed in the protonation aromatic alternant hydrocarbons can be related to loss of electron delocalization in the S_1 state (Figure 3.5), which renders the protonation at the carbon thermodynamically favorable. This basicity increase is not accompanied by appreciable proton exchange: hydrogen–deuterium exchange of deuterionaphthalenes proceeds with small quantum yields (0.01–0.04) in moderately concentrated H_2SO_4 (54–60%), with a slight preference for the β position [38]. These results indicate that, in spite of their exothermicity, simple unactivated aromatic compounds require strong acids to photoprotonate to a measurable extent, as do the corresponding endothermic ground state protonations. The small quantum yields measured, regardless of an estimated enhancement by 13–14 orders of magnitude upon excitation of the rate of protonation, are due to the short singlet lifetime in the experimental conditions employed (100 ns). In analogous conditions, the ground state exchanges show a preference for exchange at the α position [39].

The increase in acidity of benzylic protons in the excited singlet state can be assigned to an enhanced stability of the carbanions in the singlet state. The electronic charge of the benzylic carbanion can be delocalized into the aromatic rings, because they can be regarded as electron acceptors in aliphatic systems, which became stronger acceptors in the excited state. The fluorenyl carbanion has an internal cyclic array of 6 π-electrons ($4n + 2$ π-electrons, $n = 0, 1, \ldots$) and the suberenyl carbanion has 8 π-electrons ($4n$ π-electrons). In ground-state chemistry, the former has been judged 'aromatic', while the latter is given as an example of an "anti-aromatic" system. However, they have similar pK_a^* values: -8.5 to -4 were reported for fluorene and -7 for suberene [37]. Although their thermodynamic acidities cannot be distinguished by Förster cycle calculations, their kinetic acidities are remarkably different: photolysis of suberene in 50%

D$_2$O/CH$_3$CN (v/v) lead to conversion to α-deuteriosuberene with a quantum yield of 0.03, while photolysis of fluorene under similar conditions resulted in no observable deuterium incorporation [40]. This is a prime example of the difference between thermodynamic and kinetic acidities that calls for a detailed discussion of the kinetics of excited-state proton transfer reactions.

In summary, Förster cycle calculations are a valuable source of pK_a* values, but they do not guarantee that the forward or reverse proton transfer rates can both take place during the lifetime of the excited state. The excited state acidity constants thus obtained give an idea of the relative positions of the energy levels of the acidic and basic forms of the species in consideration. This should be distinguished from the connection of these electronic states by adiabatic reaction paths. If these reaction paths cannot be significantly followed in the forward and reverse directions, the excited state equilibrium is not established and the diabatic channels dominate the reactivity of the species.

5 THE INTERSECTING-STATE MODEL

The use of the rate expression given by the transition state theory (eq. 3.5), to calculate a reaction rate constant in solution, requires the knowledge of its free energy of activation. The calculation of $\Delta G^{\ddagger}$ for a chemical reaction in solution is a very difficult problem and an exact solution is inaccessible. This problem has been addressed with some success by simple theoretical models, which reduce the representation of the reaction path to one or two progress variables. Such variables, the theoretical reaction coordinate(s), are presumed to reflect the most important factors that determine $\Delta G^{\ddagger}$.

Two of the most relevant quantitative models that have been applied to proton transfer reactions are the theory of Marcus [41-44] and the intersecting-state model (ISM) [21,31,35,45-48]. Recent reviews [9,30,49] covering the application of kinetic models to excited-state proton transfer reactions have shown that the use of Marcus theory yields physically unreasonable parameters, while ISM offers an interesting link between free energies of activation and the structure of the reactants. Consequently, we recommend ISM as the kinetic model to apply for proton transfer reactions, and discuss in detail how to calculate proton transfer rates with this model.

ISM is a kinetic model designed to estimate activation energies from the knowledge of some properties of the reactants and of the energy of reaction. The theoretical basis of ISM has been discussed in detail elsewhere [50,51]. Here we will be concerned with the methodology of its application to proton transfer reactions.

The first step in the application of ISM is to identify the bonds that suffer large geometry and/or frequency changes in the course of the reaction, because they determine the reaction coordinate. In the proton transfer reaction illustrated by

Mechanism IX these are the H−A bond for the reactants and the H−B bond for the products. In some cases other bonds also suffer major changes in the course of the reaction and have to be included in the reaction coordinate. For example, in the thermal acid catalysis of alkanes, alkynes and aromatic hydrocarbons, the rate-determining step is the addition of the proton to a multiple CC bond, that becomes less unsaturated in the products. In such cases, the CC bond has to be included in the reaction coordinate [31].

The second step is to construct the energy profile of the reaction coordinate as shown in Figure 3.6. In this figure the reactants and products are represented by harmonic oscillators; this is not a requirement of the model, as they can be also represented by Morse curves. The energy of activation corresponds to the intersection of the curves representing reactants and products. The use of harmonic oscillators gives an analytical solution to this intersection, but this simplifying feature should only be used when the reactions are not very exothermic. According to Figure 3.6, the reactants bond distension to the transition state configuration, Δx_r, can be obtained from

$$\tfrac{1}{2} f_r (\Delta x_r)^2 = \tfrac{1}{2} f_p (d - \Delta x_r)^2 + \Delta G^0 \qquad \ldots (3.23)$$

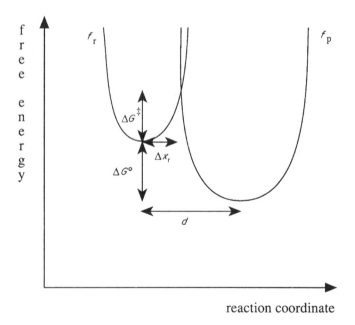

reaction coordinate

Figure 3.6. Free energy profile along the reaction coordinate defined by ISM, representing reactants and products by harmonic oscillators. The total bond extension of the reactants and products to the transition state configuration is represented by the parameter d

HOMOGENEOUS PROTON TRANSFER PHOTOCATALYSIS

and used to calculate the activation free energy,

$$\Delta G^{\ddagger} = \tfrac{1}{2} f_r (\Delta x_r)^2 \qquad \ldots (3.24)$$

In these equations, f_r and f_p are the force constants of the A—H and H—B bonds respectively and d is the total bond distension of the A—H and H—B bonds to the configuration of the transition state. If the reaction is adiabatic, ΔG^0 can be calculated from eqs. (3.9) or (3.11). The force constants and acidity constants required to perform calculations on the most common proton transfer reactions are usually available in the literature. The calculation of d requires the definition of a criterion that determines, in the spirit of this unidimensional model, the configuration of the transition state.

The following step in the calculation of a rate constant using ISM is the determination of the separation between reactants and products along the reaction coordinate, that is, the calculation of the parameter d. In order to obtain a simple expression for this parameter and understand the limitations and advantages of ISM, it is convenient to make an incursion into the theoretical foundations of this model. The reader not interested in the most theoretical aspects of this work can skip this section up to eq. (3.39).

According to the definition of d given above,

$$d = (l^{\ddagger}_{AH} - l_{AH}) + (l^{\ddagger}_{HB} - l_{HB}) \qquad \ldots (3.25)$$

where l_{AH} and l_{HB} are the equilibrium bond lengths of the AH and HB bonds respectively, and $l^{\ddagger}_{AH}$ and $l^{\ddagger}_{HB}$ are those bond lengths at the transition state configuration. The problem of calculating $\Delta G^{\ddagger}$ can now be formulated in terms of estimating the distension of the reactive bonds from their equilibrium positions to their lengths at the transition state. Pauling [52] has shown that there is an empirical relation between equilibrium bond lengths and bond orders. Extending Pauling's relation to bond lengths and bond orders at the transition state, we obtain

$$\ln n^{\ddagger}_{AH} = -\frac{1}{a}(l^{\ddagger}_{AH} - l_{AH}) \qquad \ln n^{\ddagger}_{HB} = -\frac{1}{a}(l^{\ddagger}_{HB} - l_{HB}) \qquad \ldots (3.26)$$

where $n^{\ddagger}_{AH}$ and $n^{\ddagger}_{HB}$ are the transition state bond orders of the AH and HB bonds respectively, and a is an empirical parameter adjusted by Pauling. Assuming that, like in bond-energy–bond-order model (BEBO) [53,54], the bond order is conserved along the reaction coordinate,

$$n_{AH} + n_{HB} = 1 \qquad \ldots (3.27)$$

we can define a reaction coordinate $n = n_{HB}$, such that $n = 0$ for the reactants and $n = 1$ for the products. The entropy variation along the reaction coordinate n can be simulated by a function $M(n)$ provided that, when $\Delta G^0 = 0$, certain conditions are met, namely [55]: $M(0) = 0$, $M(1) = 1$, $M(n) = M(1-n)$, $M(n) \geq 0$ and $M(n)$ has only one maximum that occurs at $n^{\ddagger} = n/2$. Now that entropy has been taken into account, the variation of the free energy along the reaction coordinate can be written

$$G(n) = \Delta E n + \lambda M(n) \qquad \ldots (3.28)$$

where λ is an energy parameter that weighs the relative contribution of the entropy variation to the variation of the free energy of the system along the reaction coordinate. The term on ΔE has been designated the purely thermodynamic contribution to the reaction free energy profile, and the term $\lambda M(n)$ the purely kinetic contribution. One possible expression for $M(n)$ is

$$M(n) = -n \ln n - (1-n)\ln(1-n) \qquad \ldots (3.29)$$

This choice of $M(n)$ makes eq. (3.28) equivalent to a form derived by Marcus [41] as a Taylor approximation to the empirical BEBO energy profile [55]. The transition state bond order, $n^{\ddagger}$, can be obtained deriving eq. (3.28), setting

$$\frac{\partial G(n)}{\partial n} = \Delta E - \lambda \ln \frac{n^{\ddagger}}{1 - n^{\ddagger}} = 0 \qquad \ldots (3.30)$$

and resolving for $n^{\ddagger}$.

At this stage, it is appropriate to reflect upon one of the approximations involved in the calculation. As presented in Figure 3.6, estimation of the activation energy neglects the effect of resonance between reactant and product configurations at the transition state. This effect may be important because the wavefunction describing this region must mix reactant and product states. As a result, there is a splitting of the surfaces at the configuration where they intersect, as shown in Figure 3.7. Resonance splitting leads to a lower energy transition state. In the framework of ISM this can be regarded as a formal enhanced bonding of the transition state, that is, as an increase of the bond order at the transition state, $n^{\ddagger}_{AH} + n^{\ddagger}_{HB} > n_{AH} + n_{HB}$. A simple expression to account for the enhancement of the bond order along the reaction coordinate, yet preserving the transition state location and respecting the asymptotic limits, is

$$n_{AH} = n^{1/m} \qquad n_{HB} = (1-n)^{1/m} \qquad \ldots (3.31)$$

where $m \geq 1$ (Figure 3.8) It is important to recall that the only energy points with thermodynamic significance along a reaction coordinate are those that correspond to equilibrium configurations, that is, configurations of mechanical equilibrium characterized by zero slopes for the potential energy [56]. In a typical potential energy vs. reaction coordinate profile there are only three such configurations: the reactants, the products and the transition state. The connection between these points is only a mathematical object, with the same value in chemical kinetics as atomic orbitals in theoretical chemistry [57].

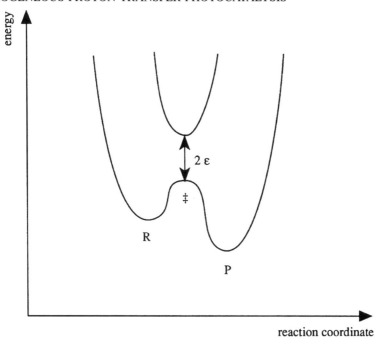

Figure 3.7. Free energy profile along the reaction coordinate, including the effect of resonance at the transition state configuration, ‡. The resonance energy at the transition state is represented by ε

Replacing n by its value at the transition state, $n^{\ddagger} = \frac{1}{2}$, one obtains the maximum for $n_{AH} + n_{HB}$,

$$n_{AH}^{\ddagger} + n_{HB}^{\ddagger} = 2^{1-1/m} \qquad \ldots(3.32)$$

The enhancement of the bonding at the transition state will depend on the value of m. This will be determined by the electronic nature of reactants and products. In principle, m should be susceptible to the flow of electrons from the nonreactive bonds of the reactants to the transition state, which can be measured by substituent parameters representing dipolar field/inductive, π-electron delocalization/hyperconjugation and polarizability effects. If m is a measure of electron count, for the large majority of cases its value should be smaller than 4. Within this limit, Figure 3.9 shows that the following approximation is valid

$$2^{1-1/m} = \sqrt{m} \qquad \ldots(3.33)$$

ISM accounts for the effect of resonance in the transition state by allowing the sum of the bond orders of the reactive bonds at the transition state to be larger than their sum at the equilibrium. In general, for a symmetric proton exchange ($\Delta E^0 =$

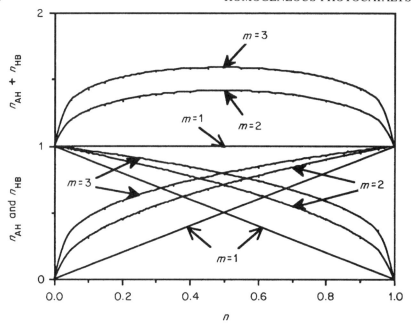

Figure 3.8. Enhancement of the bond order of the fragments AH and HB along the reaction coordinate $0 < n < 1$. The maximum of the total bond order at the transition state increases with the value of m as shown by eq. (3.32) in text

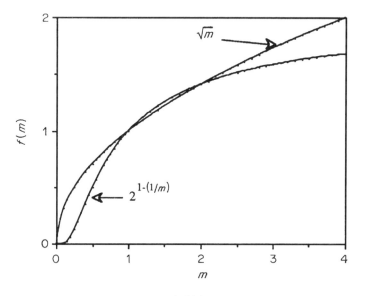

Figure 3.9. Variation of the functions $2^{1-(1/m)}$ and $\sqrt{m}$ as a function of the value of m showing their similarity in the interval $[0.5, 3]$

0), the bond order of each bond at the transition state is $n^{\ddagger}_{AH} = n^{\ddagger}_{HB} = \sqrt{m}/2$. On the other hand according to the definition of the reaction coordinate, $0 \le n \le 1$, if $\Delta E^0 = 0$ then the transition state bond order is $n^{\ddagger} = \frac{1}{2}$. We have to reconcile the resonance effect in the transition state with its bond order, without changing its location. One way of doing this is to establish the conditions that determine the value of $n^{\ddagger}$ when $n^{\ddagger}_{AH} = n^{\ddagger}_{HB} = 1/2$, (eq. 3.30), and then normalize the kinetic term (the only one subsisting when $\Delta E^0 = 0$) by the maximum bonding attained by the transition state, $\sqrt{m}$, before resolving for $n^{\ddagger}$

$$\frac{\partial G(n)}{\partial n} = \Delta E - \frac{\lambda(m)}{\sqrt{m}} \ln \frac{n^{\ddagger}}{1 - n^{\ddagger}} = 0 \qquad \ldots (3.34)$$

We can now express the location of the transition state in bond order coordinates as

$$n^{\ddagger} = \frac{1}{1 + \exp[-\sqrt{(m)}\Delta E/\lambda(m)]} \qquad \ldots (3.35)$$

In the following, we will not explicit the dependence that λ may have on the value of m. Taking the logarithms of eqs. (3.31) and (3.35) and using eq. (3.26), we obtain

$$l^{\ddagger}_{AH} - l_{AH} = \frac{a}{m} \ln \left[1 + \exp(-\sqrt{(m)}\Delta E/\lambda) \right]$$

$$l^{\ddagger}_{HB} - l_{HB} = -\frac{a}{m} \ln \left[1 - \frac{1}{1 + \exp(-\sqrt{(m)}\Delta E/\lambda)} \right] \qquad (3.36)$$

The resonance effect in the transition state, $m > 1$, is translated by ISM as a transition state configuration with a smaller bond distension. Given the definition of d, eq. (3.25), the sum of the bond distensions from the equilibrium to the transition state configuration will be given by

$$d = \frac{a}{m} \ln \frac{1 + \exp(-\sqrt{(m)}\Delta E/\lambda)}{1 - [1 + \exp(-\sqrt{(m)}\Delta E/\lambda)]^{-1}} \qquad \ldots (3.37)$$

The Pauling relation assumes that a is a constant relating to equilibrium bond lengths and chemical bond orders. It is reasonable to expect that when the equilibrium bond lengths are small (large), their distension to the transition state configuration will be small (large). Thus, d must be proportional to the equilibrium bond lengths of the reactive bonds

$$d = \frac{a'}{m}(l_{BC} + l_{AB}) \ln \frac{1 + \exp(-\sqrt{(m)}\Delta E/\lambda)}{1 - [1 + \exp(-\sqrt{(m)}\Delta E/\lambda)]^{-1}} \qquad \ldots (3.38)$$

The proportionality constant a' will be different from that determined by Pauling. It remains an empirical constant that can be obtained by fitting the distance d

to simple gas phase systems with reliable potential energy surfaces. Such fitting can best be done after expanding eq. (3.38) in a Taylor series and retaining the values to the second power of ΔE^0,

$$d = \left[\frac{2a' \ln(2)}{m} + \frac{a'}{2}\left(\frac{\Delta E}{\lambda}\right)^2\right](l_{AH} + l_{HB}) \qquad \ldots (3.39)$$

Fitting eq. (3.39) to atom transfer reactions in the gas phase gives $a' = 0.156$ [50]. These reactions also exhibit the expected linear dependence of the ratio of d by the sum of the bond lengths $(l_{AH} + l_{HB})$ on the square of the reaction energy ΔE^0. The parameters λ and m can be calculated from the slope and intersect respectively, of that plot. For simple atom transfer reactions in the absence of significant resonance effects, like in $H + H_2$ exchange, $m = 1$ is obtained; in such cases the transition state bond order is not enhanced, $n^{\ddagger} = \frac{1}{2}$. In atom transfers of the type $X + H_2 \longrightarrow XH + H$, where X is a halogen atom, $m = 2$, which means that an extra pair of electrons acquires bonding character at the transition state. Finally, in reactions of the class $H + X_2 \longrightarrow HX + X$, $m = 3$ is obtained, that is, two extra pairs of electrons contribute to the transition state bond order. In general, the values of m can be rationalized by the increased-valence structures proposed by Harcourt [58]. Given the relation between m and the electronic density on the reactive bonds at the transition state, we will use $m = n^{\ddagger}_{AH} + n^{\ddagger}_{HB}$, with the understanding that it is not the relative position of the transition state along the reaction coordinate that we are changing, but its bond order that we are setting.

The value of λ is a measure of the entropy of activation. It has been related to the topography of the potential energy surfaces and to the conversion of kinetic energy of the reactants into vibrational energy of the products. However, for reactions in solution, λ has to be treated as an empirical parameter, constant for a family of reactions. For proton transfers to/from carbon, values of λ larger than 100 kJ/mol have been obtained. Proton transfers involving only oxygen or nitrogen acids or bases gave $\lambda < 120$ kJ/mol. Given eq. (3.39), the knowledge of λ is most relevant for the calculation of rate constants of very exothermic reactions and of reactions with small λ values.

We are now in a position to proceed with the third step of the calculation of proton transfer rate constants by ISM. The distance d can be calculated from eq. (3.39), given reasonable estimates for $n^{\ddagger}$ and λ. In general, reactions involving C–H bond breaking or bond forming are not expected to involve large resonance effects, thus a reasonable estimate is $1 < m < 1.2$. On the other hand, if the transfer of a proton between two oxygen atoms is implicated in the reaction coordinate, the nonbonding electrons of the oxygen atom may acquire a large bonding character at the transition state and $1.8 < m < 2.2$ is expected. The value of $n^{\ddagger}$ for the transfer of a proton between two nitrogen atoms should be similar to that obtained for the transfer between oxygen atoms, but slightly smaller.

HOMOGENEOUS PROTON TRANSFER PHOTOCATALYSIS

In eqs. (3.38) and (3.39) ΔE can used either as the difference in potential energy between products and reactants in gas phase reactions, or as the free energy of reaction in solution. This parameter represents a well-defined thermodynamic quantity that can be used together with a transition state defined by a parameter d that accounts for the entropy variation along the reaction coordinate.

In summary, a simple calculation of a proton transfer adiabatic rate is possible if the reaction free energy is not very large. Then, the reactants and products can be represented by harmonic oscillators and eq. (3.39) can be employed. In such a calculation, it is necessary to have literature data on the equilibrium bond lengths and force constants of the reactive bonds, as well as on the acidity constants of the acid and base exchanging the proton. With these data, the ΔG^0 value of the reaction can be calculated from eq. (3.11), and, given reasonable estimates for $n^{\ddagger}$ and λ, it is possible to employ eq. (3.39) to calculate d. Next, the value of d is used in eq. (3.23) to calculate Δx_r. The value of this parameter in then used in eq. (3.24) to calculate $\Delta G^{\ddagger}$. Finally, considering that the reaction is adiabatic, $\chi_{el} = 1$, the rate constant can be obtained from eq. (3.5). Typical values of the most current bond lengths, force constants and bond dissociation energies are presented in Table 3.1. Examples of calculations with ISM will be given on pages 87–88.

If the reactions are very exothermic, the products have to be represented by a Morse oscillator:

$$E_{HB} = D_{HB}\{1 - \exp[-\beta_{HB}(d - \Delta x_r)]\}^2 + \Delta G^0 \qquad \ldots (3.40)$$

where D_{HB} is the dissociation energy of the HB bond and β_{HB} is the Morse parameter:

$$\beta_{HB} = \left(\frac{f_p}{2D_{HB}}\right)^2 \qquad \ldots (3.41)$$

The reactants can still be represented by harmonic oscillators if the barrier for the reaction is not very high. This is always true for excited-state proton transfer reactions. The other requirement of very exothermic reactions is the use of eq. (3.38).

Table 3.1. Bond lengths, force constants and bond enthalpies of bonds frequently encountered in proton transfer reactions

Bond type	Bond length[a] (pm)	Force constant[a] (J mol^{-1} pm^{-2})	Mean bond enthalpy[b] (kJ mol^{-1})
O–H	97	420	463
O–H (water)	95.8	420	492
N–H	101	380	388
C–H	107	290	412

[a] A. J. Gordon, and R. A. Ford, *The Chemist's Companion* (Wiley, New York, 1975) pp. 107, 114.
[b] P. W. Atkins, *Physical Chemistry, 5th edn* (Oxford Univ. Press, Oxford, 1994) p. C7.

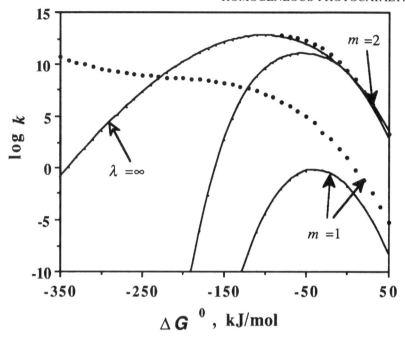

Figure 3.10. Calculation of rate constants according to ISM, as a function of the reaction free energy. The lines were calculated using harmonic oscillators to represent reactants and products and employing eqs. 23 and 39. The dots were calculated using an harmonic oscillator to represent the reactants and a Morse oscillator to simulate the products, together with eqs. 48 and 38. Unless otherwise marked, these calculations used $\lambda = 110$ kJ/mol and $m = 1$ or $m = 2$. Other parameters used: $f_r = 290$ kJ/mol, $f_p = 420$ kJ/mol, $l = 202.8$ pm and $D_e = 492$ kJ/mol

In Figure 3.10 we compare the results of calculations using the harmonic approximation and eq. (3.39) with those obtained with a harmonic and a Morse oscillator and eq. (3.38), as a function of the reaction free energy.

The calculation of diabatic factors is not within the scope of ISM. State correlation diagrams can be employed to predict the possible existence of diabatic factors, but calculations giving the order of magnitude of such factors are very difficult to perform.

6 pH AND pOH JUMP EXPERIMENTS

The fast adiabatic proton transfer taking place in the S_1 state of some aromatic alcohols and amines with appreciable ΔpK_a values, can be used to produce sudden local variations of the pH or pOH. The first studies of pH jump experiments were reported by Clark et al. [59] and by Smith et al. [27]. Typically, an aqueous solution containing a sulfonated aromatic alcohol is irradiated by a relatively intense picosecond laser pulse that excites the aromatic alcohol to its S_1 state. Such alcohols have $\Delta pK_a \approx -8$, so it is easy to find experimental conditions where the alcohol is almost completely in its neutral form in the ground state, but ionizes very rapidly and extensively in the excited state. In water, the ionization rate of sulfonated aromatic alcohols in their S_1 state is larger than 10^7 $M^{-1} s^{-1}$ (water concentration taken as 55.5 M). Thus, the ionization in the excited state can take place in less than 1 ns. The protonation of the excited state anion formed proceeds with a rate constant smaller than 10^{11} $M^{-1} s^{-1}$. If the pH of the solution drops to 3 upon electronic excitation of the aromatic alcohol, the lifetime on the anion in the S_1 state will still be 10 ns. This is also the order of magnitude of the singlet lifetimes of these molecules at room temperature when only physical decay processes are operative. Once the anion decayed to the ground state, it will reprotonate to regenerate the aromatic alcohol in the ground state. The rate of ground state thermodynamically favoured protonation of the anion to form the aromatic alcohol is diffusion controlled. Thus, with sufficiently intense and short laser pulses, it is possible to generate pH jumps with a nanosecond lifetime. The reaction controlling the decay of free proton concentration is the diffusion-controlled recombination of H^+ with the ground-state ionized proton emitter, whose rate constant is ca. 10^{11} $M^{-1} s^{-1}$. u

The pH jump technique is capable of generating 10^{-5}–10^{-4} mol of protons within a few nanoseconds. The photogenerated proton may react with any component in the solution. The reaction of the proton with any stable compounds may be followed at any timescale, but the reaction with the excited state anion is measurable only during the few nanoseconds of its lifetime. This short observation window focus the rate measurements only on those protons that may diffuse to encounter the excited state anion during its lifetime. The processes relevant in this systems are illustrated in Mechanism X.

$$\{^*ArOH(aq)\} \underset{k_p}{\overset{k_{-p}}{\rightleftarrows}} \{^*ArO^-(aq) + H^+(aq)\} \underset{k_{-s}}{\overset{k_s}{\rightleftarrows}} {^*ArO^-}(aq) + H^+(aq)$$

$$\downarrow k_{f,ic} \qquad\qquad \downarrow k_{f,ic} \qquad\qquad\qquad \downarrow k_{f,ic}$$

$$\{ArOH(aq)\} + h\nu_{AH} \quad \{ArO^-(aq) + H^+(aq)\} + h\nu_A \quad ArO^-(aq) + H^+(aq) + h\nu_{A^-}$$

Mechanism X

In a homogeneous solution the reaction volume where excited state aromatic anion can be probed for the presence of a proton is determined by a radius r, such that [28]

$$r = \sqrt{(D_{H^+} \tau_{A^-})} \qquad \ldots (3.42)$$

where D_{H^+} is the diffusion coefficient of the proton and τ_{A^-} is the lifetime of the excited state anion. The dynamics of the protons cannot be followed experimentally. However, the fluorescence of $^*ArOH$ and of $^*ArO^-$ have characteristic lifetimes, τ_{AH} and τ_{A^-} respectively. Following the fluorescence of the excited states, it is possible to obtain information concerning the local environment of the emiting aromatic alcohol or its anion. For example, if the proton emitter is placed in a microcavity, the probed volume is smaller than the reaction volume and the recombination rate will be faster, reflecting the high formal concentration of protons in the site; however, if the cavity is open to the bulk, some of the sites will irreversibly lose their protons within the lifetime of *ArO$^-$, the recombination cannot occur, these sites can be equated to the *ArO$^-$ ions that cannot be reprotonated and distinguished from the rest. The application of the pH jump technique to biochemistry has been discussed in detail by Gutman [28].

Pines and Fleming [60] have also used the pH jump technique to probe the structure of water and of the proton solvation shell in water–organic solvent mixtures. The probes utilized include aromatic alcohols of the type described above as well as the protonated 1-aminopyrene molecule.

The pOH jump technique is based on the same principles as the pH jump, but makes use of photobases (PhN) that, in the S_1 state, rapidly abstract a proton from water, producing equal amounts of a cation in the S_1 state and OH$^-$. The two molecules that have been used in pOH jump experiments are acridine (pK_a = 5.3, pK_a^* = 10.7) and 6-methoxyquinoline (pK_a = 5.0, pK_a^* = 11.8) [61,62]. The interesting feature of these experiments is that, after the return of the cation to the ground state, there is competition between its recombination with OH$^-$, its reaction with the basic form of an indicator (In$^-$) present in solution, and the reaction of the excess OH$^-$ generated with the acidic form of that indicator (HIn). These reactions, that occur in the ground state, are illustrated in Mechanism XI.

$$OH^-(aq) + PhNH^+(aq) \rightleftharpoons H_2O(l) + PhN(aq)$$

$$In^-(aq) + PhNH^+(aq) \rightleftharpoons HIn(aq) + PhN(aq)$$

$$OH^-(aq) + HIn(aq) \rightleftharpoons H_2O(l) + In^-(aq)$$

Mechanism XI

If the reaction between PhNH$^+$ and In$^-$ predominates, a transient acidification of the indicator will be observed. However, if the dominant reaction is that between OH$^-$ and HIn, then a transient alkalinization of the indicator will result. Given that $\Delta[PhNH^+] = \Delta[OH^-]$, it is the initial ratio of [HIn]/[In$^-$] that will determine whether an acidification or an alkalinization of the indicator will be observed. This ratio depends on the pK_a of the indicator and the initial pH of the solution. The competition between these reactions is observed in the microsecond time scale.

The application of ISM to estimate excited-state proton transfer rate constants can be illustrated by three simple systems: the deprotonations of 2-naphthol and of 8-hydroxy-1,3,6-pyrene trisulfonate (HPTS) in water, and the protonation of 6-methoxyquinoline also in water. HPTS and 6-methoxyquinoline were chosen for their importance in pH and pOH jump experiments, and the choice of two alcohols allows us to cover the range of rate constants observable in excited-state proton transfers. Detailed calculations on these systems have already been performed. Here we wish to approach these systems at the simplest theoretical and computational level. Thus, advantage is taken of the modest ΔG^0 changes involved in these adiabatic proton transfers to neglect the effect of the term on λ in eq. (3.39) and to use harmonic oscillators. The calculations involved are so simple that they are easily performed with a pocket calculator.

The deprotonation of the alcohols in water can be represented by the chemical equation

$$^*\text{ArOH} + \text{H}_2\text{O} \longrightarrow {}^*\text{ArO}^- + \text{H}_3\text{O}^+$$

The force constant and bond length of the reactants are that of the O–H bond of the alcohol and can be taken from Table 3.1. The O–H bond of water represents the products, and the relevant data is also available in Table 3.1. Making the approximation that $\lambda \gg \Delta G^0$ and using $m = 2$ for proton transfers to and from oxygen atoms, as previously discussed, eq. (3.39) gives $d = 20.85$ pm. The free energy of the reaction can be calculated from the equilibrium constant of the chemical equation, which is given by

$$K_{\text{eq}} = \frac{K^*_{\text{ArOH}}}{[\text{H}_2\text{O}]} \qquad \ldots (3.43)$$

Using $pK_{a^*} = 2.72$ for 2-naphthol and $pK_{a^*} = 1.28$ for HPTS, we obtain $\Delta G^0 = 25.5$ kJ/mol and 17.3 kJ/mol at 298 K respectively. For the particular case of the deprotonation of alcohols in water, $f_r \approx f_p$, and from eq. (3.23)

$$\Delta x_r = \frac{f_r d^2 + 2\Delta G^0}{2 f_r d} \qquad \ldots (3.44)$$

Using the data for 2-naphthol we obtain $\Delta x_r = 13.33$ pm, and for HPTS the data yields $\Delta x_r = 12.39$ pm. The adiabatic rate constants can now be calculated with eqs. (3.24) and (3.25). For 2-naphthol at 298 K we calculate 1.8×10^6 M^{-1} s^{-1}; the experimental rate constant is 1.8×10^6 M^{-1} s^{-1}. This perfect agreement is coincidental. It results from a cancelation of errors, because more detailed calculations show that, for these reactions, m is slightly larger than 2 and the calculated rate should be faster; however, this is compensated by the fact that $\lambda = 88$ kJ/mol and the condition $\lambda \gg \Delta G^0$ is not closely verified. The calculation for HPTS at 298 K gives 1.4×10^7 M^{-1} s^{-1}, while the experimental rate constant is 1.4×10^8 M^{-1} s^{-1}. Obviously the smaller ΔG^0 of this reaction does not favour the cancelation of errors as much as in the previous case.

For the protonation of 6-methoxyquinoline in water:

$$^*\text{RN} + \text{H}_2\text{O} \longrightarrow {}^*\text{RNH}^+ + \text{OH}^-$$

we can write the equilibrium constant as

$$K_{eq} = \frac{K_w}{[H_2O]K^*_{RNH^+}} \qquad \ldots (3.45)$$

and, using the value of pK^*_a for this compound, obtain $\Delta G^0 = 22.5$ kJ/mol. Representing the reactants by the O–H bond of water and the products by an N–H bond, we can calculate $d = 21.28$ pm using the data of Table 3.1 and eq. (3.39), and presuming that $m = 2$ and $\lambda \gg \Delta G^0$. The force constant representing the reactants is that of an OH bond, and the one representing the products is that of an NH bond. Thus $f_r - f_p$, and eq. (3.23) gives

$$\Delta x_r = \frac{f_p d \pm \sqrt{[(f_p d)^2 - (f_p - f_r)(f_p d^2 + 2\Delta G^0)]}}{(f_p - f_r)} \qquad \ldots (3.46)$$

With the relevant data for this reaction we obtain $\Delta x_r = 13.02$ pm. Now we can use eqs. (3.24) and (3.25), to calculate $k = 2.9 \times 10^6$ M^{-1} s^{-1} at 298 K; the experimental value is 5.2×10^6 M^{-1} s^{-1}.

The reasons behind the remarkable agreement between calculated and experimental rate constants can be related to two circumstances: the small ΔG^0 values involved and the "choice of a good value" for m. The first fact allows us to avoid the use of an empirical parameter, λ, in the calculation. The choice of $m = 2$ for these reactions follows from the physical meaning of this parameter. The pair of nonbonding electrons of the oxygen or nitrogen base may acquire bonding character at the transition state and enhance the transition state bond order by ca. $\frac{1}{2}$; this adds to the already existing transition state bond order of 1/2, to give a total of about unity, $n^{\ddagger}_{AH} + n^{\ddagger}_{HB} = 1$. In fact, more detailed calculations show that m is 15% higher and 15% lower than this when the proton transfer occurs between two oxygen atoms or between an oxygen and a nitrogen atom, respectively. This is related to the number of non-bonding electrons in these atoms.

7 CASE-STUDIES OF PROTON-TRANSFER PHOTOCATALYSIS

7.1 ACID CATALYSIS IN THE PHOTOHYDRATION OF STYRENES AND PHENYLACETYLENES

Yates and co-workers [63] have shown that the irradiation of nonnitro-substituted styrenes and phenylacetylenes in aqueous (5–10% H_2SO_4) acetonitrile gives the corresponding Markovnikov hydration products, with quantum yields that depend strongly on medium acidity and that reach unity in some cases. For styrene and naphthylacetylene, the variation of the fluorescence quenching quantum yield with medium acidity is complementary to the variation of the product quantum yield. It was suggested that the first step in the photohydration of styrenes and phenylacetylenes is the rate-determining protonation of their S_1 state. The following step is the rapid attack of H_2O on the intermediate carbocation, to give regiospecific products in the Markovnikov direction, Mechanism XII.

Mechanism XII

Yates and co-worker have also shown that these reactions originate curved Brönsted plots, and are a paradigmatic case of general acid photocatalysis [23]. The photoprotonation rates were found to be in the range $(0.2-6) \times 10^7$ $M^{-1} s^{-1}$. The ground-state protonation rates of these same compounds are in the range $10^{-5}-10^{-7}$ $M^{-1} s^{-1}$. There is some controversy concerning the adiabaticity of this photocatalysis. If these reactions are adiabatic, the more favorable thermodynamics of the excited state reaction does not justify by themselves the increased reactivity, and other reasons have to be sought for the driving force of the excited state reactions. The high polarization of the S_1 states (4–6 debye) of these normally nonpolar molecules has been invoked as the driving force for the reaction [64]. In order to obtain a better insight into these systems it is convenient to discuss in detail an illustrative example.

The acidity constant of the carbocation produced by photoprotonation of 4-methylstyrene was estimated to be between 1 and 10^2, based on the sigmoid behavior of fluorescence quenching and product quantum yield data [14,63]. The pK_a of 1-(4-methylphenyl)ethyl chloride in the ground state was reported to be -11.2 [65], thus $\Delta pK_a \approx -10.2$. On the other hand, the excited state protonation of 4-methylstyrene by H_3O^+ proceeds with a rate of 3.6×10^7 $M^{-1} s^{-1}$, while the rate of the ground-state protonation of the same system is 12 orders of magnitude slower. Using the Brönsted relation (eq. 3.13), to relate the increase in reactivity from the ground to the excited state reaction with the increase in the pK_a of the conjugated acid of the substrate, we obtain $\beta \approx 1.2$. This is a meaningless value, showing that the more favorable thermodynamics of the (presumed) adiabatic photoprotonation are not sufficient to justify its increase in reactivity. The insufficiency of the thermodynamic explanation for the excited state reactivity is even more evident if we recall that α coefficients close to 0.14 were calculated for the excited state reaction, and the ground state reaction has $\alpha = 0.86$ [23].

Furthermore, if these reactions are adiabatic, their reaction free energies can be estimated from eq. (3.9), where pK_{BH} now refers to the carbocation and pK_{AH} to the acid catalyst. When the acid is H_2O and using $[H_2O] = 55.5$ M, we have $pK_{AH} = 15.7$. Using $pK_{BH} = -1$, we obtain $\Delta G^0 = 95.3$ kJ/mol. For such an endothermic reaction, eq. (3.5) determines that the rate constant cannot exceed 10^{-4} $M^{-1} s^{-1}$. The observed photoprotonation rates with H_2O as the catalyst are more than seven orders of magnitude larger than this. The photohydration of 4-methylstyrene by water is the fastest one observed in water, 1.5×10^5 $M^{-1} s^{-1}$, and it is straightforward to calculate that it requires $pK_{BH} > 8.1$. It was claimed that Förster cycle calculation give acidities of this magnitude [30], but this was not substantiated by experimental data. In any case, such unusually high acidity constants would lead to other inconsistencies in the data, if these photohydrations are treated as adiabatic [31].

If the photohydrations are considered to be diabatic, proceeding from the S_1 state of the arylacetylenes to the ground state of the carbocation in one elementary step, the reaction energy has to be calculated according to

$$\Delta G^0 = -E_{S_1} + 2.303 RT (pK_{BH} - pK_{AH}) \qquad \ldots (3.47)$$

where K_{BH} represents the acidity constant of the carbocation and K_{AH} that of the acid catalyst. The energy of the singlet state of styrene is very high, $E_{S_1} \approx 404$ kJ/mol, thus all the diabatic reactions are very exothermic, and the driving force for the reaction can be mostly thermodynamic.

The treatment of these very exothermic reactions requires the use of eq. (3.38) to determine the value of the reaction coordinate d, and the use of a Morse oscillator to represent the products to estimate the value of Δx_r

$$\tfrac{1}{2} f_r (\Delta x_r)^2 = D_{HB} \{ 1 - \exp[-\beta_{HB}(d - \Delta x_r)] \}^2 + \Delta G^0 \qquad \ldots (3.48)$$

In order to calculate the rate constants of these very exothermic diabatic reactions, it is necessary to estimate the values of λ and χ_{el}. This cannot be done a priori by ISM. Therefore, it is not possible to use this model to make absolute rate calculations on these systems. ISM can still be applied to rationalize experimental results and provide more insight into the reaction mechanism. Such applications require experimental information on the rate constants of structurally related systems. It is very fortunate that general acid catalysis has been observed in the photohydration of styrenes and phenylacetylenes, because the series of catalyzed reactions provide the necessary data to use the capacity of ISM to rationalize the observed rate constants. Detailed calculations have been reported elsewhere [31]. Good fits to the experimental rate constants were obtained with $m \approx 1$, λ ranging between 100 and 150 kJ/mol and $\chi_{el} \approx 10^{-5}$. The values of m and λ are normal for this type of reactions, but the electronic factor is remarkably low. It is possible to account for the value of χ_{el} considering that the reaction is initiated in the Z_1 state of twisted ethylene that does not correlate with the ground state ion pair of the products.

No fluorescence of nitro-substituted phenyl alkynes and alkenes is observed [63]. The photohydrations of these species are regiospecific in the anti-Markovnikov direction. It was suggested that these species react via the triplet state, because the nitro group enhances the intersystem crossing from S_1. The nitro group also polarizes the molecule in the reverse manner as observed for the non-nitro-substituted compounds. A positive charge is now localized in the terminal carbon of the alkyne or alkene, facilitating the nucleophilic attack by water at this site. The lower reactivity of the triplet states is agreement with the fact that $E_{S_1} > E_{T_1}$, which is relevant in diabatic reactions.

7.2 BASE CATALYSIS IN THE PHOTOKETONIZATION OF DIBENZOSUBERENOL

The photolysis of dibenzosuberenol in water/acetonitrile solution at pH 7 gives dibenzosuberone with a quantum yield of 0.003; when 0.25 M aqueous NaOH was used, the quantum yield of product formation increased to 0.011 [66]. Fluorescence quenching studies showed that water in acetonitrile quenches the S_1 state of dibenzosuberenol with a rate constant of $k_q = 1.2 \times 10^8$ $M^{-1} s^{-1}$. Deuteration in the 5-position of dibenzosuberenol reduced k_q by a factor of 0.35, that is, a kinetic isotope effect of 2.9 in the quenching rate constants was observed. It was proposed that the measured quenching rate constants can be equated to the rates of C—H bond ionization and a reaction mechanism involving the photogeneration of a carbanion at the 5-position was suggested. This is shown in Mechanism XIII.

Mechanism XIII

Förster cycle calculations of hydrocarbons and their conjugated bases show that the benzylic protons of such compounds may become more acidic in the S_1 state by as many as 30 orders of magnitude. Thus, it is not so much remarkable that dibenzosuberenol may ionize rapidly in the S_1 state, what is fascinating is that Wan has provided numerous examples showing that, in the S_1 state, benzylic C—H bond heterolysis is only observed if a carbanion with an internal cyclic array (ICA) of $4n$ π electrons is formed [14]. If a carbanion with an ICA of electrons cannot be formed or if it is formed with $(4n + 2)\pi$ electrons, then the systems do not display a carbon acid behavior on photolysis. This is the case of fluorene.

Acidity changes from $pK_a = 23$ to $pK_a^* = -8.5$ have been proposed for fluorene [34] and from $pK_a = 32$ to $pK_a^* = -7$ to suberene [37]. Taking the fluorescence quenching rate constant of suberene ($\approx 10^8$ $M^{-1} s^{-1}$) as reference for the rate constants of reactive S_1 states proton transfer to water, and calculating the rate constant for the ground-state deprotonation of fluorene from the barrier obtained by adding the endothermicity of the deprotonation ($\approx$140 kJ/mol) to the barrier of the thermodynamically favoured reprotonation in methanol (43 kJ/mol from the rate $\approx 2 \times 10^5$ M^{-1} s^{-1}) [67], we estimate that the excited state reaction is 26–28 orders of magnitude faster than that of the ground state. It is still possible to account for this extraordinary increase in reactivity simply in terms of an unprecedented change in thermodynamic acidity: $-\Delta pK$ is in the range of 31–39 units.

The adiabatic deprotonation of the S_1 state of these species in 50% water/acetonitrile (v/v) is exothermic, and the free energies of reaction can be calculated by eq. (3.43), using [H_2O] = 27.8 M. The values obtained are $\Delta G^0 = -40.3$ kJ/mol for fluorene and $\Delta G^0 = -31.7$ kJ/mol for suberene. Thus, there is no thermodynamic constraint to prevent these reactions being adiabatic. Other criteria has to be found to assess the adiabaticity of these reactions.

We can try to reproduce the experimental rate constant using ISM and the approximation that the reactions are adiabatic. We will represent the reactants by a C–H bond and the products by an O–H bond; the data for these bonds is in Table 3.1. Due to the small free energy involved, we can calculate d according to eq. (3.39). We are dealing with carbon acids, thus we expect that $m \approx 1.0$. This gives $d = 43.9$ pm for the deprotonation of suberene, under the approximation that $\lambda \gg \Delta G^0$. With this value we obtain $k \approx 4$ M^{-1} s^{-1}. The contribution of λ to d leads to even smaller calculated rates. In fact, it is necessary to increase the value of m to 1.5 with $\lambda \approx 150$ kJ/mol to reproduce the experimental rate constant of fluorescence quenching of suberene, 1.7×10^8 M^{-1} s^{-1} [37]. Comparing this value of m with those of other hydrocarbons, we verify that it is unreasonably high, unless we postulate that an extra electron acquires bonding character at the transition state. The rate constants calculated for fluorene are of the same order of magnitude as those of suberene. The fact that no proton exchange was observed for fluorene limits the proton exchange rate constant to $<10^5$ M^{-1} s^{-1}.

An alternative calculation can be made considering the reactions to be diabatic. The free energies of reaction have to be calculated by eq. (3.47), where now K_{BH} represents the acidity constant of the carbon acid in the ground state and $pK_{AH} = -1.44$ (from [H_2O] = 27.75 M). Using $E_{s_1} = 365$ kJ/mol for suberene we obtain $\Delta G^0 \approx -175$ kJ/mol and for fluorene $\Delta G^0 \approx -230$ kJ/mol. The value of d has to be obtained from eq. (3.38), using appropriate values for m and λ. Next, Δx_r has to be estimated from eq. (3.48) using an iterative procedure. Finally, the activation free energy can be calculated from eq. (3.24) and the rate constant from eq. (3.5). Using $m = 1$, $\lambda = 110$ kJ/mol and $\chi_{el} = 1$, it is possible to reproduce the fluorescence quenching rate constant of suberene. We remark that these values are typical of carbon acids, as shown in the photohydration of styrenes and phenylacetylenes in aqueous acetonitrile. The only important difference between the proton transfer to 4-methylstyrene and the proton transfer from suberene is the value of χ_{el}. In the first system we had emphasized the low electronic factor (10^{-5}) and mentioned that it originates from the lack of correlation between reactant and product electronic states. In the present system, the calculations indicate that the conversion from the S_1 state of suberene to the ground state of the suberenyl carbanion is not electronically forbidden.

These results can be rationalized in light of the recent correlation diagrams for the deprotonation of cycloheptatriene and cyclopentadiene, published by Klessinger and co-worker [68]. The cycloheptatrienyl anion in the ground state has a cyclic array of $4n$ π electrons, so is a model for suberene. On the other hand, the cyclopentadienyl anion is a model for fluorene, because it has a ICA of $(4n + 2)\pi$ electrons in the ground state. The configuration correlation diagram shows that the first excited state of cycloheptatriene correlates with the ground state of the cycloheptatrienyl anion, as opposed to what happens in the systems with $(4n + 2)\pi$ electrons. This supports the view that the reactions are diabatic, generating the ground-state carbanions in one elementary step, and that their difference in reactivity is related to the fact that for $(4n + 2)\pi$ electron systems $\chi_{el} \ll 1$, whereas $4n$ π electron systems have $\chi_{el} \approx 1$.

7.3 STATIC AND DYNAMIC CATALYSIS IN THE PHOTOTAUTOMERIZATION OF LUMICHROME

Excited-state intramolecular proton-transfer reactions of aromatic molecules were recently reviewed [10]. A few examples of photocatalysis can be identified when the proton donor group is distal from the proton acceptor, and a concerted biprotonic transfer mediated by a molecular companion takes place. Generally, the mediator is a solvent molecule and in such cases the proton transfer cannot be considered a catalytic process. However, in some instances, solvent cannot act as a mediator and other molecular species are required to promote the biprotonic transfer. In such cases the mediator is a true bifunctional acid–base catalyst of a photochemical process.

An illustrative example is the excited-state intramolecular proton transfer in lumichrome (7,8-dimethylalloxazine), which exhibits two fluorescence emission maxima at 447 and 545 nm in mixtures of acetic acid–ethanol, with an isoemissive point at 488 nm indicating two excited-state species in equilibrium [69–71]. This reaction is shown in Mechanism XIV.

In absolute ethanol there is no emission at 540 nm. Such an emission grows with an increase in acetic acid concentration, at the expense of the 447 nm emission. This is clear evidence that acetic acid catalyzes the phototautomerism in ethanol. There is also evidence for the formation of a ground-state complex between lumichrome and acetic acid, therefore the processes described above are the result of a static quenching mechanism. In fact, the phototautomerization persists at 77 K in rigid-solvent matrices, and the rise time of the tautomers is very short (<10 ps at 293 K) [71].

lumichrome:acetic acid complex

flavin tautomer:acetic acid complex

Mechanism XIV

Pyridine appears to have a similar effect on the emission of lumichrome in dioxane as solvent. The emission bands are now at 440 and 540 nm. However, all proton transfer is blocked in low-temperature rigid matrices and the rise time of tautomer formation is now larger than 50 ps at 293 K, indicating a solvent relaxation mechanism. From these observations, Kasha suggested that the reaction mechanism involves dynamic catalysis via an ion-pair intermediate.

8 CONCLUSION

The definition of proton-transfer photocatalysis is not as straightforward as the definition of the corresponding thermal reactions. In this account the term was applied to reactions involving, at or before the rate-determining step, a proton-transfer step from an acid and to a base, one of them being in the excited state, and excluding the solvent as the proton donor or acceptor.

Understanding the mechanisms of the photochemical proton transfer reactions is a key step to controlling these reactions and optimizing the conditions to increase their yields. The use of models to calculate thermodynamic and kinetic acidities can be very helpful in establishing reaction mechanisms. The use of the Förster cycle to estimate the thermodynamic acidities of excited state species and of the ISM to calculate proton-transfer rate constants, is recommended for their simplicity, but one must be aware of their limitations.

The adiabaticity of some proton transfers remains controversial. It is shown how important insight into these reactions can be obtained from thermodynamic reasoning, kinetic modelling and orbital correlation diagrams. Presently, it seems that proton transfers involving only oxygen or nitrogen atoms are predominantly adiabatic. On the other hand, when proton transfer to/from carbon atoms is involved, the models indicate that the reactions are most likely diabatic.

REFERENCES

[1] C. F. Brønsted and K. Pederson, Z. Phys. Chem. **108**, 185 (1924).
[2] R. P. Bell, *Acid-Base Catalysis* edited by (Oxford University Press, London, 1941).

[3] H. Kisch, in *Photocatalysis*, N. Serpone and E. Pelizetti, (Wiley, New York, 1989), p. 1.
[4] H. Hennig, R. Billing and H. Knoll, in *Photosensitization and Photocatalysis Using Inorganic and Organometallic Compounds*, edited by K. Kalyanasundaram and M. Grätzel (Kluwer, Dordrecht, 1993).
[5] G. G. Wubbels, Acc. Chem. Res., **16**, 285 (1983).
[6] S. J. Formosinho and L. G. Arnaut, Adv. Photochem., **16**, 67 (1991).
[7] G. G. Wubbels, B. R. Sevetson and H. Sanders, J. Am. Chem. Soc., **111**, 1018 (1989).
[8] G. G. Wubbels and B. R. Sevetson, J. Phys. Org. Chem. **2**, 177 (1989).
[9] L. G. Arnaut and S. J. Formosinho, J. Photochem. Photobiol., A: Chem., **75**, 1 (1993).
[10] S. J. Formosinho and L. G. Arnaut, J. Photochem. Photobiol. A: Chem., **75**, 21 (1993).
[11] P. Wan, K. Yates and M. K. Boyd, J. Org. Chem., **50**, 2881 (1985).
[12] T. Förster, Pure Appl. Chem. **24**, 443 (1970).
[13] L. M. Tolbert and J. E. Haubrich, J. Am. Chem. Soc., **116**, 10593 (1994).
[14] P. Wan and D. Shukla, Chem. Rev., **93**, 571 (1993).
[15] H. Shizuka, Acc. Chem. Res., **18**, 141 (1985).
[16] R. Pollard, S. Wu, G. Zhang and P. Wan, J. Org. Chem., **58**, 2605 (1993).
[17] A. Streitwieser, Jr., W. B. Hollyhead, A. H. Pudjaatmaka, P. H. Owens, T. L. Kruger, P. A. Rubenstein, R. A. MacQuarrie, M. L. Brokaw, W. K. C. Chu and H. M. Niemeyer, J. Am. Chem. Soc., **93**, 5088 (1971).
[18] J. E. Leffler, Science, **117**, 340 (1953).
[19] F. G. Bordwell, W. J. Boyle, Jr., J. A. Hautala and K. C. Yee, J. Am. Chem. Soc., **91**, 4002 (1969).
[20] S. J. Formosinho, J. Chem. Soc. Perkin Trans. 2, 839 (1988).
[21] L. G. Arnaut, J. Phys. Org. Chem., **4**, 726 (1991).
[22] F. G. Bordwell and W. J. Boyle, Jr., J. Am. Chem. Soc., **94**, 3907 (1972).
[23] J. McEwen and K. Yates, J. Am. Chem. Soc., **109**, 5800 (1987).
[24] A. Argile, A. R. E. Carey, G. Fukata, M. Harcourt, R. A. More O'Ferrall and M. G. Murphy, Isr. J. Chem. **26**, 303 (1985).
[25] C. J. Murray and W. P. Jencks, J. Am. Chem. Soc., **110**, 7561 (1988).
[26] M. Eigen, Angew. Chem. Int. Ed. Engl., **3**, 1 (1964).
[27] K. K. Smith, K. J. Kaufmann, D. Huppert and M. Gutman, Chem. Phys. Lett., **64**, 522 (1979).
[28] M. Gutman, Methods Biochem. Anal., **30**, 1 (1984).
[29] H. Z. Cao, M. Allavena, O. Tapia and E. M. Evleth, J. Phys. Chem. **89**, 1581 (1985).
[30] K. Yates, J. Phys. Org. Chem., **2**, 300 (1989).
[31] L. G. Arnaut and S. J. Formosinho, J. Photochem. Photobiol. A: Chem., **69**, 41 (1992).
[32] T. Z. Förster, Elektrochem., **54**, 43 (1950).
[33] Z. R. Grabowski and A. Grabowska, Z. Phys. Chem. N. F. (Wiesbaden), **101**, 197 (1976).
[34] J. F. Ireland and P. A. H. Wyatt, Adv. Phys Org. Chem., **12**, 131 (1976).
[35] L. G. Arnaut and S. J. Formosinho, J. Phys. Chem., **92**, 685 (1988).
[36] R. Daudel, Adv. Quantum Chem., **5**, 1 (1970).
[37] D. Budac and P. Wan, J. Org. Chem., **57**, 887 (1992).
[38] C. G. Stevens and S. J. Strickler, J. Am. Chem. Soc., **95**, 3922 (1973).
[39] C. G. Stevens and S. J. Strickler, J. Am. Chem. Soc., **95**, 3918 (1973).
[40] P. Wan, E. Krogh and B. Chak, J. Am. Chem. Soc., **110**, 4073 (1988).

[41] R. A. Marcus, J. Phys. Chem., **72**, 891 (1968).
[42] A. O. Cohen and R. A. Marcus, J. Phys. Chem., **72**, 4249 (1968).
[43] R. A. Marcus, J. Am. Chem. Soc., **91**, 7224 (1969).
[44] R. A. Marcus, Faraday Symp. Chem. Soc., **10**, 60 (1975).
[45] S. J. Formosinho, J. Chem. Soc., Perkin Trans. 2, 61 (1987).
[46] S. J. Formosinho and V. M. S. Gil, J. Chem. Soc., Perkin Trans. 2, 1655 (1987).
[47] L. G. Arnaut and S. J. Formosinho, J. Phys. Org. Chem., **3**, 95 (1990).
[48] L. G. Arnaut, in *Proton Transfer in Hydrogen-Bonded Systems*, edited by T. Bountis, (Plenum Press, New York, 1992) p. 281.
[49] K. Yates, J. Am. Chem. Soc., **108**, 6511 (1986).
[50] A. J. C. Varandas and S.J. Formosinho, J. Chem. Soc., Faraday Trans. 2, **82**, 953 (1986).
[51] S. J. Formosinho, in *Theoretical and Computational Models for Organic Chemistry*, edited by S. J. Formosinho, I. G. Csizmadia and L. G. Arnaut, (Kluwer, Dordrecht, 1991) p. 159.
[52] L. Pauling, J. Am. Chem. Soc., **69**, 542 (1947).
[53] H. S. Johnston, Adv. Chem. Phys., **3**, 131 (1960).
[54] H. S. Johnston and C. A. Parr, J. Am. Chem. Soc., **85**, 2544 (1963).
[55] N. Agmon, Int. J. Chem. Kinet., **13**, 333 (1981).
[56] E. Grunwald, Prog. Phys. Org. Chem. **17**, 55 (1990).
[57] I. G. Csizmadia, in *Theoretical and Computational Models for Organic Chemistry*, edited by S. J. Formosinho, I. G. Csizmadia and L. G. Arnaut, (Kluwer, Dordrecht, 1991) p. 1.
[58] R. D. Harcourt, *Qualitative Valence-Bond Descriptions of Electron-Rich Molecules: Pauling '3-Electron Bonds' and 'Increased-Valence' Structures* (Springer-Verlag, Berlin, 1982)
[59] J. H. Clark, S. L. Shapiro, A. J. Campillo and K. R. Winn, J. Am. Chem. Soc., **101**, 746 (1979).
[60] E. Pines and G. R. Fleming, J. Phys. Chem., **95**, 10448 (1991).
[61] E. Pines, D. Huppert, M. Gutman, N. Nachliel and M. Fishman, J. Phys. Chem., **90**, 6366 (1986).
[62] E. Nachliel, Z. Ophir and M. Gutman, J. Am. Chem. Soc., **109**, 1342 (1987).
[63] P. Wan, S. Culshaw and K. Yates, J. Am. Chem. Soc., **104**, 2509 (1982).
[64] H. K. Sinha, P. C. P. Thomson and K. Yates, Can. J. Chem., **68**, 1507 (1990).
[65] J. P. Richard and W. P. Jencks, J. Am. Chem. Soc., 106, 1373 (1984).
[66] P. Wan, D. Budac and E. Krogh, J. Chem. Soc., Chem. Commun., 255 (1990).
[67] F. Hibbert, in *Comprehensive Chemical Kinetics*, edited by C. H. Bamford and C. F. H. Tipper, (Elsevier, Amsterdam, 1977) p. 97.
[68] H. -M. Steuhl and M. Klessinger, Angew. Chem. Int. Ed. Engl., **33**, 2431 (1994).
[69] P. S. Song, M. Sun, A. Koziolowa and J. Koziol, J. Am. Chem. Soc., **96**, 4319 (1974).
[70] J. D. C. Choi, R. D. Fugate and P. -S. Song, J. Am. Chem. Soc., **102**, 5293 (1980).
[71] M. Kasha, J. Chem. Soc., Faraday Trans. 2, **82**, 2379 (1986).

4 Principles and Organic Synthetic Applications of Photoinduced Electron Transfer Photosensitization

J. SANTAMARIA[‡] and C. FERROUD
Ecole Supérieure de Physique et de Chimie Industrielles de Paris, Paris France

1	Introduction	97
2	General considerations	98
3	Photosensitized electron-transfer reaction	99
	3.1 Reaction from alkenes	99
	3.1.1 Isomerizations and rearrangements.	100
	3.1.2 Cycloadditions and dimerizations.	100
	3.1.3 Nucleophilic additions	105
	3.2 Reactions from aromatic compounds	105
	3.2.1 Reactions from arene radical cations	106
	3.2.2 Reactions from arene radical anions	107
	3.3 Reactions from amines	108
	3.4 Dissociation reactions	112
	3.5 Reactions of strained ring systems	115
4	Electron-transfer Photooxygenations	117
	4.1 Selective benzylic oxidations	119
	4.2 Synthesis of dioxanes, dioxetanes and related compounds	120
	4.3 Synthesis of ozonides	122
	4.4 Selective photocatalytic amine oxidations	123
5	Conclusion and prospects	127
	References	128

1 INTRODUCTION

It is surprising to note that although the theory of photoinduced electron transfer (PET) has been known for a long time [1], it only started to be considered by photochemists and to become a research subject about 20 years ago.

[‡] Deceased June 7th 1996 in Paris

Homogeneous Photocatalysis. Edited by M. Chanon
© 1997 John Wiley & Sons Ltd

Publications on eximers and exciplexes [2] were the first to show the importance of electron transfer in organic photochemistry. Then, the rapid development of photophysics and photochemistry of coordination complexes made it possible to impose definitively this concept [3] whose theoretical basis was established by Taube [4], Libby [5] and Marcus [6]. Subsequently, the concept of electron transfer reactions became one of the main subjects of scientific inquiry in the 1980s and was discussed in several reviews [7].

In the 1970s, photochemical applications in organic synthesis mainly relied on energy transfer processes [8]. The concept of electron transfer reactions has made possible, since 1980, the emergence of new mild and/or selective reactions which cover a large variety of structures and provide new methodologies in organic synthesis.

2 GENERAL CONSIDERATIONS

The photochemical excitation of electron–donor (D) or electron–acceptor (B) compounds leads to a radical ion pair formation by one electron transfer from the donor to the acceptor:

$$D + A \xrightarrow{h\nu} D^{+\bullet} + A^{-\bullet}$$

The evidence for these radical ion intermediates has been obtained by means of various photochemical methods such as fluorescence [9], flash photolysis [10], chemically induced nuclear polarization (CIDNP) [11], RPE [12], Raman resonance [13] or photoinduced charge-recombination luminescence [14].

The thermodynamic feasibility of such an electron transfer may be established from the free energy change (ΔG) of the donor–acceptor pair. This free energy change may be evaluated using the well-known Weller equation [15a,b] whose diverse parameters appear in data tables or may be easily accessible experimentally. This relation can be used for systems whose ΔG is included between -120 and $+20$ kJ mole^{-1}.

$$\Delta G = E_{(D^{+\circ}/D)} - E_{(A/A^{-\circ})} - e^2/\varepsilon d - E_{AA^*}$$

where $E_{(D^{+\circ}/D)}$ and $E_{(A/A^{-\circ})}$ terms represent oxidation and reduction potentials of the donor and the acceptor respectively. The $e^2/\varepsilon d$ term represents the coulombic interaction energy of the ion pair separated by a distance d in a solvent of ε dielectric constant; E_{AA^*} stands for the electronic activation energy of the acceptor.

A description of quenching dynamics by electron transfer must take into account the positions and motions of the reactants in a fluid medium. Scheme 4.1 represents the charge-transfer species preceding and following electron transfer and their relation with the solvent polarity.

If $\Delta G < 0$, the electron transfer becomes exergonic and should proceed at a nearly diffusion-controlled rate [15c]. Rehm and Weller have estimated that this is realized for reactions such as $\Delta G > -40$ kJ mole^{-1}. In these conditions the

$$D + A^* \longrightarrow {}^1(D^{\delta+}A^{\delta-}) \underset{2}{\overset{1}{\rightleftarrows}} (D^{+\bullet}A^{-\bullet}) \underset{2}{\overset{1}{\rightleftarrows}} (D^{+\bullet} + A^{-\bullet}) \overset{1}{\underset{2}{\longrightarrow}} D^{+\bullet} + A^{-\bullet}$$
$$\quad\quad\quad\quad\quad\quad\text{Ex} \quad\quad\quad\quad \text{CIP} \quad\quad\quad\quad \text{SSIP} \quad\quad\quad\quad \text{FI}$$

Scheme 4.1. (1) Polar solvent; (2) non-polar solvent; (Ex) exciplex; (CIP) contact radical ion pair; (SSIP) solvent-separated radical ion pair; (FI) free ions

radical ions can be formed as a CIP, SSIP or FI. Farid recently reported a correlation between the nature of these intermediates and the solvent polarity [16]. The use of polar solvents, e.g. acetonitrile, facilitates the generation of radical ions and their reactions with other substrates. The efficiency of these radical ion reactions may be diminished, especially in solvents of lower polarity, by the possibility of the reverse electron transfer, which regenerates the starting materials in their ground state. This latter process can be facilitated in the particular case of very exergonic reactions (Marcus-inverted region [17] of electron transfer). Only fast chemical reactions or the use of specific conditions such as the photooxygenation (see Section 4) or the "special salt effect" [18] may overcome this restriction.

In case of an endergonic electron transfer ($\Delta G > 0$) more or less polar exciplex intermediates may be involved. Here a reaction between the donor and the acceptor is preferred, which is often a cycloaddition or a substitution. These processes are often cage reactions.

Products formation in PET reactions is often governed by the secondary processes of these initially formed radical ions since these serve as precursors for various reaction intermediates: neutral radicals and ions (by loss of nucleofugal or electrofugal groups), nitrones, ylids, iminium ions, ... required to initiate chemical reactions.

This chapter is limited to photosensitized electron transfer reactions of synthetic importance initiated by photoexcitation of donor–acceptor pairs.

Over the years, reviews have been written on the mechanistic and synthetic aspects of this subject [7]. Recently, PET reactions of carbanions and carbocations have been discussed in detail by Wan and Krogh [19].

3 PHOTOSENSITIZED ELECTRON TRANSFER REACTION

3.1 REACTION FROM ALKENES

The study of reactions from PET-generated alkene radical cations has been one of the important fields of research over the years and several reviews have been written on this subject [7b,d,g]. Recently, a vivid summary of this topic has been provided by Mattay [7h,20]. The reactivity of alkene radical ions can be illustrated according to the classification of Mattay as shown in Scheme 4.2:

In this section, we notify only the results of synthetic interest, focusing on cycloadditions and dimerizations. The photooxygenation aspects will be developed in Section 4.

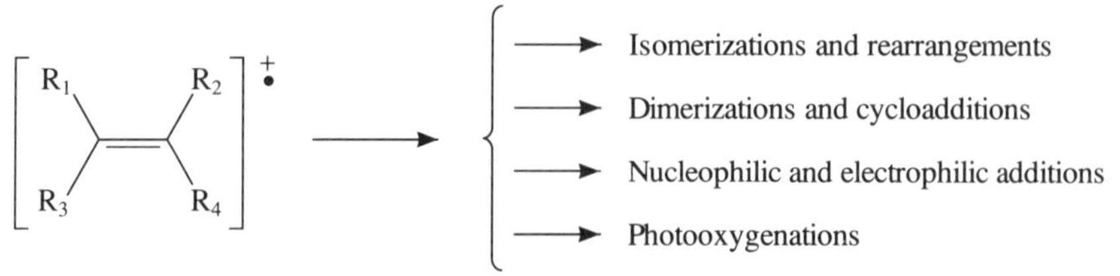

HOMOGENEOUS PHOTOCATALYSIS

[diagram: alkene radical cation $[R_1R_3C=CR_2R_4]^{+\bullet}$ leading to:]
- Isomerizations and rearrangements
- Dimerizations and cycloadditions
- Nucleophilic and electrophilic additions
- Photooxygenations

Scheme 4.2

3.1.1 Isomerizations and Rearrangements

Due to the diminution of the double bond character by removing one electron, alkene radical cations are subject to isomerization and rearrangement. Among these are *cis-trans* isomerizations, which have been intensively studied by several groups [7a–d]. For rearrangements of alkenes we will also refer to Farid and Mattes's excellent review [7b].

3.1.2 Cycloadditions and Dimerizations

After Arnold's [21] and Bauld's [22a] reports concerning cyclobutane formation and Diels–Alder dimerization, the enhanced reaction rates and regioselectivities (head-to-head) in the alkene dimerization via radical cation catalysed reactions have to led numerous studies in this direction [7b,h,22b,23] (Scheme 4.3):

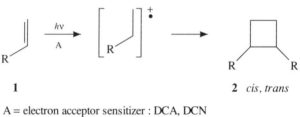

1 → **2** *cis, trans*

A = electron acceptor sensitizer : DCA, DCN
R = carbazole   62%
R = OPh   42%
R = Ar   10–30%

Scheme 4.3

The head-to-head regiochemistry of these dimerizations have been explained in terms of the addition of the radical cation to a neutral molecule giving the stabilised 1,4-radical cation.

Mizuno et al. [24] have reported an interesting application of this method to synthesize macrocycles of the type 2,*m*-dioxabicyclo [*m*-1,2,0] ring (Scheme 4.4).

After the first dimerizations of alkenes, there soon followed examples of mixed cycloadditions [7h,20] such as the heterodimer **7** formation from the photosensitization by 1,4-dicyanobenzene (DCB), of two electron-rich alkenes **5** and **6** [25] (Scheme 4.5). The same group also reported on photoinduced alkylations of five-membered heteroaromatic compounds via electron-transfer

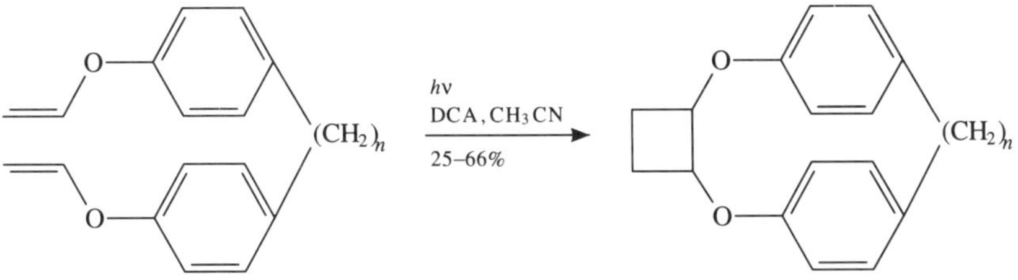

PHOTOINDUCED ELECTRON TRANSFER PHOTOSENSITIZATION

Scheme 4.4

$n = 4–7, 12, 20$
DCA = 9, 10-dicyanoanthracene

reactions [26], such as those with furans **8** and alkenes **9** photosensitized by 9,10-dicyanoanthracene (DCA), 1,4-dicyanonaphthalene (DCN) or 1,4-dicyanobenzene (DCB) (Scheme 4.5):

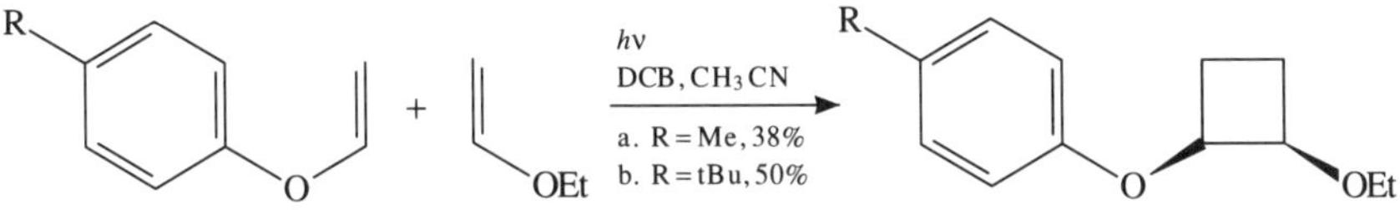

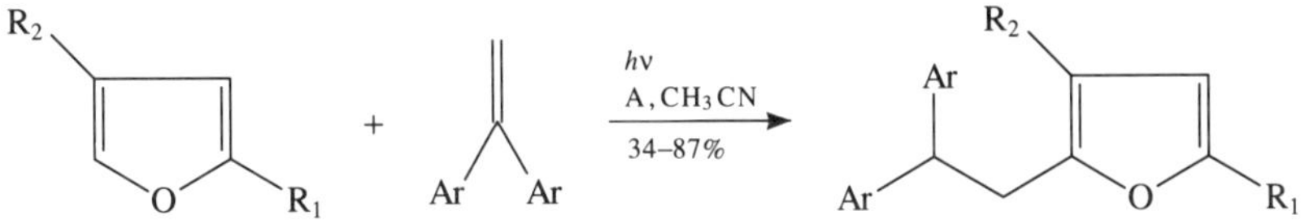

$R_1, R_2 = H, CH_3$
$Ar = 4-X-C_6H_4, X = H, MeCl, Br, OMe$
$A = DCA, DCN, 1-4$ dicyanobenzene

Scheme 4.5

The key step involved the nucleophilic attack of alkene radical cations on furan similar to the dimerization mechanism [27], but the tendency of regenerating the aromatic ring systems resulted in acyclic additions. More examples of this type have been reported by Arnold [28], Pac [29] and Mizuno [30].

In addition to the (2 + 2)-cyclodimerizations, the study of radical cation catalysed Diels–Alder cycloadditions (4 + 2 cyclodimerizations) have provided a new and significant adjuvant to the conventional Diels Alder synthetic methodology [7b,20–22] especially where both the dienes and dienophiles are electron-rich.

Extensive studies of 1,3-cyclohexadiene dimerization [7b,20,22,31–33], which show small reaction rate and poor yields under thermal conditions, have provided deep mechanistic and synthetic understanding of this methodology. Synthetically good yields of both endo and exo cycloadducts have been obtained. Some examples are given in Scheme 4.6:

11 12 13

A = DCA, DCN, TPP^+, chloranil
R = H, OAc, OMe, Me

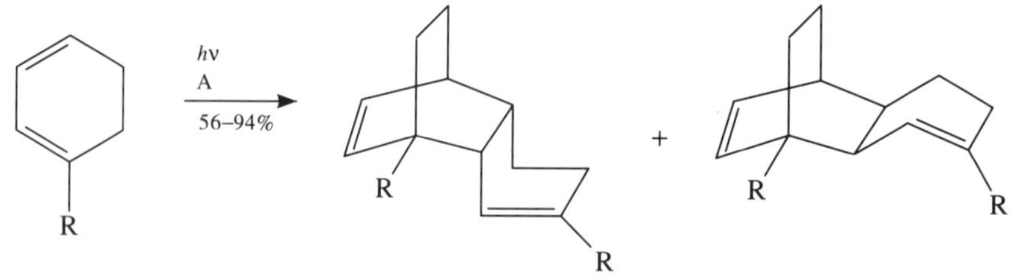

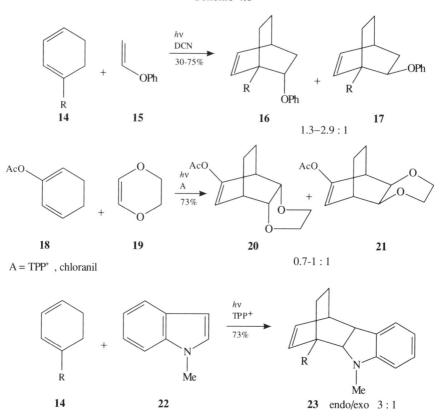

Scheme 4.6

Scheme 4.7

Radical-cation catalysed Diels–Alder cycloadditions using various electron-transfer photosensitizers such as DCA, DCN, $TPP^+2BF_4^-$ or chloranil, especially by using the "special salt effect" [18,34], have been extensively studied [7b,20,31–33,35]. Generally, the endo-isomer **12** is formed with high preference, although the endo–exo ratio depends on the reaction conditions [31].

Scheme 4.7 shows some typical examples of mixed cycloadditions. Radical-cation catalysed mixed (4 + 2)-cycloadditions between the 1,3-cyclohexadiene **14** and electron-rich dienophiles such as **15** have also been studied in detail [22a,31]. An even higher selectivity is observed in the reaction of 2-acetoxy-1,3-cyclohexadiene **18** and 2,3-dihydro-1,4-dioxin **19** photosensitized by TPP^+ or chloranil [31] (Scheme 4.7).

Recently, the (4 + 2)-cycloaddition between the indole **22** and 1,3-cyclohexadiene **14** have been reported to produce adduct **23** in 70% yield [36].

The efficiency of the selectivity in PET-initiated Diels–Alder methodology has been demonstrated by carrying out the formal total synthesis of (−)-β-selenine **27**, a sesquiterpene using Diels–Alder cycloadduct **26** as precursor (Scheme 4.8) [37]:

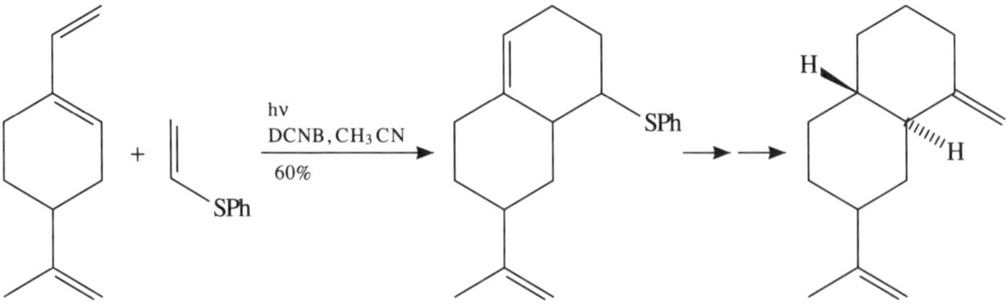

Scheme 4.8

Bauld et al. [22b] studied the influence of substituents on the cyclobutane/Diels–Alder periselectivity. Cycloaddition reaction of diene **28** with electron-rich olefines **29** gives mainly cyclobutane adduct **31** (Scheme 4.9).

This strategy has been recently extended by the same group for the Diels–Alder cycloadduct **34** via the cyclobutane **33** from the PET reaction of diene **14** with enamide **32** (Scheme 4.10).

A variation of great potential is Schuster's "triplex-Diels–Alder reaction [38]. The exciplex formed by the excited singlet state of the 2,6,9,10-tetra-cyanoanthracen (TCA) and the dienophile **35** is trapped by the 1,3-cyclohexadiene **14** to form selectively the cycloadduct **36** in good yield. The intramolecular cycloaddition of phenyl-substituted alkenes to substituted 1,3-cyclohexadiene **37** gives the cycloadduct **38** in 88% yield [39] (Scheme 4.11):

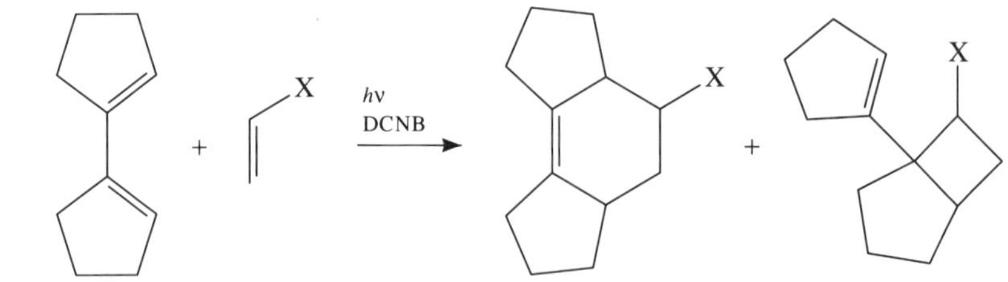

28 **29** **30** **31**

X	**30 : 31**
NMeAc	0 : 100
OR	2–18 : 98–82
SPh	31 : 69

Scheme 4.9

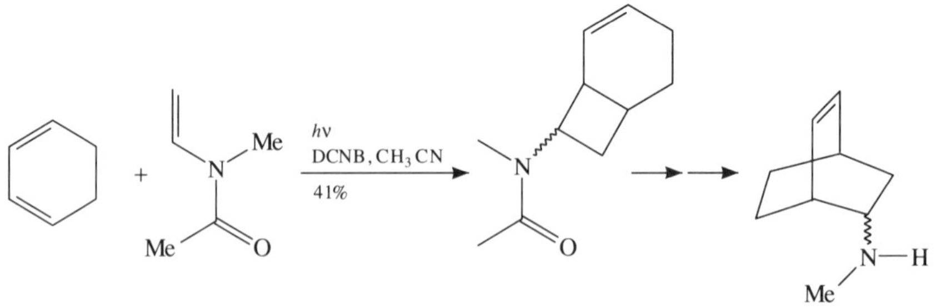

14 **32** **33** **34**

anti/syn 1,3 : 1

Scheme 4.10

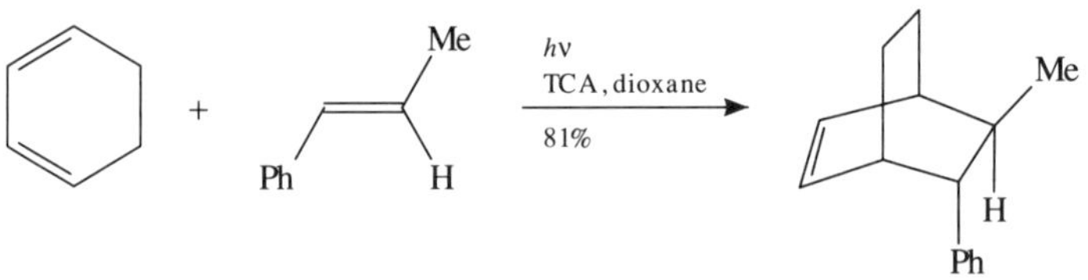

14 **35** **36**

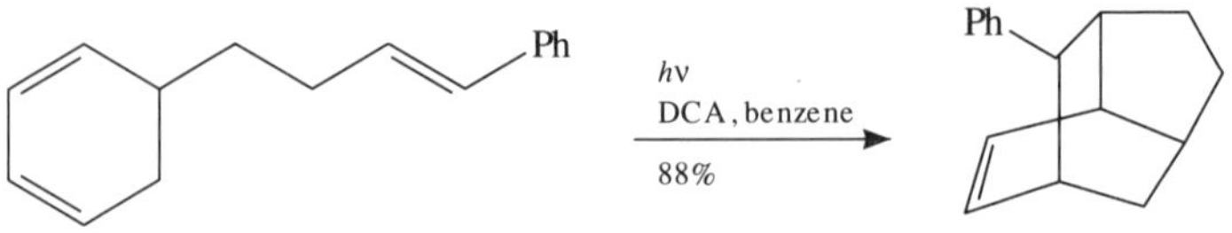

37 **38**

Scheme 4.11

An extension of the concept has been described by Kim and Schuster [40] for the formation of diastereomeric exciplex from chiral arene **39** and a prochiral dienophile **40** with the 1,3-cyclohexadiene **14** to achieve enantioselectivity in Diels–Alder adduct **41**. The observed induction was poor (ee = 15%) (Scheme 4.12):

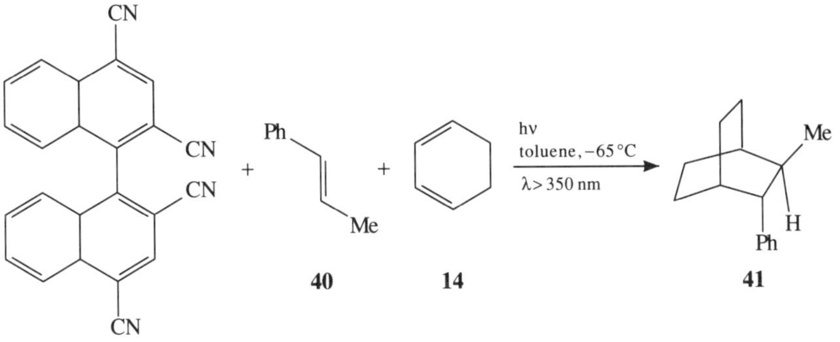

Scheme 4.12

The involvement of ternary interactions in various photoreactions have been reported by Jones [41], Mattay [42], Farid [43] and Mizuno [44].

3.1.3 Nucleophilic Additions

Nucleophilic addition reaction is one of the most common reactions involving organic radical cations. One of the first photochemical examples was reported by Arnold [45,46] as a general method for synthetizing anti-Markovnikov addition products from alkene radical cations. This is illustrated by the addition of alcohol or cyanide ion to conjugated alkene **42** sensitized by 1-cyanonaphthalene or methyl *p*-cyanobenzoate (Scheme 4.13).

Gassman et al. [47] have extended this reaction to the intramolecular addition of carboxylic acid to alkene radical cation for synthesizing lactones **46** and **47** (Scheme 4.13).

Pac and Sakurai [29], Tazuke and Kitamura [48] and Yamamoto and co-workers [49] have developed an indirect procedure giving high yields, which may be called "redox photosensitization". More examples can be found in Farid and co-worker [7b] and Mariano's and co-worker [7d] reviews.

3.2 REACTIONS FROM AROMATIC COMPOUNDS

Aromatic ions formed by photoinduced electron transfer can react with neutral or charged molecules to lead essentially to substitution and reduction reactions. Nucleophilic photosubstitutions have been treated by various authors: Havinga and Cornelisse [50], Bunnett [51] and also Julliard and Chanon [7a], Chanon and Tobe [52], Rossi [53], Todres [54] and Eberson and Radner [55]. Electrophilic

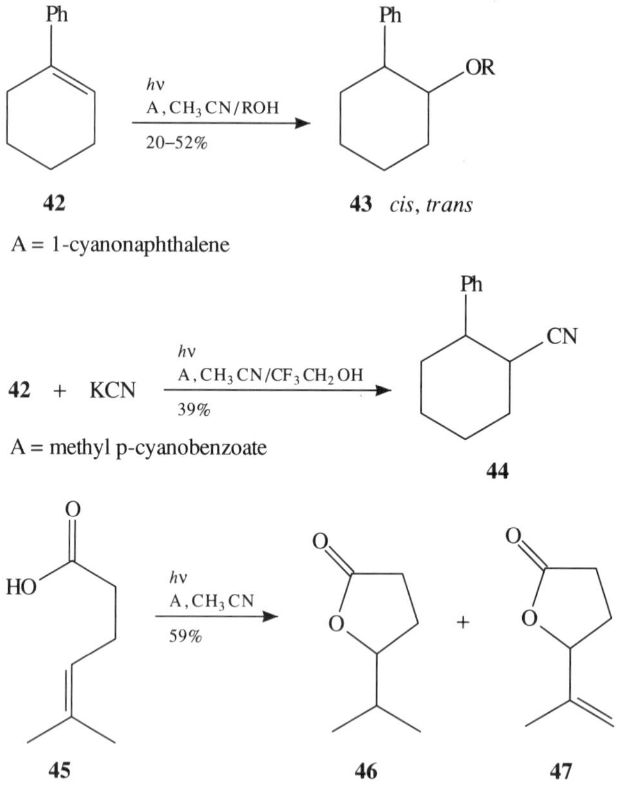

Scheme 4.13

substitutions have been reviewed mainly by Julliard and Chanon [7a] and Eberson and Radner [55].

3.2.1 Reactions from Arene Radical Cations

Aromatic radical cations give rise to nucleophilic substitutions via σ complexes (Scheme 4.14):

$$ArY \xrightarrow{-e^-} ArY^{+\bullet} \xrightarrow{Nu^-} ArNuY^{\bullet} \longrightarrow ArNu$$

Scheme 4.14

One of the most representative examples of this kind is the photocyanation of aromatic compounds [56,57]. Marquet and co-workers [58] have reported recently photosubstitution of 4-nitroveratrole. Yasuda et al. [59] have carried out aromatic photoamination with ammonia and primary amines via the arene radical cation produced by irradiating arenes in the presence of cyanoaromatic acceptors (Scheme 4.15):

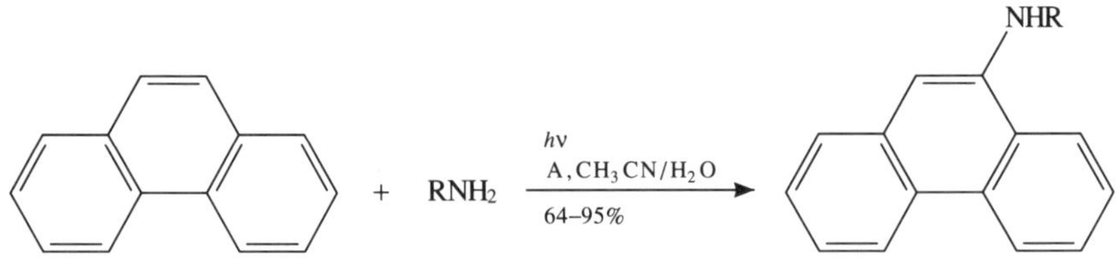

Scheme 4.15

A = DCA, 1,3-dicyanobenzene
R = H, Me, Et, Bu

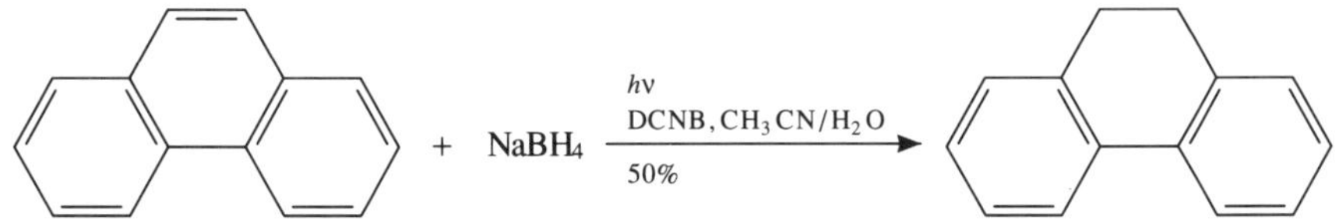

The same authors reported on Birch-like aromatic photoreduction sensitized by DCN in the presence of sodium borohydride [60].

Pandey et al. [61,62] have provided a novel application of PET-generated arene radical cations for preparing a variety of oxygen, nitrogen and carboxylic aromatic compounds by intramolecular nucleophilic cyclizations. So, coumarins **53** are synthesized directly from photosensitization (DCN) of corresponding cinnamic acids [61] (Scheme 4.16).

2-Alkylated dihydrobenzofurans [62] and precocenes-I [63], a potent anti-juvenile hormone compound, have been synthesized using this methodology. Substituted dihydroindoles **55** have also been produced by efficient and regiospecific cyclization of β-arylethylamines **54** (Scheme 4.16).

The combination of two electron-transfer reactions allowed the latter group [64,65] to make a remarkable "one-pot" synthesis of benzopyrrolidines **58** related to the mitomycin skeleton **59** starting from aromatic amine **56** (Scheme 4.17).

3.2.2 Reactions from Arene Radical Anions

Aromatic halides and related compounds are good electron acceptors for a variety of donors such as amines, alkenes or arenes to give arene radical anions. The latter can undergo expulsion of nucleofugal groups leading to aryl radicals, which are captured by nucleophiles to give substitution ($S_{RN}1$) (cq. 4.1) or H-donors to give reduction products (eq. 4.2) (Scheme 4.18). It should be noted, however, that these reactions are not photosensitized reactions but electron-transfer-induced chain reactions.

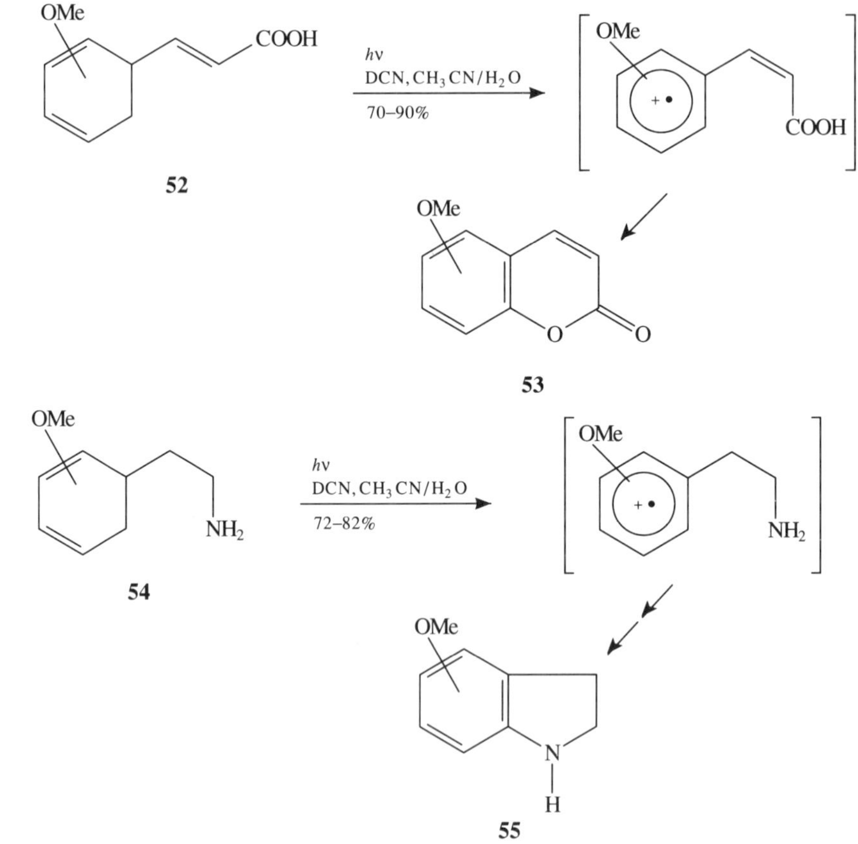

Scheme 4.16

These aspects of arene radical anion chemistry have been reviewed [7d,h,20]. Dehalogenation of aryl halides have been exploited for various organic syntheses [7d,h,66].

Arene radical anions bearing no nucleofugal group have been utilized for Birch-type reductions of aromatic hydrocarbons [67]. The reductive photocarboxylation of aromatic hydrocarbons, was reported by Tazuke et al. [68] (Scheme 4.19).

3.3 REACTIONS FROM AMINES

The studies related to the interactions of electronically excited arene molecules with tertiary amines have provided a basis for the present understanding of exciplexes and radical ion-pair phenomena [69,70].

PET reaction from excited electron acceptor sensitizer (A*)-amine complexes produces a radical anion and a planar amine radical cation. The rate constants

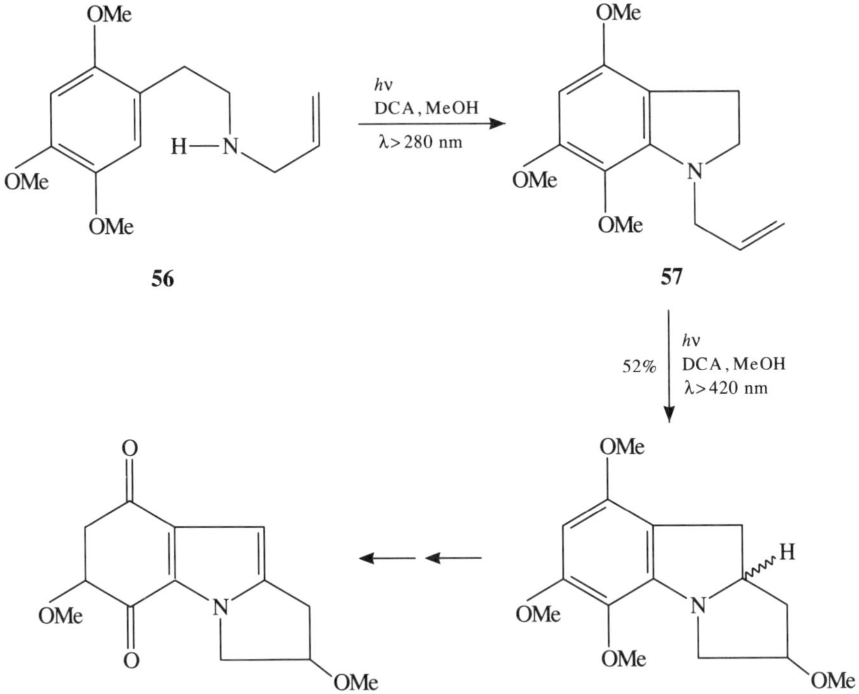

56 **57** **59** **58**

Scheme 4.17

$$\mathrm{ArX} \xrightarrow{+e^-} \mathrm{ArX}^{\bullet -} \xrightarrow{-X^-} \mathrm{Ar}^{\bullet} \xrightarrow{Nu^-} \mathrm{ArNu}^{\bullet -} \xrightarrow{-e^-} \mathrm{ArNu} \qquad \ldots(4.1)$$

$$\mathrm{Ar}^{\bullet} + \mathrm{RH} \longrightarrow \mathrm{ArH} + \mathrm{R}^{\bullet} \qquad \ldots(4.2)$$

Scheme 4.18

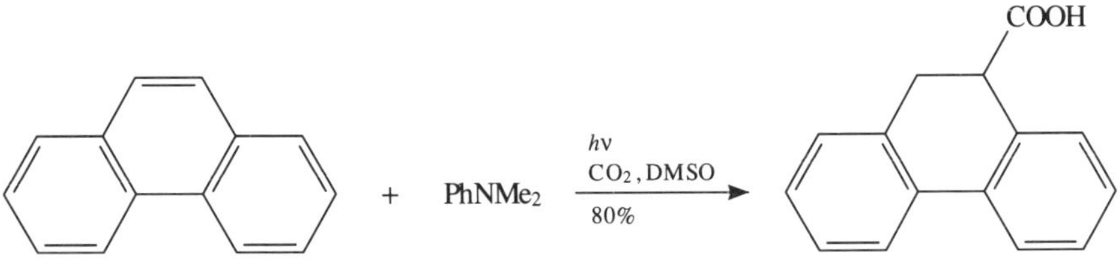

48 **60** **61**

Scheme 4.19

are sensitive to electronic energies and reduction potentials of the sensitizers and oxidation potentials of amines, as well as the structures of both sensitizers and amines. The most general pathway available to these reactive amine radical cation intermediates is probably the fragmentation to ions and neutral radicals. Proton loss following the ionization of simple tertiary amines to generate an α-amino alkyl radical is a common feature. The latter can react in two ways: (1) either by cross-coupling between radical pairs of donor–acceptors [71,72] (eq. 4.5). (2) or by electron loss, because of its reduced ionization potential [73], giving an iminium ion which is subsequently trapped with an internal nucleophile (Nu^-) allowing the obtention of products (eq. 4.6 Scheme 4.20):

$$R_1R_2N{-}CH_2R_3 + A^* \longrightarrow R_1R_2\dot{N}^{+}{-}CH_2R_3 + \dot{A}^{-} \qquad \ldots(4.3)$$

$$R_1R_2\dot{N}^{+}{-}CH_2{-}R_3 + \dot{A}^{-} \xrightarrow{-H^+} R_1R_2N{-}\dot{C}H{-}R_3 + AH^{\bullet} \qquad \ldots(4.4)$$

$$R_1R_2N{-}\dot{C}H{-}R_3 + AH^{\bullet} \longrightarrow R_1R_2N{-}C(AH)R_3 \qquad \ldots(4.5)$$

$$R_1R_2N{-}\dot{C}H{-}R_3 + A^* \xrightarrow{-e^-} R_1R_2\overset{+}{N}{=}CH{-}R_3 + \dot{A}^{-} \xrightarrow{Nu^-} \text{products} \qquad \ldots(4.6)$$

Scheme 4.20

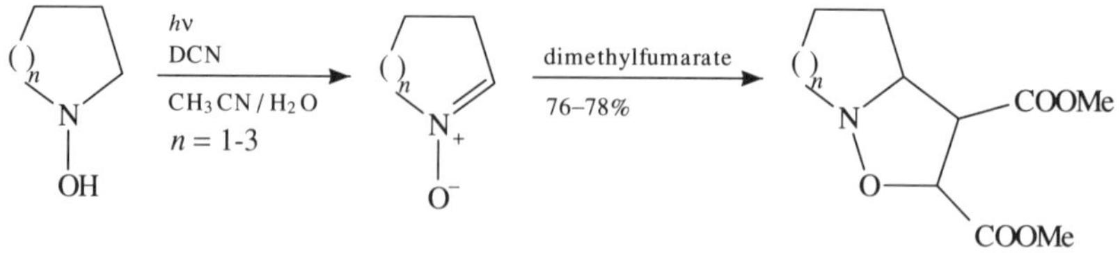

62 63 64

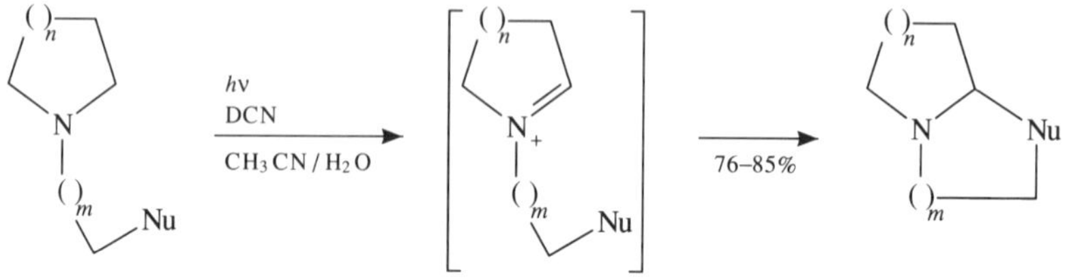

65 66

$n = 1$-2
$Nu = CH_2$, O, $CH(CO_2Et)_2$

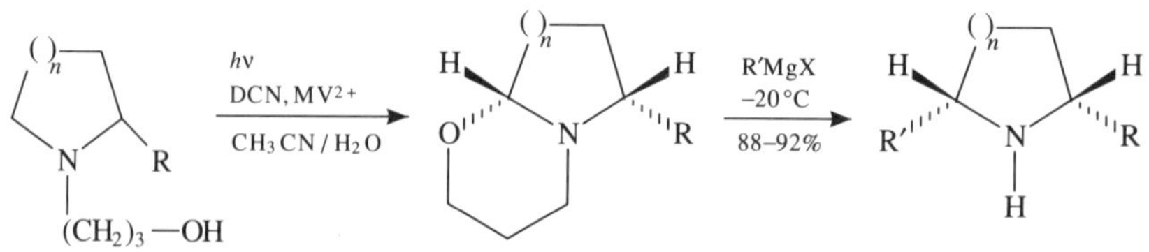

67 68 69 *cis*

$n = 1,2$
R = Me, nBu

Scheme 4.21

The mechanistic aspect of these reactions has been extensively studied by Lewis et al. [74,75].

From these mechanistic considerations, Pandey et al. [76] have designed a large number of interesting synthetic applications. Thus, the irradiation of cyclic hydroxylamines **62** in the presence of DCN provide high yields of nitrones. Trapping of nitrones by dimethylfumarate as dipolarophile gave corresponding cycloadducts **64** in good yields (Scheme 4.21).

This sequence is further extended for generating regiospecific iminium cations. Iminium cations thus formed may be trapped with an internal nucleophile to yield heterocyclic ring systems **66** [77] (Scheme 4.21). In addition, regiospecific iminium cations and their *in situ* cyclization from **67** have been used to synthesize diastereoselective tetrahydro-1,3-oxazines **68** which have been utilized to prepare *cis*-α,α'-dialkylated piperidines and pyrrolidines **69** by a nucleophilic ring opening reaction [78].

In situ trapping of iminium cation by allyltrimethylsilane has been utilised by the same group for direct $-C-C-$ bond formation at the α position of tertiary amines. An interesting application of PET-generated amine radical cations have been demonstrated to produce α-amino radicals from efficient desilylation of α-methyl silylamine **70**. High yields of both pyrrolidines and piperidines **71** are obtained by the intramolecular cyclisation to π bonds [79] (Scheme 4.22). The stereochemical aspect of these cyclizations [80] is to be associated with Beckwith's work [81].

Mariano and co-workers [82] recently reported an identical approach for the heterocyclization reaction of α-methyl silylamine **72** (Scheme 4.22):

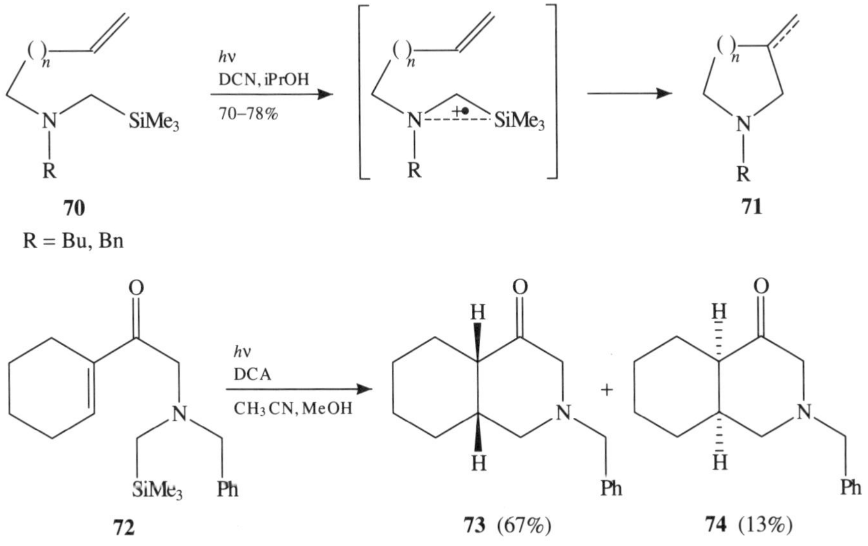

Scheme 4.22

Lewis used the intramolecular photochemical alkene–amine additions as for the preparation of these nitrogen heterocycles [83] (Scheme 4.23). The efficient desilylation from amine radical cation has been extended by Pandey's group [84] for sequential double desilylation reaction from amine **78** to generate the azomethine ylide **79** which, upon cyclization with different dipolarophiles, gives a stereoselective pyrrolidine ring system **80** (Scheme 4.23):

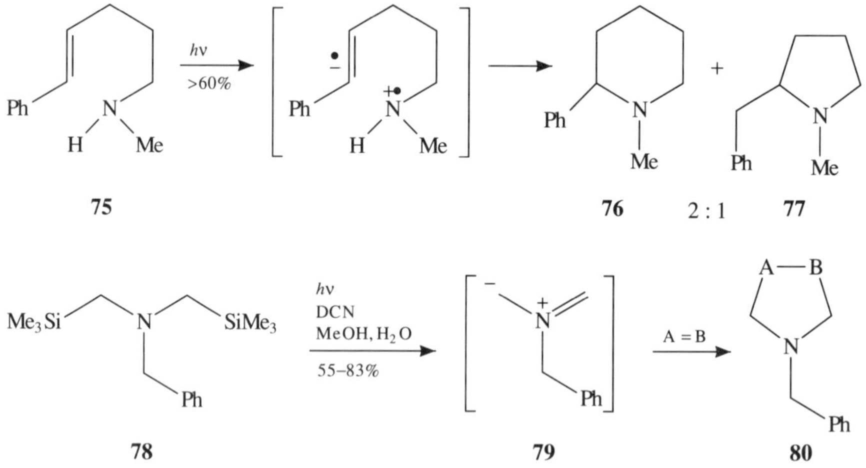

A = B : benzophenone, ethyl acrylate, ...

Scheme 4.23

3.4 DISSOCIATION REACTIONS

One of the most general evolution process of radical ions is their fragmentation to ions and neutral radicals which may serve as key intermediates in various synthetic reactions [85,86] (Scheme 4.24):

$$D-Y + S^* \longrightarrow [D-Y^{+\bullet}] \longrightarrow \dot{D} + Y^+ + S^{-\bullet}$$

$$A-X + S^* \longrightarrow [A-X^{-\bullet}] \longrightarrow \dot{A} + X^- + S^{+\bullet}$$

S*: excited photosensitizer

Scheme 4.24

Due to sharp and selective reduction in bond dissociation energy of radical cations, the cleavage of $-C-C-$ or $-C-$ heteroatom bonds has been observed during PET reactions. More than 10 years ago, Arnold and Maroulis [87] described the photosensitized cleavage of β-phenylethyl ether in the presence of 1,4-dicyanobenzene (Scheme 4.25). Similar cleavage of bibenzyls and pinacols may also be mentioned [88,89]. In an extensive study Whitten and co-workers

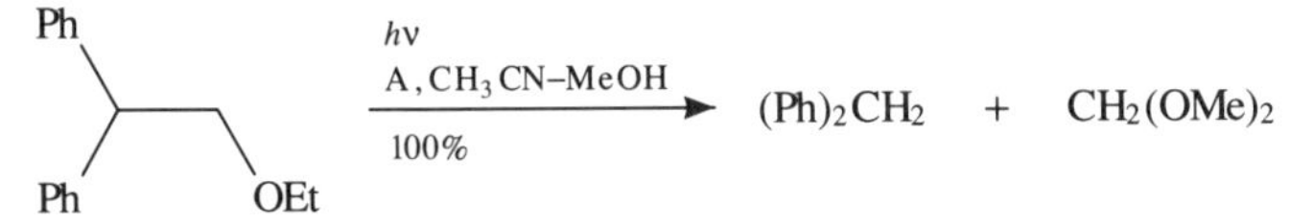

PHOTOINDUCED ELECTRON TRANSFER PHOTOSENSITIZATION

$$\text{Ph}_2\text{CH-CH}_2\text{OEt} \xrightarrow[\text{100\%}]{h\nu, \text{A, CH}_3\text{CN-MeOH}} (\text{Ph})_2\text{CH}_2 + \text{CH}_2(\text{OMe})_2$$

81   **82**   **83**

A = 1,4-dicyanobenzene

Scheme 4.25

[90,91] have reported clean –C–C– bond cleavage from the reaction of β-aminoalcohols and vinilogous aminoalcohols (Scheme 4.25).

Pandey et al. reported on the efficient debenzylations-debenzylations of benzylic alcohols **84** [92] and benzylamines **87** [93] and also *N*-demethylation of tertiary amines **89** [94] in the presence of 1,4-dicyanonaphthalene (Scheme 4.26):

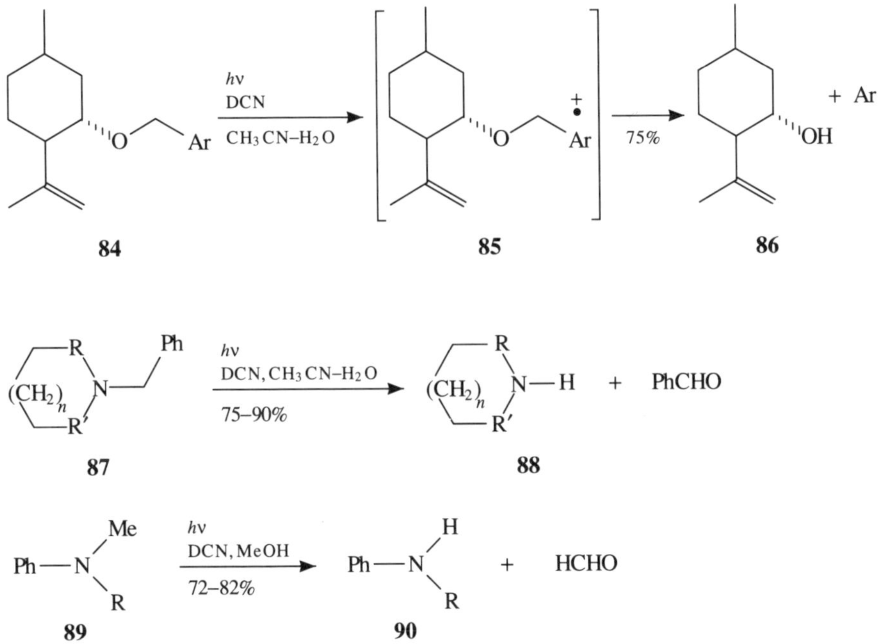

Scheme 4.26

R = Me, Et, Bu

Mizuno et al. [95], Gassman and Boottorff [96] as well as Ohga and Mariano [97], and recently Pandey et al. [98] have reported that, in the case of group 4 organometallics, PET reaction leads to the metal carbon and metal–metal bond dissociation. This strategy has been jointly utilised for a selective desilylation of trimethylsilyl enol ethers **91** [96] and a "one-pot" deselenylation of organoselenium compound **93** [98] (Scheme 4.27):

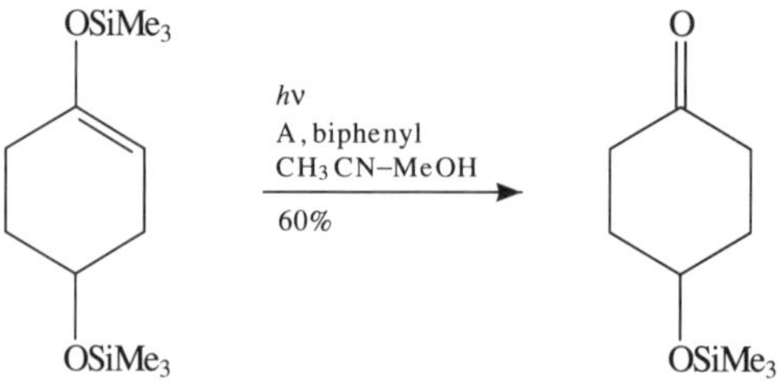

91 **92**

A = 1-cyanonaphthalene

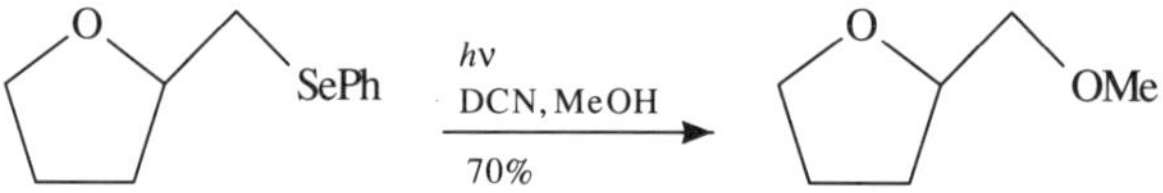

93 **94**

Scheme 4.27

Similar heterolytic –Si–Si– and –Se–Se– bond cleavage reactions have been recently reported by Sakurai and co-workers [99] and Pandey et al. [100].

Analogous to the cleavage reactions from radical cations, a number of synthetically useful reaction from radical anions are also derived. For example, Yonemitsu and co-workers [101] and Masnovi et al. [102] reported on the reductive cleavage of sulfonamides and sulfonates, respectively (Scheme 4.28).

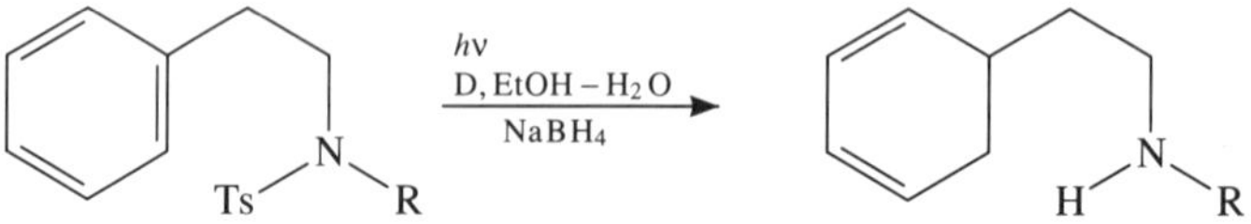

95 **96**

D = dimethoxybenzene
Ts = p-$MeC_6H_4SO_2$

R = H 89%
R = Me 87%

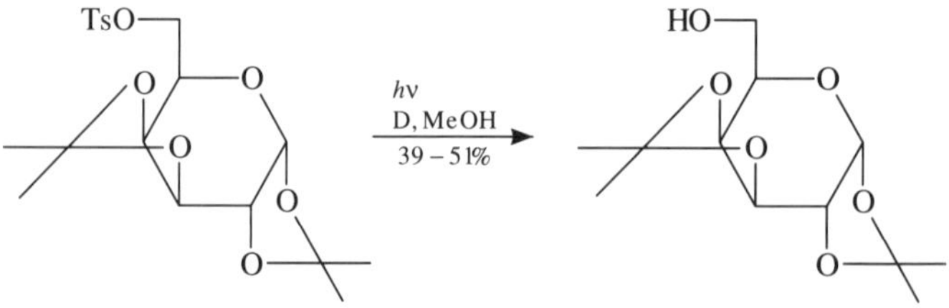

97 **98**

D = DABCO, Et_3N

Scheme 4.28

For these cases, electron donors transfer an electron to the sulfonic acid derivatives yielding radical anions, which then undergo heterolytic cleavage and protonation to the unprotected substrates (Scheme 4.28).

Saito et al. [103] have also carried out selective photosensitized deoxygenation of secondary alcohols via benzoates esters in good yields. Pete et al. [104] reported on photoreduction of carboxylic esters **99** to corresponding alkanes **100** by irradiating **99** in the presence of HMPT at 254 nm (Scheme 4.29):

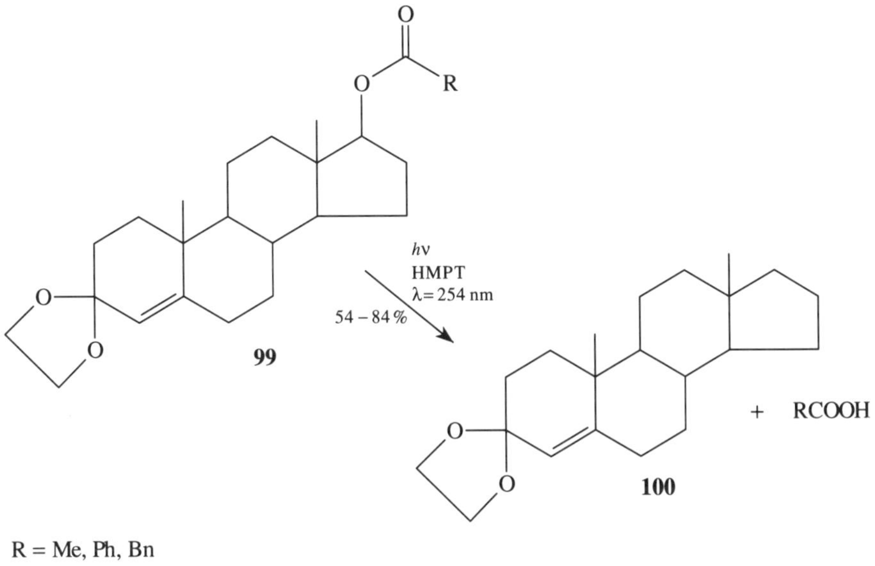

R = Me, Ph, Bn

Scheme 4.29

A review on photoremovable protecting groups has been recently published by Binkley and Flechtner [105].

3.5 REACTIONS OF STRAINED RING SYSTEMS

Strained cyclic organic molecules are efficient quenchers of excited electron-transfer and energy-transfer photosensitizers, especially if they are activated by electron-donating substituents. Since ionization leads to reduced bond strengths, ring opening followed by isomerization, rearrangement, or scavenging of the intermediates is the most common process. This has been observed for three- and four-membered carbocycles and heterocycles. Gassman recently reviewed this topic [106].

Arnold discovered the *cis-trans* isomerization of 1,2-diphenylcyclopropanes photoinduced with good effectiveness in the presence of cyanoaromatic electron acceptors [107] (Scheme 4.30). The potential of this method for organic synthesis was also demonstrated by the preparation of tetrahydrofurans **105** and **106** [108].

Mechanistic aspects of organic radical cation rearrangements have been reported by Roth [109] and are dealt with in a recent review [110] (Scheme 4.30):

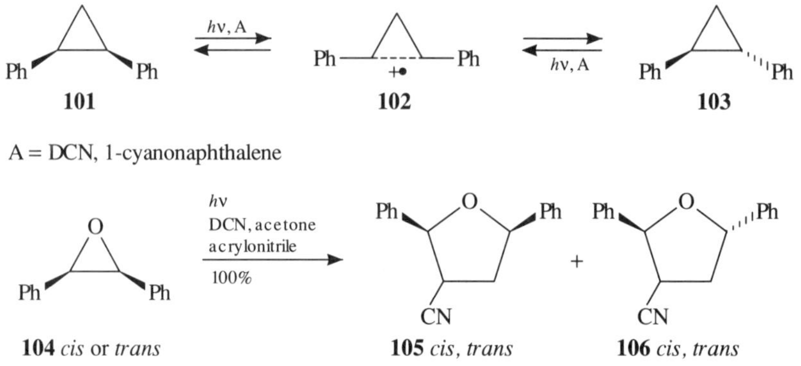

Scheme 4.30

An interesting application of PET-mediated bond cleavage reaction from azirine **107** was reported by Mattay and co-worker [111] for synthesizing N-substitued imidazoles **109** via (3 + 2) cycloaddition reaction of 2-azaallenyl radical cation with imines **108** (Scheme 4.31). Synthesis of pyrrolophane 3,4-dimethyl ester is also feasible using the same concept [112].

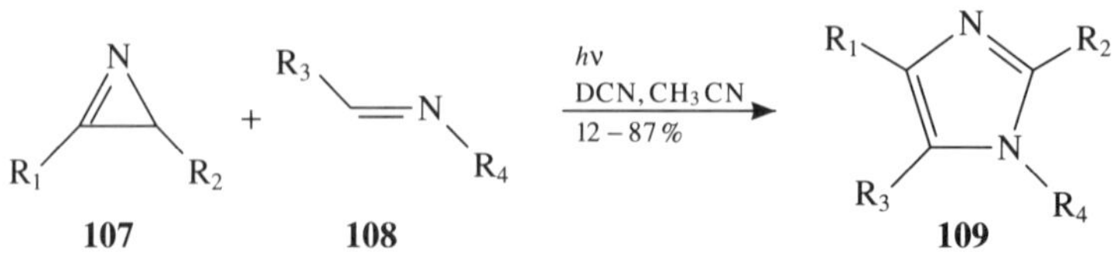

Scheme 4.31

Recently, Mizuno and co-workers [113] reported on PET reaction of cyclopropanes for preparing 3,5-diaryl-2-isoxazolines **111** in good yields by direct NO insertion into the most substituted C–C bond of 1,2-diarylcyclopropane radical cation (Scheme 4.32):

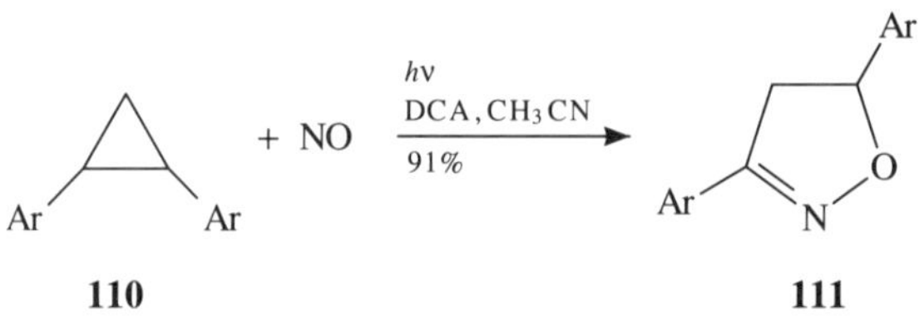

Scheme 4.32

The synthetic potential of these methodologies have been further demonstrated by the preparation of cycloaddition product from vinyl ether [114] and by the ring expansion of the cyclopropanes [115]. Recently, Whitting and co-workers extended this method to other derivatives [116].

4 ELECTRON-TRANSFER PHOTOOXYGENATIONS

It has been known for long time that, under the conjugated influence of light, oxygen and a photosensitizer, many organic compounds undergo oxidation.

In 1939, Kautsky [117] suggested that the reactive chemical entity in photooxygenations was an excited form of dioxygen, which later was found to be dioxygen in an excited singlet state, $^1O_2^*$. This singlet state is formed by electronic energy transfer from photosensitizer-excited triplet state $^3S^*$ towards triplet dioxygen. After numerous controversies, this singlet dioxygen mechanism was definitively established in the 1960s, and photooxygenations were generally considered to proceed by an energy transfer process.

A certain number of these reactions could not, however, be explained by such a mechanism. Over the past few years the emergence of the photoinduced electron transfer concept allowed the rationalization of various mechanistic aspects of these photooxygenations [7i,118,119].

A typical example involves the photooxygenation of electron-rich olefins conjugated to aromatic moieties [120], such as tetraphenylethylene **112**, with catalytic amounts of 9,10-dicyanoanthracene (DCA) (Scheme 4.33):

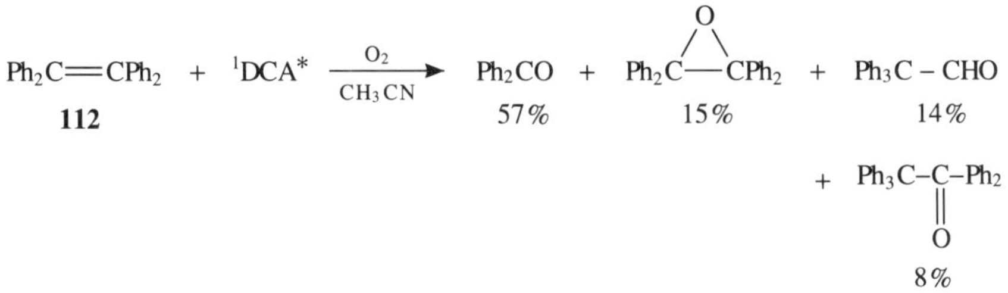

$$Ph_2C=CPh_2 + {}^1DCA^* \xrightarrow[CH_3CN]{O_2} Ph_2CO + Ph_2C\overset{O}{-\!\!\!-\!\!\!-}CPh_2 + Ph_3C-CHO$$

112   57%   15%   14%

$$+ \ Ph_3C-\underset{O}{\overset{\|}{C}}-Ph_2$$

8%

Scheme 4.33

The mechanism proposed by Foote and co-workers [120] implicates the radical anion $DCA^{-\cdot}$ generated by a first electron transfer from the donor D to excited DCA ($^1DCA^*$). From the respective reduction potentials of ground-state oxygen and DCA $[E(O_2/O_2^{-\cdot}) = -0.78$ V, $E(DCA/DCA^{-\cdot}) = -0.89$ V vs SCE] the radical anion $DCA^{-\cdot}$ should be reoxidized by molecular oxygen to produce, by a second exergonic process, the superoxide anion $O_2^{-\cdot}$. The combination between the radical cation $D^{+\cdot}$ and the superoxide anion would lead to oxidation products (Scheme 4.34).

A similar mechanism had already been proposed by Davidson and co-workers [121] for the photooxygenation of tertiary amines sensitized by aromatic compounds. Results corroborating this electron-transfer mechanism were then reported by other groups [122,123].

$$\text{DCA} \xrightarrow{h\nu} {}^1\text{DCA}^*$$
$$\text{D} + {}^1\text{DCA}^* \rightleftharpoons \text{D}^{+\bullet} + \text{DCA}^{-\bullet}$$
$$\text{DCA}^{-\bullet} + {}^3\text{O}_2 \longrightarrow \text{DCA} + \text{O}_2^{-\bullet}$$
$$\text{D}^{+\bullet} + \text{O}_2^{-\bullet} \longrightarrow \text{Oxidation products}$$

Scheme 4.34

Various sensitizers were found to induce the photooxygenation of a great variety of compounds such as tertiary amines [121,124], 1,2-diarylcyclopropanes [125], arylolefins [120,126], arylsubstituted epoxides [127], aziridines [128], alkynes [129], sulfides [120,130], and aromatic compounds [9a,c,131].

These photooxygenations sensitized by cyanoaromatic compounds are, however, really preparative reactions only for a limited number of substrates such as aryl substituted olefins which are unreactive towards singlet oxygen.

DCA-sensitized photooxygenations of alkylaromatic hydrocarbons, have provided chemical evidence for the simultaneous involvement of both singlet dioxygen-mediated and electron-transfer mechanisms [131b]. This result was then confirmed when Foote [132] demonstrated the involvement of singlet dioxygen by its 1270 nm emission in flash photolysis [133], and more recently, when Davidson and Pratt [134] and Foote and co-workers [135] reported solvent isotopic effects. Fluorescence [14,131g] and phosphorescence [14] studies of donor–acceptor pairs formed by DCA with various compounds sensitive or not sensitive to singlet dioxygen, have given evidence for the effect of the solvent polarity on this reaction mechanism [18,131e,g].

These results indicate that, usually, the photooxygenation mechanism using cyanoaromatic sensitizers is complex and difficult to predict. This duality of cyanoaromatic sensitizers seems to be a general behaviour, and is even sometimes met in classical sensitizers used to generate singlet dioxygen such as methylene blue, bengal rose and eosin. More aspects on singlet dioxygen formation can be found in the recent Lopez review [119].

Dioxygen, even in its singlet state, may act as an electron acceptor, yielding the superoxide anion, which either reacts directly with the substrate (neutral or as radical cation), or regenerates singlet dioxygen if a rapid back electron transfer within a donor radical cation–superoxide pair [14,136] occurs.

Molecular oxygen can also react with radical cations of electron-rich compounds by a chain mechanism (Barton-type process) [137]. Thus, in the presence of Lewis acids such as $Ph_3C^+BF_4^-$ or $(pBr-Ph)_3N^+BF_4^-$, in methylene chloride, dienes [138], alkenes [122,139], 1,2-diarylcyclopropanes [125b] and arylsubstituted epoxides [140] react with dioxygen, via their radical cations, to give the same oxygenated products as those obtained by reaction of singlet dioxygen. This mechanism involving molecular oxygen may also compete with that of Foote.

In terms of synthetic applications, the subsequent reduction of molecular oxygen to superoxide ion (Scheme 4.34) is of great importance:

1. It generates a more reactive oxygen species than molecular oxygen;
2. It avoids, or at least, reduces, the back electron-transfer process in the primary step;
3. It allows, with the regeneration of the sensitizer, the building up of catalytic processes illustrative of electron-transfer photocatalysis.

4.1 SELECTIVE BENZYLIC OXIDATIONS

One the first typical examples of benzylic photooxygenations is due to Wie and Adelman [141] who have obtained a mixture of benzylic aldehyde and alcohol from direct irradiation of toluene in the presence of dioxygen (Scheme 4.35):

$$Ph-CH_3 \xrightarrow{h\nu} [Ph-CH_2OOH] \longrightarrow Ph-CHO + Ph-CH_2OH$$
$$\quad\; 113 \qquad\qquad\;\; 114 \qquad\qquad\;\; 115 \qquad\;\; 116$$

Scheme 4.35

The alkylaromatic photooxygenations, particularly those of methyl aromatics, have been explored by numerous groups which have used various electron acceptor sensitizers: cyanoaromatics [9c,131,142,143], cerium IV salts [144], semiconductors such as TiO_2 [145] or nitroanilines supported over silica [146]. These various systems give different results in benzylic oxidations, but none of them offer an actual preparative method. Thus sensitizers which absorb the visible light allow selective oxidations, but have a limited oxidation power. For instance, excited DCA oxidizes hydrocarbons whose oxidation potential is around to 2 V. Sensitizers which absorb UV light are more oxidant, but, in the case of methylaromatics, they yield mixtures of benzaldehydes and of corresponding carboxylic acids.

A sequence of electron-transfer–proton-transfer steps are proposed to rationalize these benzylic oxidations, and an alkylperoxide radical and a hydroperoxide are probable intermediates in the product formation.

Recently, a selective and mild photochemical procedure has been reported for benzylic oxidations with DCA, a usual electron acceptor, in the presence of methyl viologen (MV^{2+}), an electron relay. Benzylic methyl and methylene groups are selectively oxidized in good to excellent yields to the corresponding primary and secondary hydroperoxides **118, 120** which can be isolated as such [147] (Scheme 4.36). With polymethyl substrates, only one methyl group was oxidized in these conditions, and in the naphthalene series, oxidation occurs selectively on a methyl group in the α position (Scheme 4.36):

The same group [148] reported a selective photocatalytic procedure for converting polyalkylated aromatic hydrocarbons into monoaldehydes or

Scheme 4.36

$$Ar-CH_2-R \xrightarrow[\text{visible } \lambda]{\substack{h\nu/O_2 \\ DCA + MV^{2+}}} Ar-CH-OOH-R$$

117 R = H 60-95% **118**
119 R = Alkyl 95% **120**

$$Ar-CH_2-R \xrightarrow[\text{visible } \lambda]{\substack{h\nu/O_2 \\ DCA + MV^{2+}}} [Ar-CH-OOH-R] \xrightarrow[60-90\%]{Fe(II)} Ar-CO-R$$

117, 119 **118, 120** **121**
R = H, R = Alkyl

$$Ar-CH_2-R \xrightarrow[\text{visible } \lambda]{\substack{h\nu/O_2 \\ DCA + MV^{2+}}} [Ar-CH-OOH-R] \xrightarrow[60-90\%]{red.} Ar-CH(OH)-R$$

117, 119 **118, 120** **122**
R = H, R = Alkyl

Scheme 4.36

monoketones **121** in good yields. These carbonyl derivatives arise from an iron(II)-photocatalyzed decomposition (Fenton type) of the corresponding hydroperoxides, **118** and **120**, selectively generated *in situ* by a PET oxidation, using catalytic amounts of DCA in the presence of MV^{2+} and $FeCl_2$ (Scheme 4.36). Similarly, the corresponding benzylic alcohols **122** arise from a "one-pot" reduction of these hydroperoxides generated *in situ*.

Adam et al. [149] have developed a convenient "one-pot" synthesis of epoxy alcohols via photooxygenation of alkenes in the presence of titanium(IV) catalyst. Several examples demonstrate the wide applicatibility of this method. In addition, the authors succeeded in achieving relatively high enantioselectivities in oxirane formation such as **124**. The corresponding Sharpless oxidation of the hindered olefin **123** leads to the same oxirane, with a slightly higher enantioselectivity (ee = 85%) (Scheme 4.37):

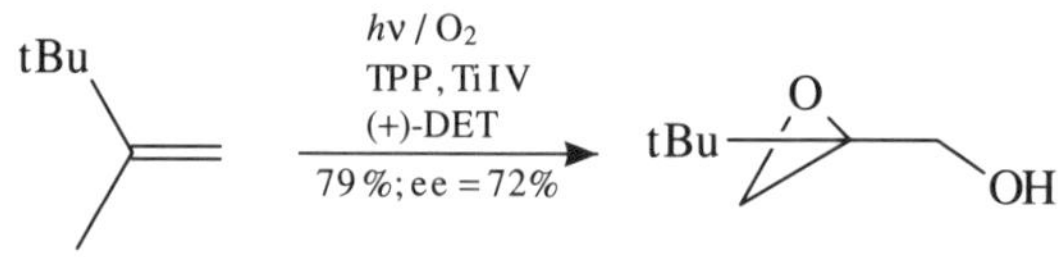

$$\text{tBu-alkene} \xrightarrow[79\%; ee = 72\%]{\substack{h\nu / O_2 \\ TPP, Ti^{IV} \\ (+)-DET}} \text{tBu-epoxide-OH}$$

123 **124**

Scheme 4.37

4.2 SYNTHESIS OF DIOXANES, DIOXETANES AND RELATED COMPOUNDS

The photooxygenation of 1,1- diarylethylenes in the presence of various electron acceptor sensitizers gives 1,2-dioxanes in good yields [7b,h,120,150,151]. Gollnick and Schanatterer [150] have reported, in contrast with Foote's results

[120], that the irradiation of diarylethylene **125** in the presence of DCA yields the corresponding 1,2-dioxane **126** along with small amounts of the corresponding carbonyl compound (Scheme 4.38):

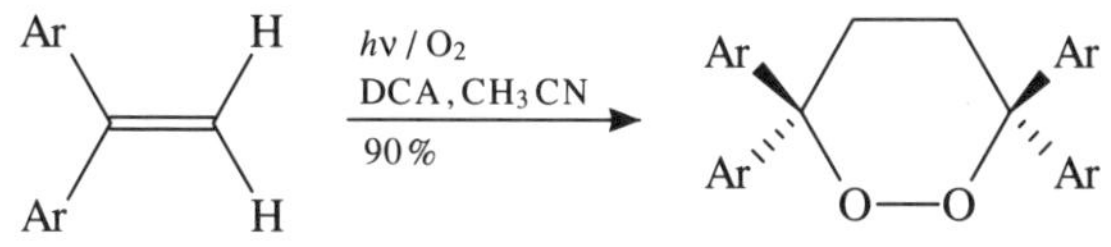

Scheme 4.38

This method has also been applied to the synthesis of 1,2-dioxetanes and dioxetenes from olefins or alkynes, respectively [7b,d]. The latter, finally leads to α-diketones. Thus, Foote and Liang [131d] have prepared in modest yields the *trans,trans*-dimethylmuconate **130** via the dioxetane **128** (Scheme 4.39):

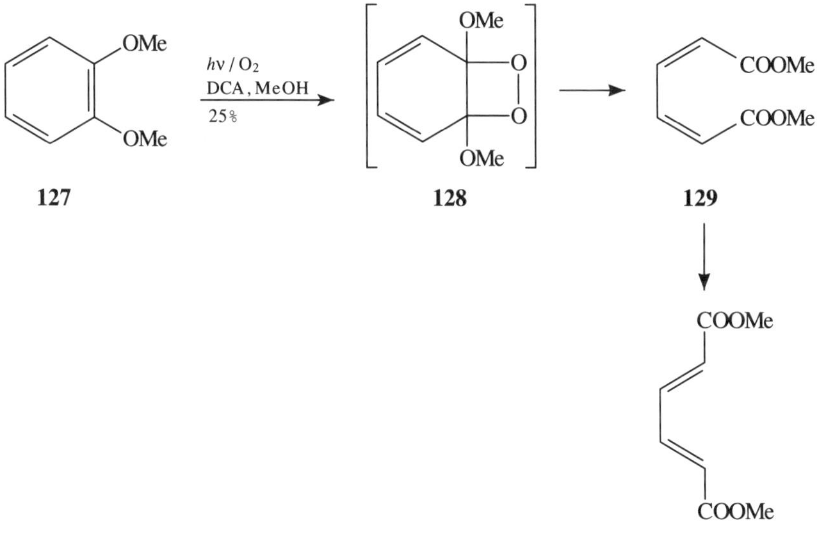

Scheme 4.39

Recently, Lopez and Troisi [152a] have prepared thermally stable dioxetanes such as **132** from various aromatic enol ethers **131** (Scheme 4.40).

Schaap et al. [152b,153] have used the same methodology to obtain, in moderate yields, 1,2-dioxolanes from diarylcyclopropanes. Independently, Mizuno et al. reported the photooxygenations of several substituted 1,2-diarylcyclopropanes **133** affording the corresponding 1,2-dioxolanes **134**, **135** in high yields [125b]. These photoreactions are greatly accelerated by the addition

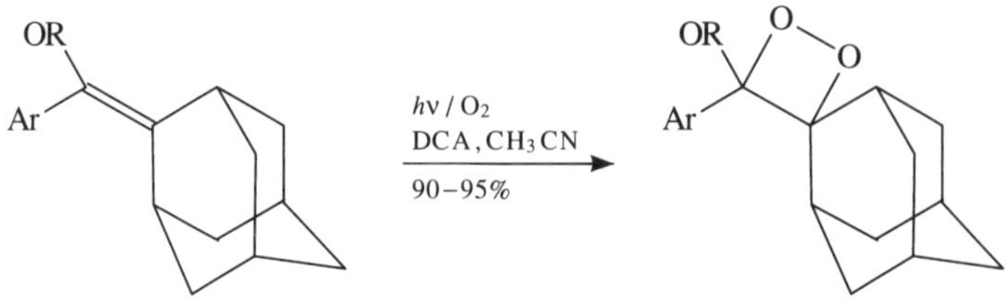

131 **132**

R = Me, Bn
Ar = Ph, $C_{10}H_7$, $C_{12}H_{10}$

Scheme 4.40

of certain aromatic hydrocarbons such as biphenyl, or by metal salts, such as $Mg(ClO_4)_2$ and $LiBF_4$ (Scheme 4.41):

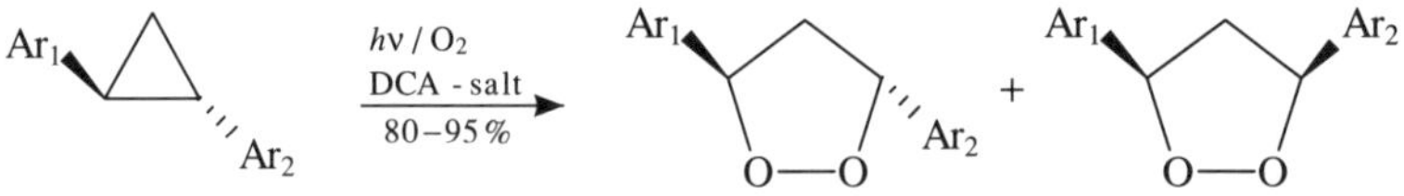

133 **134** **135**

$Ar_1 = Ar_2$ = 4–MeO–C_6H_4, 4–Me-C_6H_4, 4–Me_2N–C_6H_4

Scheme 4.41

Schaap et al. [128,154] have also applied this method to the synthesis of 1,2,4-dioxazolidines from aziridines. The yields, however, are usually not as high as in the preceding case.

4.3 SYNTHESIS OF OZONIDES

Some years ago, Schaap et al. [155] developed a method by which compounds that do not quench the fluorescence of the singlet DCA sensitizer may nevertheless be rapidly oxidized. Thus, epoxides **136**, unreactive under standard DCA-sensitized conditions, can be readily converted into the corresponding ozonides **137** in high yields by use of a nonlight-absorbing aromatic hydrocarbon such as biphenyl (BP) as a cosensitizer in conjunction with DCA (Scheme 4.42):

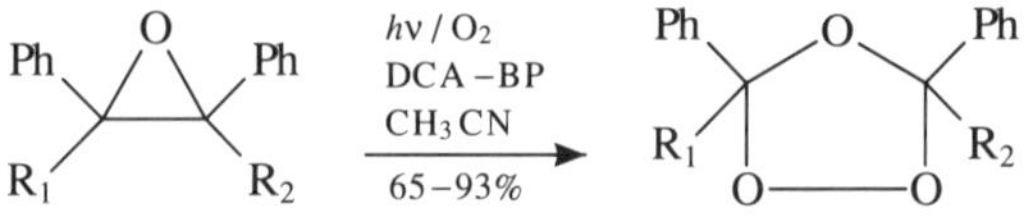

136 **137**

$R_1 = R_2$ = Ph, Me, H

Scheme 4.42

More interestingly, the DCA/BP cosensitized photooxygenations of *cis* and/or *trans* epoxides, as well as the DCA-sensitized photoreaction on several *cis* and *trans*-2,3-dinaphthyloxiranes, not requiring the BP cosensitization, afforded exclusively the corresponding cis-ozonides [155c]. Further examples have been reported by different groups [127b,156–158]. These results are explained in terms of a mechanism which involves the formation of a carbonyl ylide intermediate. Griffin and co-workers [159] and Miyashi et al. [156] have confirmed that ylides can be trapped by dipolarophiles affording substituted tetrahydrofurans with an identical stereochemical course.

The photooxygenation of disubstituted furans **138** gives the corresponding endoperoxides **139**, which, according to their ozonide structure, can be easily converted to unsaturated 1,4-dicarbonyl compounds **140** [160,161] (Scheme 4.43):

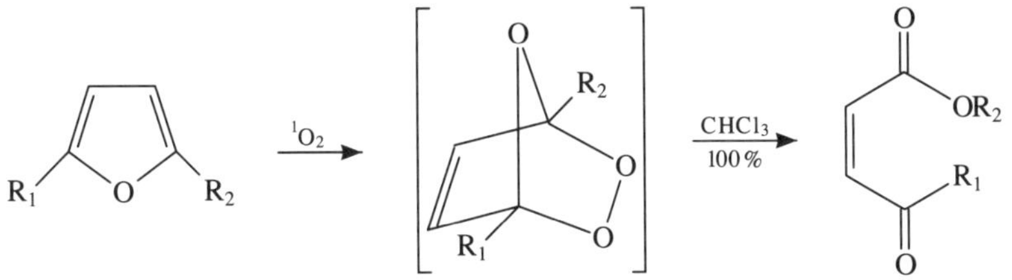

138   **139**   **140**

$R_1 = R_2 = CH_2OMe, CH_2OH, ...$

Scheme 4.43

4.4 SELECTIVE PHOTOCATALYTIC AMINE OXIDATIONS

The reduction of molecular oxygen to superoxide ion according to Scheme 4.34 (p. 118) is also applicable to the electron-transfer photooxidation of amines. The regeneration of the sensitizer makes it possible to design catalytic processes.

Chemical N-demethylation of tertiary amines with chloroformate reagents leads generally to limited yields and the toxicity of required reagents is a real drawback; an alternative method would, therefore, be welcome. In addition, oxidative N-demethylation of tertiary amines is relevant as an enzymatic model for the cytochrome P-450 specific reaction.

A mild and useful alternative photochemical method for N-demethylation of some alkaloids which leads to important intermediates for the preparation of biologically active analogues has been reported [124c]. This method is based on the interesting finding that added salts can greatly affect PET reactions [18]. Thus, the irradiation of an acetonitrile solution of the amine **141** in the presence of a catalytic amount of DCA, under dioxygen bubbling, provides both nor and N-formyl derivatives respectively **142** and **143**, in variable yields. In the presence of added salt such as $LiClO_4$ or $Mg(ClO_4)_2$, with an optimum of concentration (0.25 to 0.5 equiv.), the nor derivative is obtained highly efficiently (Scheme 4.44).

From these results a specific sensitizer which could mimic cytochrome P-450 selective reactions: the dication N,N'-dimethyl-2,7-diazapyrenium (DAP^{2+}, $2BF_4^-$) has been synthesized. The latter is an electron acceptor which can be

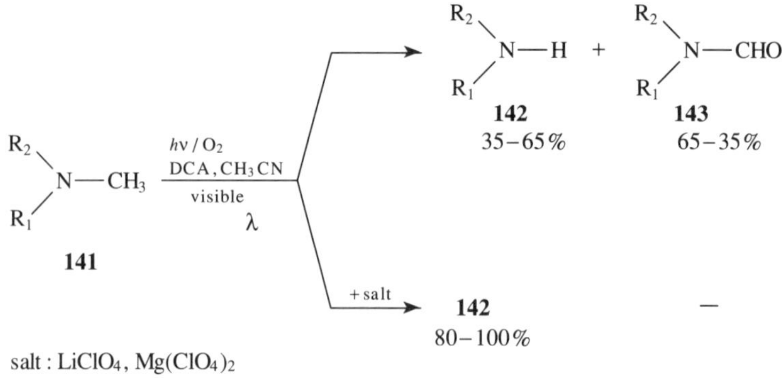

Scheme 4.44

$$\underset{\mathbf{141}}{\overset{R_2}{\underset{R_1}{>}}N-CH_3} \xrightarrow[\text{visible } \lambda]{\substack{h\nu / O_2 \\ DCA, CH_3CN}}$$

→ $\underset{\mathbf{142}}{\overset{R_2}{\underset{R_1}{>}}N-H}$ (35–65%) + $\underset{\mathbf{143}}{\overset{R_2}{\underset{R_1}{>}}N-CHO}$ (65–35%)

+ salt → **142** (80–100%)

salt: LiClO$_4$, Mg(ClO$_4$)$_2$

Scheme 4.44

excited by visible light, not absorbed by amines. Thus, the PET oxidation of the preceding *N*-methylated alkaloid models in the presence of catalytic amount of DAP^{2+} gave exclusively nor products in excellent yields [124b] (Scheme 4.45). The specificity of this sensitizer is probably connected to both its powerful electron acceptor properties and its salt character (Scheme 4.45):

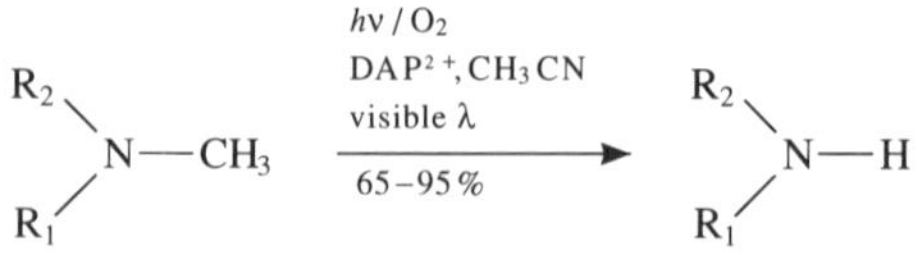

$$\underset{\mathbf{141}}{\overset{R_2}{\underset{R_1}{>}}N-CH_3} \xrightarrow[\substack{\text{visible } \lambda \\ 65-95\%}]{\substack{h\nu / O_2 \\ DAP^{2+}, CH_3CN}} \underset{\mathbf{142}}{\overset{R_2}{\underset{R_1}{>}}N-H}$$

Scheme 4.45

In a similar photochemical *N*-demethylation using methylene blue or bengal rose the nor products were also obtained but in poorer yields [162,163].

From these results, an efficient and mild photoinduced cyanation of various alkaloids by trapping the iminium intermediate with CN$^-$ as nucleophile, has been reported [124d]. Thus, the irradiation of tertiary amines and alkaloids **144** in the presence of trimethylsilyl cyanide (TMSCN) and a catalytic amount of DAP^{2+}, yields regioselectively and often stereoselectively α-aminonitriles **145** in good yields (75–95%) (Scheme 4.46). For example, the irradiation of the eburnamonine **146** leads regio- and stereoselectively to the α-aminonitrile **147** (Scheme 4.46).

In this reaction, TMSCN is a better nucleophile source than the alkali cyanides because it does not require addition of another solvent, water for example, which could modify the reaction medium.

This methodology has been extended to the regioselective synthesis of 2-cyano-1,2,5,6-tetrahydropyridines **149**, versatile synthetic equivalents of 5,6-dihydropyridinium salts, which are powerful synthons for the preparation of functionalized piperidines [164]. In contrast to the unstable 5,6-dihydropyridinium salts, which can exist only under a very restricted set of conditions, i.e. acidic media, their cyanoadducts are readily isolable entities (Scheme 4.47):

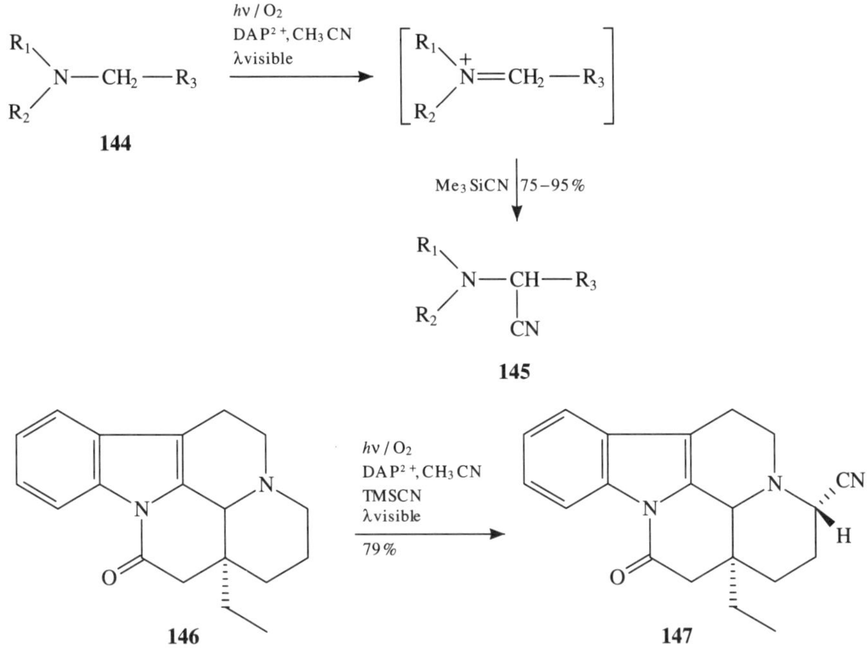

PHOTOINDUCED ELECTRON TRANSFER PHOTOSENSITIZATION

Scheme 4.46

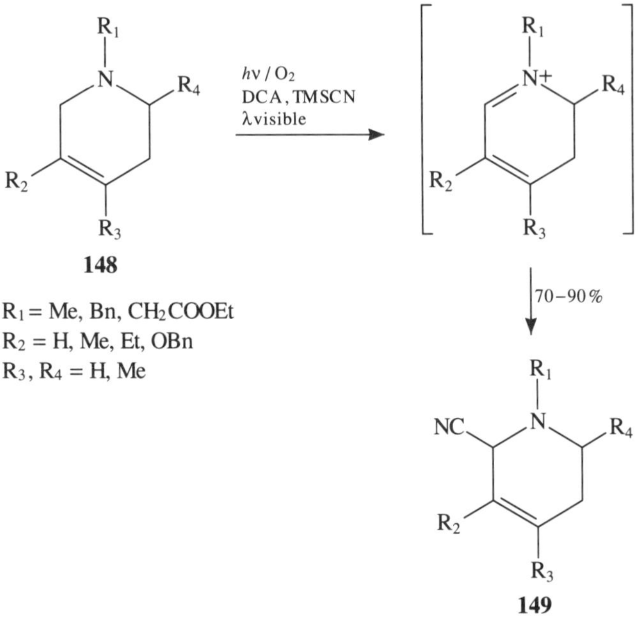

$R_1 = $ Me, Bn, CH_2COOEt
$R_2 = $ H, Me, Et, OBn
$R_3, R_4 = $ H, Me

Scheme 4.47

These results have been extended to regioselective synthesis of 2-cyano-*N*-tryptophyl-Δ^3-piperidines **151** which can be used as synthons for the construction of complex alkaloids. In indole series, TMSCN is a superior trapping agent for iminium ions and it also displays a suitable protection towards enamine moieties allowing the α-aminonitrile formation in good yields [165,166] (Scheme 4.48):

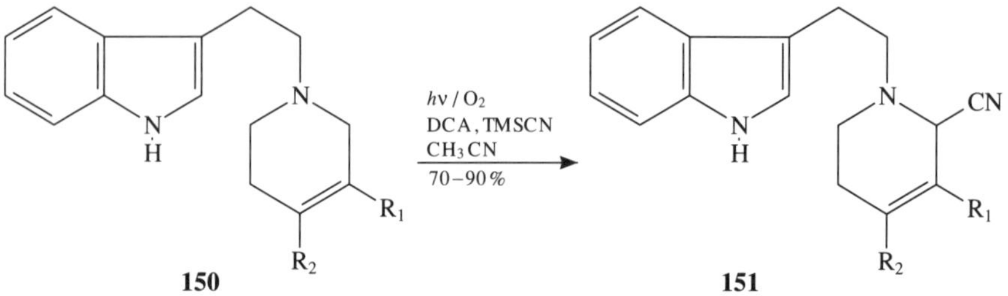

150 **151**

R_1 = Me, Et, OBn, CH_2COOEt, ...
R_2 = H, Me

Scheme 4.48

A mild and efficient "one-pot" synthesis of some indoloquinolizidine alkaloids through a Pictet–Spengler reaction applied to these α-amino nitriles **151** selectively generated *in situ* was carried out [165] (Scheme 4.49). This methodology, applied to the tetrahydropyridine **153** leads directly, in two steps, to the pentacyclic eburnane type alkaloid **154** (Scheme 4.49):

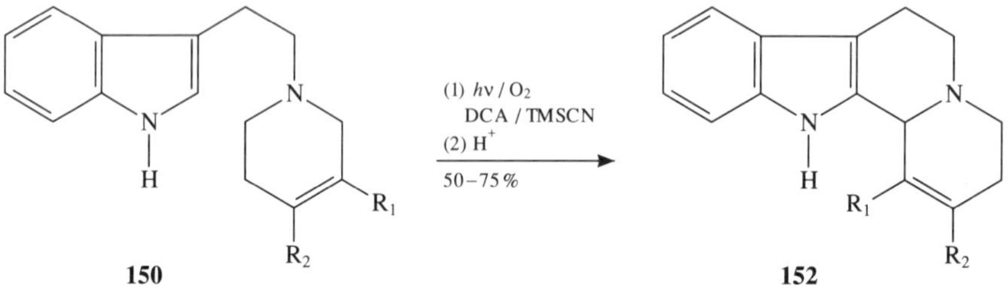

150 **152**

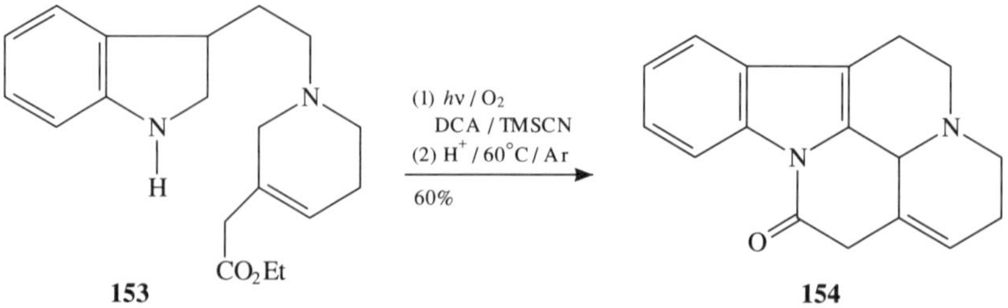

153 **154**

Scheme 4.49

Recently, a new efficient synthesis of the *cis*-eburnamonine **158**, a cerebrovascular agent, has been described [167] through a tetracyclic intermediate **157** obtained from 2-cyano-3-ethylidenepiperidines **156** generated *in situ* by a regioselective ET photocatalysis (Scheme 4.50):

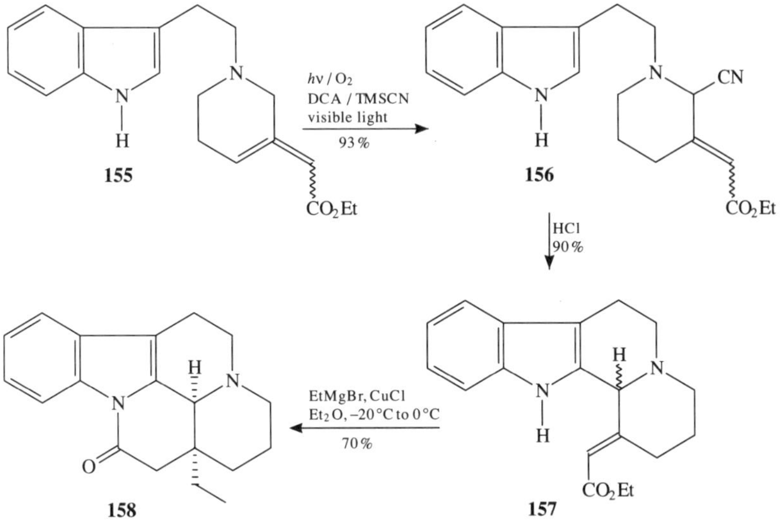

Scheme 4.50

These results demonstrate the development, from the photocatalytic process using a PET, of a new concept for the synthesis of reactive iminium cations, which has proved to be useful in the synthesis of biologically active alkaloids. These new methodologies will find a variety of applications in the synthetic design of alkaloids and other heterocycles.

5 CONCLUSION AND PROSPECTS

During the past decade, important progress have been made in the field of photoinduced electron-transfer reactions, particularly in terms of their applications in organic synthesis. This evolution is linked to the intense activity which has been developed for the exploration of several aspects of electron-transfer reactions. It has been marked by the presentation of the Nobel prize in chemistry to A. Marcus in 1992.

Thus, photophysics, organic electrochemistry and time-resolved spectroscopies, made possible the identification and the study of radical ions, fleeting intermediates, photochemically or thermally generated. These results led to a better comprehension of the thermodynamic and kinetic aspects of these reactions needed to design new methodologies for organic synthesis.

Electron-transfer photosensitization reactions still present a wide potential which is linked to a better knowledge of the nature of the reaction medium:

solvent and salt effects, pH influence .... It is the same for sensitizers concerning their photophysics and redox properties and their stability. The development of new sensitizers, in particular supported ones [146], should allow the development of a fully fledged preparative photocatalysis.

REFERENCES

[1] F. M. Penning, Naturwiss, **15**, 281 (1927).
[2] T. Förster, in *The Exciplex*, (Academic Press, New York, 1975).
[3] (a) V. Balzani, L. Maggi, M. F. Manfin and F. Bolleta, Coord. Chem. Rev., **15**, 321 (1975); (b) V. Balzani, F. Bolleta, M. T. Gandolfi and M. Maestri, Topics Current Chem., **10**, 19 (1978); (c) T. J. Meyer, Isr. J. Chem, **15**, 200 (1977); (d) D. G. Whitten, Acc. Chem. Res., **13**, 83 (1980).
[4] (a) H. Taube, in *Electron Transfer Reactions of Complex Ions in Solution*, (Academic Press, New York, 1970); (b) H. Taube, Angew. Chem., **23**, 329 (1984).
[5] W. F. Libby, Annu. Rev. Phys. Chem., **28**, 105 (1977).
[6] R. A. Marcus, Annu. Rev. Phys. Chem, **15**, 155 (1964).
[7] (a) M. Julliard and M. Chanon, Chem. Rev., **83**, 425 (1983); (b) S. L. Mattes and S. Farid in *Organic Photochemistry*, (Marcel Dekker, New York, 1986) pp. 6, 233; (c) M. A. Fox in *Adv. Photochemistry*, (Wiley, New York, 1986) pp. 13, 237; (d) P. S. Mariano and J. L. Stavinoha, in *Synthetic Organic Photochemistry*, (Plenum Press, New York, 1984) p. 145; (e) G. J. Karnavos and N. J. Turro, Chem. Rev., **86**, 401 (1986); (f) L. Eberson, in *Electron Transfer Reactions in Organic Chemistry* (Springer-Verlag, Berlin, 1987); (g) M. A. Fox and M. Chanon, (editors), *Photoinduced Electron Transfer Reactions*, (Elsevier, Amsterdam, 1988) Part A-D; (h) J. Mattay, Synthesis, **4**, 233 (1989); (i) M. Chanon, M. Julliard, J. Santamaria and F. Chanon, New. J. Chem., 16 (1992); (j) M. Chanon and L. Eberson in *Photoinduced Electron Transfer Reactions*, ref. [7g].
[8] A. Albini, Synthesis, 249 (1981).
[9] (a) J. Eriksen and C. S. Foote, J. Phys. Chem., **82**, 2659 (1978); (b) E. F. Hilinski and P. M. Rentzepis, Acc. Chem. Res., **16**, 224 (1983); (c) J. Santamaria and R. Ouchabane, Tetrahedron, **42**, 5559 (1986).
[10] (a) R. F. Bartholomew, D. R. G. Brimage and R. S. Davidson, J. Chem. Soc., Chem. Commun., 3482 (1971); (b) T. Hino, H. Akazawa, H. Mashara and N. Mataga, J. Phys. Chem., **80**, 33 (1976).
[11] (a) H. D. Roth and M. L. Manion, J. Am. Chem. Soc., **97**, 6896 (1975); (b) R. Kaptein, Adv. Free Radical Chem., **5**, 319 (1975).
[12] (a) A. P. Schaap, K. A. Zaklika, B. Kaskae and L. W. M. Fung, J. Am. Chem. Soc., **102**, 389 (1980).
[13] M. Forster and R. E. Hester, Chem. Phys. Lett., **85**, 287 (1982).
[14] T. B. Truong and J. Santamaria, J. Chem. Soc., Perkin Trans. II, 1 (1987).
[15] (a) D. Rehm and A. Weller, Isr. J. Chem., **8**, 259 (1970); (b) A. Z. Weller, Phys. Chem. Neue Folge, **133**, 93 (1982); (c) A. Z. Weller, Phys. Chem. Neue Folge, **130**, 129 (1982).
[16] I. R. Gould, D. Ege, J. E. Moser and S. Farid, J. Am. Chem. Soc., **112**, 4290 (1990).
[17] (a) R. A. Marcus and P. Spiders, J. Phys. Chem., **86**, 622 (1982); (b) R. A. Marcus, Nouv; J. Chem., **11** 79 (1987).
[18] J. Santamaria, in *Photoinduced Electron Transfer Reactions*, edited by M. A. Fox and M. Chanon, (Elsevier, Amsterdam, 1988), Part B, p. 483.

[19] E. Krogh and P. Wan, Topics in Current Chemistry, **156**, 93 (1990).
[20] J. Mattay, Angew. Chem., Int. Ed. Engl., **26**, 825 (1987).
[21] R. A. Neunteufel and D. R. Arnold, J. Am. Chem. Soc., **95**, 4080 (1973).
[22] (a) D. J. Bellville, D. D. Wirth and N. L. Bauld, J. Am. Chem. Soc., **103**, 718 (1981). (b) N. L. Bauld, D. J. Bellville, B. Harirchian, K. T. Lorentz, R. A. Pabon, Jr., D. W. Reynolds, D. D. Wirth, H. S. Chiou and B. K. Marsh, Acc. Chem. Res., **20**, 37 (1987).
[23] (a) H. Al-Ekabi and P. De Mayo, Tetrahedron, **42**, 6277 (1986); (b) H. Al-Ekabi and P. De Mayo, J. Org. Chem., **52**, 4756 (1987).
[24] K. Mizuno, H. Kagano and Y. Otsuji, Tetrahedron Lett., **24**, 3849 (1983).
[25] K. Mizuno, H. Veda and Y. Otsuji, Chem. Lett., 1237 (1981).
[26] K. Mizuno, M. Ishii and Y. Otsuji, J. Am. Chem. Soc., **103**, 5570 (1981).
[27] N. Heinrich, W. Koch, J. C. Morrow and H. Schwarz, J. Am. Chem. Soc., **110**, 6332 (1988).
[28] (a) A. J. Maroulis and D. R. Arnold, J. Chem. Soc., Chem. Commun., 351 (1979); (b) D. R. Arnold, R. M. Bory and A. Albini, J. Chem. Soc., Chem. Commun., 138 (1981).
[29] T. Majima, C. Pac, A. Nakasone and H. Sakurai, J. Am. Chem. Soc., **103**, 4499 (1981).
[30] K. Mizuno, R. Kaji, H. Okadu and Y. Otsuji, J. Chem. Soc., Chem. Commun., 594 (1978).
[31] J. Mattay, G. Trampe and J. Runsink, Chem. Ber., **121**, 1991 (1988).
[32] J. Mattay, J. Gersdorf and J. Mertes, J. Chem. Soc., Chem. Commun., 1088 (1985).
[33] (a) J. Mlcoch and E. Steckhan, Angew. Chem., Int. Ed. Engl., **24**, 412 (1985); (b) J. Mlcoch and E. Steckhan, Tetrahedron Lett., **28**, 1081 (1987).
[34] A. Loupy and B. Tchoubar, *Effets de Sel en Chimie Organique et Organométallique*, (Bordas, Paris, 1988).
[35] M. Kojima, H. Sakurai and K. Tokumaru, Tetrahedron Lett., **22**, 2889 (1981).
[36] A. Gieseler, E. Steckhan, O. Wiest and F. Knoch, J. Org. Chem., **56**, 1405 (1991).
[37] (a) B. Harirchian and N. L. Bauld, J. Am. Chem. Soc., **111**, 1826 (1989); (b) N. L. Bauld, B. Harirchian, D. W. Reynolds and J. C. White, J. Am. Chem. Soc., **110**, 8111 (1988).
[38] (a) G. C. Calhoun and G. B. Schuster, J. Am. Chem. Soc., **106**, 6870 (1984); (b) G. C. Calhoun and G. B. Schuster, Tetrahedron Lett., **27**, 911 (1986); (c) G. C. Calhoun and G. B. Schuster, J. Am. Chem. Soc., **108**, 8021 (1986); (d) N. Akbulut and G. B. Schuster, Tetrahedron Lett., **29**, 5125 (1988); (e) D. Hartsough and G. B. Schuster, J. Org. Chem., **54**, 3 (1989); (f) N. Akbulut, D. Hartsough, J. I. Kim and G. B. Schuster, J. Org. Chem., **54**, 2549 (1989).
[39] I. Wolfe, S. Chan and G. B. Schuster, J. Org. Chem., **56**, 7313 (1991).
[40] J. I. Kim and G. B. Schuster, J. Am. Chem. Soc., **112**, 9635 (1990).
[41] C. R. Jones, B. J. Allman, A. Morring and B. Spahic, J. Am. Chem. Soc., **105**, 652 (1983).
[42] H. Leismann, J. Mattay and H. D. Scharf, J. Am. Chem. Soc., **106**, 3985 (1984).
[43] S. L. Mattes and S. Farid, J. Am. Chem. Soc., **108**, 7356 (1986).
[44] (a) K. Mizuno, T. Hashizume and Y. Otsuji, J. Chem. Soc., Chem. Commun., 772 (1983); (b) K. Mizuno, K. Nakanishi and Y. Otsuji, J. Am. Chem. Soc., 90 (1991).
[45] (a) Y. Shigemitsu and D. R. Arnold, J. Chem. Soc., Chem. Commun., 407 (1975); (b) A. J. Maroulis and A. J. Arnold, Synthesis, 819 (1979).
[46] A. J. Maroulis, Y. Shigemitsu and D. R. Arnold, J. Am. Chem. Soc., **100**, 535 (1978).

[47] (a) P. G. Gassman and K. J. Bottorff, J. Am. Chem. Soc., **109**, 7547 (1987). (b) P. G. Gassman and S. A. De Silva, J. Am. Chem. Soc., **113**, 9870 (1991).
[48] S. Tazuke and N. Kitamura, J. Chem. Soc., Chem. Commun., 515 (1977).
[49] T. Asanuma, T. Gotoh, A. Tsuchida, M. Yamamoto and A. Nishijima, J. Chem. Soc., Chem. Commun., 485 (1977).
[50] (a) J. Cornelisse and E. Havinga, Chem. Rev., **75**, 353 (1975); (b) E. Havinga and J. Cornelisse, Pure Appl. Chem., **47**, 1 (1976).
[51] J. F. Bunnett, Acc. Chem. Res., **11**, 413 (1978).
[52] (a) M. Chanon and M. L. Tobe, Angew. Chem., Int. Ed. Engl., **21**, 21 (1982); (b) M. Chanon, Acc. Chem. Res. **20**, 214 (1987).
[53] R. A. Rossi, Acc. Chem. Res., **15**, 164 (1982).
[54] Z. W. Todres, Tetrahedron, **41**, 2771 (1985).
[55] L. Eberson and F. Radner, Acc. Chem. Res., **20**, 53 (1987).
[56] J. Den Heijer, O. B. Shadid, J. Cornelisse and E. Havinga, Tetrahedron, **33**, 779 (1977).
[57] (a) K. Mizuno, C. Pac and H. Sakurai, J. Chem. Soc., Chem. Commun., 553 (1975); (b) M. Yasuda, C. Pac and H. Sakurai, J. Chem. Soc., Perkin Trans. I, 746 (1981); (c) N. J. Bunce, J. P. Bergsma and A. Schmidt, J. Chem. Soc., Perkin Trans. II, 713 (1981).
[58] A. Cantos, J. Marquet, M. Moreno-Manas and A. Castello, Tetrahedron, **44**, 2607 (1988).
[59] M. Yasuda, Y. Yamashita, K. Shima and C. Pac, J. Org. Chem., **52**, 753 (1987).
[60] M. Yasuda, C. Pac and H. Sakurai, J. Org. Chem., **46**, 788 (1981).
[61] G. Pandey, A. Krishna and J. M. Rao, Tetrahedron Lett., **27**, 4075 (1986).
[62] G. Pandey, A. Krishna and U. T. Bhalearo, Tetrahedron Lett., **30**, 1867 (1989).
[63] G. Pandey and A. Krishna, J. Org. Chem., **53**, 2364 (1988).
[64] G. Pandey, M. Sridhar and U. T. Bhalerao, Tetrahedron Lett., **31**, 5373 (1990).
[65] G. Pandey and G. Kumaraswamy, Tetrahedron Lett., **29**, 4153 (1988).
[66] A. Lablache-Combier, *Photoinduced Electron Transfer Reactions*, edited by M. A. Fox and M. Chanon, (Elsevier, Amsterdam, 1988) Part C, p. 134.
[67] (a) J. A. Barltrop, Pure Appl. Chem., **33**, 179 (1973); (b) V. R. Rao and V. Ramakrishnan, J. Chem. Soc., Chem. Commun., 971 (1971).
[68] S. Tazuke, S. Kazama and N. Kutamura, J. Org. Chem., **51**, 4548 (1986).
[69] K. Mizuno, H. Ueda and Y. Otsuji, Chem. Lett., 1237 (1981).
[70] H. Mashuara and M. Mataga, Acc. Chem. Res., **14**, 312 (1981).
[71] J. A. Barltrop, Pure Appl. Chem., **33**, 179 (1973).
[72] (a) M. Bellas, D. B. Bryce-Smith, M. T. Clarke, A. Gilbert, G. Klunkin, S. Krestonosich, C. Manning and S. Wilson, J. Chem. Soc., Perkin Trans. I, 2571 (1977); (b) N. C. Yang, D. M. Shold and B. Kim, J. Am. Chem. Soc., **98**, 6587 (1976).
[73] D. Griller and F. P. Lossing, J. Am. Chem. Soc., **103**, 1586 (1981).
[74] (a) F. D. Lewis, T. I. Ho and J. T. Simpson, J. Am. Chem. Soc., **104**, 1924 (1982); (b) W. Hub, S. Schneider, F. Dorr, J. T. Simpson, J. D. Oxman and F. D. Lewis, J. Am. Chem. Soc., **104**, 2044 (1982).
[75] (a) F. D. Lewis, Acc. Chem. Res., **79**, 401 (1986); (b) F. D. Lewis, T. I. Ho and J. T. Simpson, J. Org. Chem., **46**, 1077 (1981).
[76] G. Pandey, G. Kumaraswamy and A. Krishna, Tetrahedron Lett., **28**, 2649 (1987).
[77] (a) G. Pandey and G. Kumaraswamy, Tetrahedron Lett., **29**, 4153 (1988); (b) G. Pandey, G. Kumaraswamy and P. Y. Reddy, Tetrahedron, 48, 8295 (1992).
[78] G. Pandey, P. Y. Reddy and U. T. Bhalerao, Tetrahedron Lett., **32**, 5147 (1991).
[79] (a) G. Pandey, G. Kumaraswamy and U. T. Bhalerao, Tetrahedron Lett., **30**, 6059 (1989). (b) G. Pandey, Synlett, 546 (1992).
[80] G. Pandey and P. Y. Reddy, Tetrahedron Lett., **33**, 6533 (1992).

[81] A. L. J. Beckwith, Tetrahedron, **37**, 3073 (1981).
[82] Y. T. Jeon, C. P. Lee and P. S. Mariano, J. Am. Chem. Soc., **113**, 8847 (1991).
[83] F. D. Lewis and P. Y. Reddy, Tetrahedron Lett., **37**, 5293 (1990).
[84] G. Pandey, G. Lakshmaiah and G. Kumaraswamy, J. Chem. Soc., Chem. Commun., 1313 (1992).
[85] F. D. Saeva, Topics in Current Chemistry, **156**, 59 (1990).
[86] D. F. Eaton, Pure Appl. Chem., **56**, 1191 (1984).
[87] D. R. Arnold and A. J. Maroulis, J. Am. Chem. Soc., **98**, 5931 (1976).
[88] A. Sulpizio, A. Albini, N. d'Alessandro, E. Fasani and S. Pietra, J. Am. Chem. Soc., **111**, 5773 (1989).
[89] S. Sankararaman, S. Perrier and J. K. Kochi, J. Am. Chem. Soc., **111**, 6448 (1989).
[90] X. Ci and D. G. Whitten, *Photoinduced Electron Transfer Reactions*, edited by M. A. Fox and M. Chanon, (Elsevier, Amsterdam 1988) Vol. C, p. 553.
[91] (a) X. Ci and D. G. Whitten, J. Am. Chem. Soc., **109**, 7215 (1987); (b) X. Ci and D. G. Whitten, J. Am. Chem. Soc., **111**, 3459 (1989); (c) W. R. Bergmark and D. G. Whitten, J. Am. Chem. Soc., **112**, 4042 (1990); (d) X. Ci, M. A. Kellett and D. G. Whitten, J. Am. Chem. Soc., **113**, 3893 (1991); (e) L. Y. C. Lee, C. Giannotti and D. G. Whitten, J. Am. Chem. Soc., **108**, 175 (1986); (f) K. S. Schanze, L. Y. C. Lee, C. Giannotti and D. G. Whitten, J. Am. Chem. Soc., **108**, 2646 (1986).
[92] G. Pandey and A. Krishna, Synthetic Commun., **18**, 2309 (1988).
[93] G. Pandey, K. Sudha Rani and U. T. Bhalerao, Tetrahedron Lett., **31**, 1199 (1990).
[94] G. Pandey, P. Y. Reddy and U. T. Bhalerao, Tetrahedron Lett., **32**, 5147 (1990).
[95] K. Mizuno, M. Ikeda and Y. Otsuji, Tetrahedron Lett., **26**, 461 (1985).
[96] P. G. Gassman and K. J. Boottorff, J. Org. Chem., **53**, 1097 (1988).
[97] (a) K. Ohga and P. S. Mariano, J. Am. Chem. Soc., **104**, 617 (1982); (b) K. Ohga, U. C. Yoon and P. S. Mariano, J. Org. Chem., **49**, 213 (1984).
[98] G. Pandey, B. B. V. Soma Sekhar and U. T. Bhalerao, J. Am. Chem. Soc., **112**, 5650 (1990).
[99] Y. Nakadaira, A. Sekiguchi, Y. Funada and H. Sakurai, Chem. Lett., 327 (1991).
[100] (a) G. Pandey, V. J. Rao and U. T. Bhalerao, J. Chem. Soc., Chem. Commun., 416 (1989); (b) G. Pandey, B. B. V. Soma Sekhar, J. Org. Chem., **57**, 4019 (1992).
[101] T. Hamada, A. Nishida, O. Yonemitsu, J. Am. Chem. Soc., **108**, 140 (1986).
[102] J. Masnovi, D. J. Koholic, R. J. Berki and R. W. Binkley, J. Am. Chem. Soc., **109**, 2851 (1987).
[103] I. Saito, H. Ikehira, R. Kasatani, M. Watanabe and T. Matsuura, J. Am. Chem. Soc., **108**, 3115 (1986).
[104] C. Portella, H. Deshayes, J. -P. Pete and D. Scholler, Tetrahedron, **40**, 3635 (1984).
[105] R. W. Binkley and W. Flechtner, in *Synthetic Organic Photochemistry*, edited by Horspool William M. (Plenum Press, New York, 1984) p. 375.
[106] P. G. Gassman, in Photoinduced Electron Transfer Reactions, edited by M. A. Fox and M. Chanon, Eds. (Elsevier, Amsterdam, 1988) Vol C, p. 70.
[107] P. C. Wong and D. R. Arnold, Tetrahedron Lett., **20**, 2101 (1979).
[108] A. Albini and D. R. Arnold, Can. Journ. Chem., **56**, 2985 (1978).
[109] H. D. Roth, Acc. Chem. Res., **20**, 343 (1984).
[110] J. P. Dinnocenzo, W. P. Todd, T. R. Simpson and I. R. Gould, J. Am. Chem. Soc., **112**, 2462 (1990).
[111] F. Muller and J. Mattay, Angew. Chem., Int. Ed. Engl., **30**, 1336 (1991).
[112] F. Muller and J. Mattay, Angew. Chem., Int. Ed. Engl., **31**, 209 (1992).
[113] N. Ichinose, K. Mizuno, K. Yoshida and Y. Otsuji, Chem. Lett., 723 (1988).
[114] H. Tomioka, D. Kobayashi, A. Hashimoto, S. Murata, Tetrahedron Lett., **30**, 4685 (1989).

[115] P. G. Gassman and S. J. Burns, J. Org. Chem., **53**, 5576 (1988).
[116] P. Clawson, P. Lunn and D. A. Whitting, J. Chem. Soc., Chem. Commun., 134 (1984).
[117] H. Kautsky, Trans Faraday Soc., 35, 216 (1939).
[118] J. Saito and T. Matsuura, Tetrahedron, **41**, 2037 (1985).
[119] L. Lopez, *Topics in Current Chemistry*, Springer-Verlag, Berlin Heidelberg 156, p. 117 (1990).
[120] J. Eriksen, C. S. Foote,; T. L. Parker, J. Am. Chem. Soc., **99**, 6455 (1977).
[121] R. F Bartholomew, D. R. G Brimage and R. S Davidson, J. Chem. Soc. Chem. Comm., **34**, 3482 (1971).
[122] A. P. Schaap, K. A. Zaklika, B. Kaskar and L. W. M. Fung, J. Am. Chem. Soc., **102**, 389 (1980).
[123] L. T. Spada and C. S. Foote, J. Am. Chem. Soc., **102**, 391 (1980).
[124] (a) G. Pandey and G. Kumaraswamy, Tetrahedron Lett., **29**, 4153 (1988); (b) J. Santamaria and R. Ouchabane J. Rigaudy, Tetrahedron Lett., **30**, 2927 (1989); (c) J. Santamaria, R. Ouchabane and J. Rigaudy, Tetrahedron Lett., **30**, 3977 (1989); (d) J. Santamaria, M. T. Kaddachi and J. Rigaudy, Tetrahedron Lett., **31**, 4735 (1990).
[125] (a) K. Mizuno, N. Kamiyama and Y. Otsuji, Chem. Lett., 477 (1983); (b) K. Mizuno, N. Kamiyama, N. Ichinose and Y. Otsuji, Tetrahedron Lett., **41**, 2207 (1985).
[126] J. Eriksen and C. S. Foote, J. Am. Chem. Soc., **102**, 6083 (1980).
[127] (a) S. Futamara, S. Kusunose, H. Otha and Y. Kamiya, J. Chem. Soc., Chem. Commun., 1223 (1982); (b) S. Futamura, S. Kusunose, H. Otha and Y. Kamiya, J. Chem. Soc., Perkin Trans. I, 15 (1985); (c) A. P. Schaap, L. Lopez and S. D. Gagnon, J. Am. Chem. Soc., **105**, 663 (1983); (d) A. P. Schaap, S. Siddiqui and S. D. Gagnon, J. Am. Chem. Soc., **105**, 5149 (1983).
[128] A. P. Schaap, G. Prasad, S. Siddiqui, Tetrahedron Lett., **25**, 3035 (1984).
[129] (a) N. Berenjian, P. de Mayo, F. M. Phonix and A. C. Weeden, Tetrahedron Lett., 4179 (1979); (b) S. L. Mattes and S. Farid, J. Chem. Soc., Chem. Commun., 457 (1980).
[130] (a) T. Akasaka, W. Ando, Tetrahedron Lett., 5049 (1985); (b) W. Ando, T. Nagashima, K. Saito and S. Kohmoto, J. Chem. Soc., Chem. Commun., 154 (1979).
[131] (a) I. Saito, K. Tamoto and T. Matsuura, Tetrahedron Lett., **31**, 2889 (1979); (b) J. Santamaria, Tetrahedron Lett., **22**, 4511 (1981); (c) S. L. Mattes and S. Farid, J. Am. Chem. Soc., **104**, 1454 (1982); (d) J. J. Liang and C. S. Foote, Tetrahedron Lett., **23**, 3039 (1982); (e) J. Santamaria, P. Gabillet and L. Bokobza, Tetrahedron Lett., 2139 (1984). (f) K. Mizuno, N. Ichinose, T. Tamai and Y. Otsuji, Tetrahedron Lett., 5823 (1985); (g) L. Bokobza and J. Santamaria, J. Chem. Soc., Perkin Trans. II, 269 (1985); (h) M. Julliard, A. Galadi and M. Chanon, J. Photochem. Photobiol., A: Chem., **54**, 79 (1990).
[132] D. S. Steichen and C. S. Foote, J. Am. Chem. Soc., **103**, 1855 (1981).
[133] D. C. Dobrowolski and P. O. Ogilby, C. S. Foote, J. Phys. Chem., **87**, 2261 (1983).
[134] R. S. Davidson and J. E. Pratt, Tetrahedron, **40**, 999 (1984).
[135] Y. Araki, D. C. Dobrowolski, T. E. Goyne, D. C. Hanson, Z. Q. Jiang, K. J. Lee and C. S. Foote, J. Am. Chem. Soc., **106**, 4570 (1984).
[136] E. A. Mayeda and A. J. Bard, J. Am. Chem. Soc., **95**, 6223 (1973).
[137] D. H. R. Barton, R. K. Haynes, G, L.; Magnus, P. D.; I. D. Menzies, J. Chem. Soc., Perkin Trans. I, 2055 (1975).
[138] R. Tang, H. J. Yue, J. F. Wolf and F. Mares, J. Am. Chem. Soc., **100**, 5248 (1978).

[139] (a) S. F. Nelsen and R. Akaba, J. Am. Chem. Soc., **103**, 2096 (1981); (b) E. L. Clennan, W. Simmons and C. W. Almgren, J. Am. Chem. Soc., **103**, 2098 (1981).
[140] S. Futamura, S. Kusunose, H. Ohta and Y. Kamuya, J. Chem. Soc., Chem. Commun., 462 (1983).
[141] K. S. Wie and A. H. Adelman, Tetrahedron Lett., **10**, 3297 (1969).
[142] F. D. Lewis and J. Petisce, Tetrahedron, **42**, 6207 (1986).
[143] (a) E. Baciocchi, A. Piermattei, C. Rol, R. Ruzziconi and G. V. Sebastiani, Tetrahedron, **45**, 7049 (1989); (b) A. Albini and S. Spreti, Z. Naturforsch, **41b**, 1286 (1986).
[144] E. Baciocchi, T. Del Giacco, T. Rol and G. V. Sebastiani, Tetrahedron Lett., **26**, 3353 (1985).
[145] P. Pichat, J. Disdier, J. M. Hermann and P. Vaudano, Nouv. J. Chem., **10**, 545 (1986).
[146] M. Julliard, C. Legris and M. Chanon, Photochem. Photobiol., A: Chem, **61**, 137 (1991).
[147] J. Santamaria, R. Jroundi and J. Rigaudy, Tetrahedron Lett., **30**, 4677 (1989).
[148] J. Santamaria and R. Jroundi, Tetrahedron Lett., **32**, 4291 (1991).
[149] (a) W. Adam, A. Griesbeck, E. Staab, Tetrahedron Lett., **27**, 2839 (1986); (b) W. Adam and E. Staab, Tetrahedron Lett., **29**, 531 (1988).
[150] K. Gollnick and A. Schanatterer, Tetrahedron Lett., **25**, 185–2735 (1984).
[151] R. K. Haynes, M. K. S. Probert and I. D. Wilmot, Aust. J. Chem., **31**, 1737 (1978).
[152] (a) L. Lopez and L. Troisi, Tetrahedron Lett., **30**, 485 (1989); (b) A. P. Schaap, L. Lopez, S. D. Anderson and S. D. Gagnon, Tetrahedron Lett., 5493 (1982).
[153] A. P. Schaap, S. Siddiqui, G. Prasad, E. Palomino and L. Lopez, Photochem., **25**, 167 (1984).
[154] A. P. Schaap, G. Prasad and S. D. Gagnon, Tetrahedron Lett., **24**, 3047 (1983).
[155] (a) A. P. Schaap, S. Siddiqui, S. D. Gagnon and L. Lopez, J. Am. Chem. Soc., **105**, 5149 (1983); (b) A. P. Schaap, S. Siddiqui, G. Prasad, E. Palomino, M. Sandison and Tetrahedron, **41**, 2229 (1985); (c) A. P. Schaap, S. Siddiqui, G. Prasad and A. F. M. Raham, J. P. Olivier, J. Am. Chem. Soc., **106**, 6087 (1984).
[156] (a) T. Miyashi, M. Kamata and T. Mukai, J. Chem. Soc., Chem. Commun., 1577 (1986); (b) T. Miyashi, M. Kamata and T. Mukai, J. Am. Chem. Soc., **108**, 2755 (1986).
[157] M. Sakuragi, H. Sakuragi, Chem. Lett., 1017 (1980).
[158] G. W. Griffin, G. P. Kirschenheuter, C. Vaz, P. P. Umrigar, D. C. Lankin and S. Christensen, Tetrahedron, **41**, 2069 (1985).
[159] J. P. K. Wong, A. A. Fahmi, G. W. Griffin and N. S. Bhacca, Tetrahedron, **37**, 3354 (1981).
[160] B. L. Feringa, Rec. Trav. Chim. Pays-Bas, **106**, 469 (1985).
[161] K. Gollnick and A. Griesbeck, Tetrahedron, **41**, 2057 (1985).
[162] (a) F. Khuong-Huu and D. Herlem, Tetrahedron Lett., 3649 (1970); (b) D. Herlem, Y. Hubert-Brierre, F. Khuong-Huu, Tetrahedron Lett., 4173 (1973).
[163] J. H. E. Lindler, H. J. Khun and K. Gollnick, Tetrahedron Lett., 1705 (1972).
[164] J. Santamaria and M. T. Kadachi, Synlett, 739 (1991).
[165] J. Santamaria, M. T. Kadachi and C. Ferroud, Tetrahedron Lett., **33**, 781 (1992).
[166] C. Ferroud, E. L. Cavalcanti de Amorim, L. Dallery and J. Santamaria, Synthesis, 291 (1994).
[167] A. Da Silva Goes, C. Ferroud and J. Santamaria, Tetrahedron Lett., **36**, 2235 (1995).

5 Transition Metal Complexes and Homogeneous Photocatalytic Transformations of Organic Substrates

CHARLES KUTAL
Department of Chemistry, The University of Georgia, Athens, Georgia, USA

1 Introduction 135
2 Homogeneous thermal catalysis by transition metal complexes 137
3 Homogeneous photocatalysis by transition metal complexes 139
   3.1 General considerations 139
   3.2 Mechanisms of photocatalysis 140
      3.2.1 Reactions catalytic in photons 140
      3.2.2 Reactions not catalytic in photons 143
   3.3 A caveat concerning mechanistic labels 148
4 Properties of excited states of transition metal complexes 148
5 Selected examples of photocatalysis 153
   5.1 Hydrogenation and hydrosilation 153
   5.2 Cycloaddition 157
   5.3 C–H bond activation 162
6 Final remarks 165
Acknowledgements 166
References 166

1 INTRODUCTION

Homogeneous transition-metal catalysis of organic reactions has been the focus of extensive research activity during the past 40 years [1–3]. Practical interest in soluble catalysts arises from the desire to develop systems that are more active, per metal atom, and more selective than their heterogeneous counterparts. Higher activity allows the catalyst to function under milder conditions, thereby conserving energy, while greater selectivity results in the more efficient use of chemical feedstocks. From a fundamental perspective, homogeneous systems are

---

Homogeneous Photocatalysis. Edited by M. Chanon
© 1997 John Wiley & Sons Ltd

amenable to examination by a variety of spectroscopic techniques that can yield important structural and kinetic information relevant to the mechanism of catalytic behavior. This information, in turn, facilitates the rational design of new catalysts exhibiting improved performance.

Most studies of transition-metal catalysis have dealt with thermally activated reactions. In recent years, however, there has been a growing recognition of the benefits of using light rather than heat to activate a system. *Photocatalysis* is the generally accepted term for a process in which light and a catalyst bring about or accelerate a chemical reaction [4–7]. The type of photocatalytic transformation of interest in this chapter can be represented generically by eq. (5.1):

$$O \xrightarrow[M]{h\nu} P \qquad \ldots (5.1)$$

where O and P denote the organic reactant(s) and product(s), respectively, and M is a soluble mononuclear or polynuclear (containing two or more metal atoms) transition metal catalyst or catalyst precursor. Photocatalysis has been observed for a diverse assortment of metal/substrate combinations. Table 5.1 lists a sampling of systems that will be discussed in later sections. A more extensive, annotated compilation of examples appears in a recent review by Chanon and Chanon [5].

Table 5.1. Examples of transition-metal photocatalysis

Substrate(s)	Metal compound	Transformation	Ref.
1-pentene	$Fe(CO)_5$	Double-bond migration and *cis-trans* isomerization (Scheme 5.4)	[12, 13]
α-cyanoacrylate	*trans*-$Cr(NH_3)_2(NCS)_4^-$	Anionic polymerization (Scheme 5.5)	[14]
endo-dicyclopentadiene	$Cu(O_3SCF_3)$	Intermolecular cycloaddition (Scheme 5.6)	[15]
Norbornadiene	$Ir(bipy)_2(bipy')^{2+}$ [a]	Valence isomerization (Scheme 5.11)	[17]
1-methylcyclooctene + O_2	$O[Fe(TPP)]_2^b$	Oxygenation (Scheme 5.13)	[18]
α, β-unsaturated ketone + O_2	$CuSO_4$	Dimeric lactone formation (Scheme 5.16)	[19]
Norbornadiene + H_2	$Cr(CO)_6$	Hydrogenation (Scheme 5.19)	[31,32]
Alkene + silane	$Pt(acac)_2^c$	Hydrosilation (Scheme 5.22)	[36]
n-pentane	$RhCl(CO)(PMe_3)_2$	Carbonylation (Scheme 5.26)	[48]

[a] bipy is 2,2'-bipyridine; bipy' is 2,2'-bipyrid-3-yl-C^3,N'.
[b] TPP is the tetraphenylporphyrinato dianion.
[c] acac is the 2,4-pentanedianato anion.

This chapter will focus upon the mechanistic aspects of homogeneous transition-metal photocatalysis of organic reactions. The tripartite interactions of M, O, and light implied by eq. (5.1) yield a multiplicity of possible reaction pathways for this process. We shall find that many of the principles that have emerged from studies of transition-metal-catalyzed thermal reactions of organic compounds (interaction of M and O; see eq. (5.2)) and inorganic/organometallic photochemistry (interaction of M and light) will be extremely helpful in unraveling the mechanistic complexities of photocatalytic behavior. Accordingly, the contents of the chapter have been organized along the following lines. In the next section we review some important features of the thermal catalysis of organic reactions by transition metal complexes. This is followed by a general discussion of photocatalysis that includes a useful classification of possible mechanisms. Thereafter, we consider the excited-state reactivity patterns of metal complexes and their relationship to photocatalytic behavior. Lastly, some selected classes of photocatalyzed reactions are surveyed. The treatment of these topics is intended to be didactic rather than exhaustive. Readers interested in additional information can consult several of the comprehensive sources cited as references.

2 HOMOGENEOUS THERMAL CATALYSIS BY TRANSITION METAL COMPLEXES

Homogeneous thermal catalysis of organic reactions by transition metals can be represented by eq. (5.2):

$$O \xrightarrow[M]{\Delta} P \qquad \ldots (5.2)$$

where Δ denotes thermal activation. While elevated temperatures may be required to generate the actual catalyst from an inactive precursor, it is important to note that all species participating in the transformation of O to P reside in their electronic ground states. The generality of eq. (5.2) can be attributed to the chemical versatility of the d-block and f-block elements. Transition metals form complexes with a wide range of σ-bonding and π-bonding ligands, including nearly every class of organic compounds. Coordination to a metal alters the electron density of stable molecules and ions (e.g. CO, alkenes, CN^-) and thereby renders them susceptible to reaction. Moreover, the metal can serve as a template for assembling the substrates that subsequently react to form the desired product(s). As shown in Scheme 5.1 for the rhodium-catalyzed hydroformylation of an alkene to an aldehyde [2], the metal must accommodate the alkene molecule, carbon monoxide, and hydride ion in its first coordination sphere. Other coordinated ligands (e.g. triphenylphosphine), which do not participate directly in the reaction, nonetheless can influence the outcome via steric and/or electronic effects. This ability to modify reactivity by changes in non-participating ligands allows the design of catalysts that yield the targeted product with high selectivity. Other

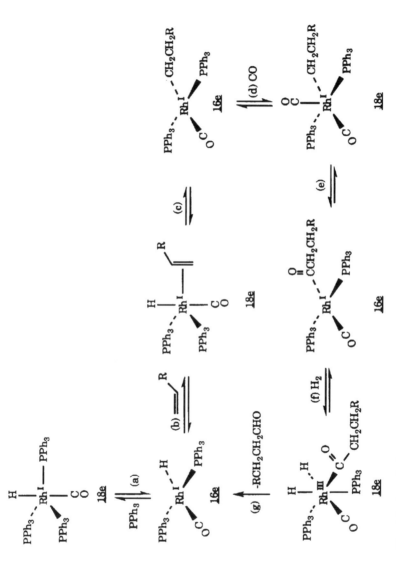

Scheme 5.1. Simplified catalytic cycle for the thermally activated hydroformylation of an alkene to an aldehyde. The oxidation state (Roman numeral) and number of metal valence electrons (16e or 18e) are indicated. (From reference [2], reproduced by permission of John Wiley & Sons.)

characteristics of transition metals pertinent to catalysis are variable coordination number, stereochemistry, and oxidation state. This variability facilitates the formation of different metal-containing intermediates during the course of the catalytic process.

Catalytic reactions typically consist of several elementary steps connected in a cyclic scheme. A journey around the cycle results in the conversion of reactants to products, via one or more intermediates, with the regeneration of the catalytically active species. Referring again to the rhodium-catalyzed hydroformylation reaction in Scheme 5.1, we note that the catalyst precursor, $HRh(PPh_3)_3(CO)$ (PPh_3 is triphenylphosphine), undergoes thermally activated dissociation of a PPh_3 ligand to yield the coordinatively unsaturated intermediate, $HRh(PPh_3)_2(CO)$ (step a). Thereafter, the following succession of reactions occurs: coordination of an alkene molecule (step b), insertion of the alkene into the Rh−H bond (step c), coordination of a CO molecule (step d), insertion of CO into the metal alkyl bond (step e), oxidative addition of dihydrogen (step f), and reductive elimination of aldehyde with the regeneration of the active catalyst, $HRh(PPh_3)_2(CO)$ (step g). During this catalytic cycle, the central metal changes its coordination number, geometry, and oxidation state while activating the various substrates (alkene, CO, H_2) that yield the final product.

One of the most useful paradigms of organometallic chemistry is the 18 valence-electron rule, which can be stated as follows: a d-block transition metal will tend to form complexes in which the number of electrons in its valence orbitals (one s, three p, and five d) equals 18. This rule recognizes the tendency to achieve an inert gas electron configuration around the central metal, much as first-row elements strive to complete an octet of electrons in their valence shell. Although exceptions to the rule exist, the structures and reactivities of most organometallic complexes can be rationalized in terms of the number of metal valence electrons. Thus, complexes containing 16 or 14 electrons are coordinatively unsaturated and tend to bind additional ligands via simple complexation or oxidative addition. Paramagnetic 17-electron complexes display radical character and undergo reactions such as dimerization and atom abstraction. Not surprisingly, complexes containing 17, 16, or fewer metal valence electrons are often postulated as reactive intermediates in catalytic cycles (e.g. Scheme 5.1).

3 HOMOGENEOUS PHOTOCATALYSIS BY TRANSITION METAL COMPLEXES

3.1 GENERAL CONSIDERATIONS

As specified by eq. (5.1), transition-metal photocatalysis of an organic reaction requires both light and a metal-containing catalyst or catalyst precursor. Absorption of visible or ultraviolet radiation by one or more components in the system results in the population of electronic excited states. These energy-rich species possess geometries, metal–ligand bond orders, acid–base strengths, and redox potentials that can differ significantly from the corresponding properties of the ground state [8,9]. Consequently, excited states may introduce reaction pathways that either do not occur in conventional thermal catalysis or occur only at elevated temperatures. Endoergic reactions, for example, now become feasible, since a portion of the absorbed photon energy can be used to overcome the unfavorable change in free energy. Other potential consequences of photoactivation are greater temporal control of a reaction through regulation of light intensity and greater spatial control through the use of patterned light. Judicious choice of the irradiation wavelength allows the selective population of different excited states within a molecule or of different compounds within a mixture; both outcomes provide a means of controlling the nature of the primary photochemical process. Finally, activation of catalysis by light rather than heat permits reactions to be conducted at ambient or lower temperatures, thereby prolonging the lifetimes of thermally sensitive catalysts and facilitating the isolation of thermally sensitive products.

3.2 MECHANISMS OF PHOTOCATALYSIS
3.2.1 Reactions Catalytic in Photons

Transition-metal photocatalysis of organic reactions can be divided into two broad mechanistic categories on the basis of the role played by light [5–7,10,11]. The first category comprises reactions that are catalytic in photons. Such behavior occurs in systems where light generates an active ground-state catalyst, C, from M (Scheme 5.2) or M and O (Scheme 5.3). In a subsequent thermal cycle, C catalyzes the conversion of the organic substrate to product. Since C is regenerated in each cycle, the overall reaction is also catalytic in one substance. Transformations of this type have been labeled *photogenerated catalysis* by Salomon [10] (an equally descriptive term suggested by Hennig [6] is *photoinduced catalytic reaction*). Photogenerated catalysis frequently occurs with an induction period, during which the active catalyst is formed, and continued reaction after irradiation. Moreover, since the catalyst formed by one photon can transform several substrate molecules, the observed quantum efficiency of product

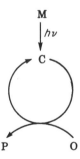

Scheme 5.2. Reproduced from reference [7]. Copyright 1993 American Chemical Society

TRANSITION METAL COMPLEXES

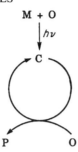

Scheme 5.3. Reproduced from reference [7]. Copyright 1993 American Chemical Society

formation, $\emptyset_{ob}$, can exceed unity. This chemical amplification of the initial photochemical act can be exploited in the design of highly photosensitive systems.

Scheme 5.4 depicts a simplified mechanism for the photogenerated catalysis of alkene isomerization in the presence of Fe(CO)$_5$ [12,13]. In the specific example considered, 1-pentene is isomerized to an equilibrium mixture of *cis* and *trans* 2-pentenes. Upon near-ultraviolet irradiation, the metal carbonyl complex undergoes successive loss of two CO ligands to yield the active catalyst, Fe(CO)$_3$ (1-pentene). In the ensuing thermal catalytic cycle, the coordinated alkene isomerizes via a π-allyl hydride intermediate. Substitution of the isomerized product by the parent substrate then regenerates the catalyst. Occurrence of an induction period, inhibition of catalytic behavior by added CO and strongly coordinating dienes, and continuation of isomerization after the cessation of irradiation are characteristics of the system that support the proposed mechanism. Here $\emptyset_{ob}$ equals the quantum efficiency of catalyst formation, $\emptyset_{cat}$, multiplied by the catalyst turnover number, η_{cat} (moles of product per mole of C). The observation that $\emptyset_{ob} > 1$ confirms the assignment of the mechanism as photogenerated catalysis. In those systems where $\emptyset_{ob} < 1$ owing to a low value of $\emptyset_{cat}$, the more general criterion, $\emptyset_{ob} > \emptyset_{cat}$ (i.e. $\eta_{cat} > 1$) can be used to infer the operation of a photogenerated catalytic cycle.

Another process, termed *photoinitiation*, also is catalytic in photons. A transformation of this type begins with the light-induced production of an initiator, I, from M (eq. 5.3) or M and O (eq. 5.4). In a subsequent step, I is consumed while initiating a chain reaction such as polymerization (eq. 5.5). Thus, unlike photogenerated catalysis, photoinitiation is not catalytic in any substance. Apart from this distinction, however, the two processes display many of the same characteristics.

$$M \xrightarrow{h\nu} I \qquad \ldots (5.3)$$

$$M + O \xrightarrow{h\nu} I \qquad \ldots (5.4)$$

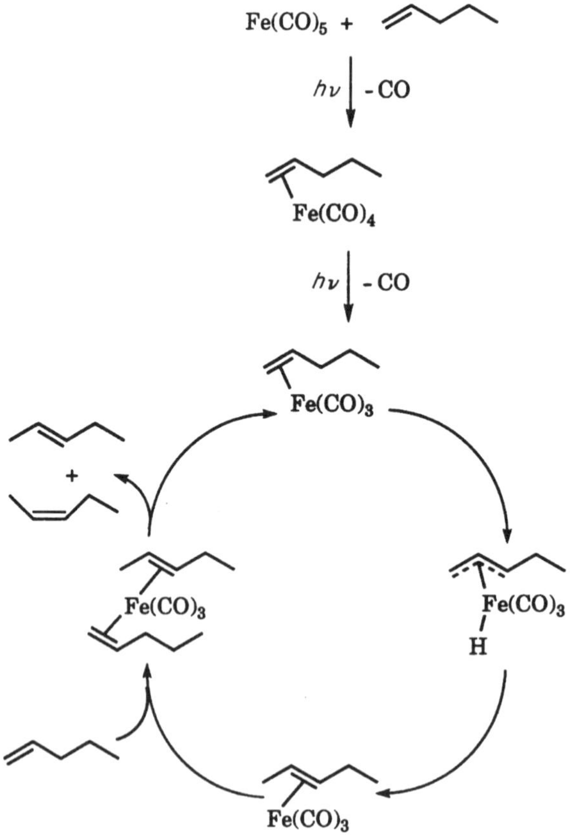

Scheme 5.4. Reproduced from reference [7]. Copyright 1993 American Chemical Society

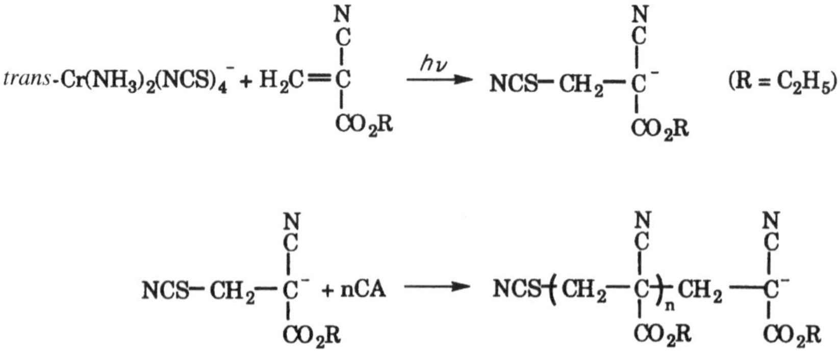

$$trans\text{-}Cr(NH_3)_2(NCS)_4^- + H_2C=\underset{CO_2R}{\underset{|}{\overset{\overset{N}{\underset{|}{C}}}{C}}} \xrightarrow{h\nu} NCS-CH_2-\underset{CO_2R}{\underset{|}{\overset{\overset{N}{\underset{|}{C}}}{C^-}}} \quad (R = C_2H_5)$$

$$NCS-CH_2-\underset{CO_2R}{\underset{|}{\overset{\overset{N}{\underset{|}{C}}}{C^-}}} + nCA \longrightarrow NCS\text{-}(CH_2-\underset{CO_2R}{\underset{|}{\overset{\overset{N}{\underset{|}{C}}}{C}}})_n CH_2-\underset{CO_2R}{\underset{|}{\overset{\overset{N}{\underset{|}{C}}}{C^-}}}$$

Scheme 5.5

$$O + I \rightarrow P_1 \xrightarrow{O} P_2 \xrightarrow{xO} P_x \qquad \ldots (5.5)$$

Scheme 5.5 summarizes the mechanism for the photoinitiated anionic polymerization of ethyl α-cyanoacrylate (CA) in the presence of trans-$Cr(NH_3)_2(NCS)_4^-$ [14]. Visible-light irradiation of the complex dissolved in neat CA results in the release of a thiocyanate ion. Subsequent addition of NCS^- to the carbon–carbon double bond of the acrylate monomer yields a resonance-stabilized carbanion. Polymerization then proceeds in the dark by the repetitive addition (average chain length $>10^4$) of monomer units to the active anionic center. This example reminds us that an initiator (or a catalyst) need not contain a metal atom.

3.2.2 Reactions not Catalytic in Photons

The second mechanistic category of photocatalysis, which bears the generic label *catalyzed photolysis* [10], encompasses reactions that are noncatalytic in photons. This type of process begins with the absorption of light by M, O, or a preformed M–O complex (eqs. 5.6–5.8) (an asterisk denotes electronic excitation). The resulting excited state or its primary photoproduct then participates in a cycle that yields the final product and regenerates M. While the overall reaction is thus catalytic with respect to M, at least one photon is required per product molecule formed and, consequently, $\emptyset_{ob}$ never exceeds unity. Furthermore, neither an induction period nor continued reaction in the dark is observed.

$$M \xrightarrow{h\nu} M^* \xrightarrow{O} P + M \qquad \ldots (5.6)$$

$$O \xrightarrow{h\nu} O^* \xrightarrow{M} P + M \qquad \ldots (5.7)$$

$$M + O \rightleftharpoons (M\text{–}O) \xrightarrow{h\nu} (M\text{–}O)^* \begin{cases} \rightarrow P + M & \ldots(5.8a) \\ \rightarrow M + O^* \rightarrow P & \ldots(5.8b) \end{cases}$$

The metal catalyst can play several roles in catalyzed photolysis. For example, formation of a ground-state complex between M and O typically shifts the absorption spectrum of the system to longer wavelengths and may introduce a sterically or electronically favorable pathway to the product that is unavailable to the uncoordinated substrate. Following photoexcitation, the complex may yield product directly (eq. 5.8a) or via an excited state of O (eq. 5.8b).

Scheme 5.6 illustrates the marked influence of ground-state complexation upon the photochemistry of *endo*-dicyclopentadiene [15]. Irradiation in the presence of an organic triplet sensitizer favors intramolecular cycloaddition via the diene triplet state, whereas a completely different reaction, intermolecular cycloaddition, occurs in the presence of $Cu(O_3SCF_3)$. The latter pathway reflects the

Scheme 5.6. From reference [11]. Reproduced by permission of Elsevier Science

photoreactivity of a 1 : 2 copper(I)-diene complex in which the metal, serving as a template, brings the two reacting substrate molecules into close proximity.

Photoexcitation of the metal catalyst followed by energy transfer (Scheme 5.7) or electron transfer (Scheme 5.8) provides another common pathway for activating the organic substrate. Historically, this subcategory of catalyzed photolysis has been called *photosensitization*. Energy transfer yields an excited state of O, which subsequently can react to form product. In contrast, electron transfer generates the radical cation ($O^{\overset{+}{\cdot}}$) or anion ($O^{\overset{-}{\cdot}}$), both of which are usually highly reactive. Often, a sacrificial reagent (sac) is required to regenerate M from its oxidized or reduced form via a thermal redox process (Scheme 5.9).

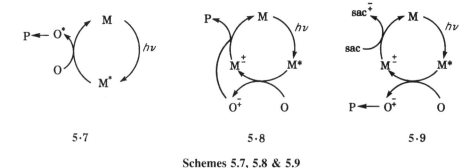

Schemes 5.7, 5.8 & 5.9

Photosensitization of an endoergic reaction is shown in Scheme 5.10 [16]. Visible-light irradiation of Ru(bipy)$_3^{2+}$ (bipy is 2,2'-bipyridine) in acetonitrile induces the characteristic orange-red phosphorescence from the lowest triplet excited state of the complex. Substituted norbornadiene, **I**, quenches this luminescence via a dynamic, collisional process and, in turn, undergoes valence isomerization to the highly strained quadricyclene, **II**. This behavior has been attributed to energy transfer from Ru(bipy)$_3^{2+}$ to **I**, the triplet state of which then rearranges to **II**. A turnover number of 2 for Ru(bipy)$_3^{2+}$ confirms its catalytic role in the reaction.

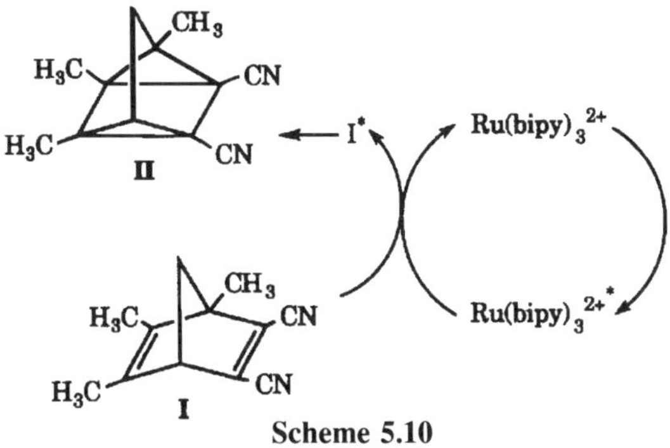

Scheme 5.10

Scheme 5.11 depicts a related example in which the Ir(III) complex, **III**, functions as a photosensitizer for the valence isomerization of unsubstituted norbornadiene, **IV**, to the corresponding quadricyclene, **V** [17]. Sensitization via energy transfer was discounted on energetic grounds and, instead, it was proposed that the key interaction was partial electron transfer from **IV** to the photoexcited metal complex. The resulting redistribution of electron density generates a polarized and structurally distorted organic substrate that is favorably disposed toward rearrangement to **V**.

Rather than interacting directly with O (eq. 5.6), M* may yield a ground-state species, C′, that assists the transformation of substrate to product and, in the process, reverts to M (Scheme 5.12). This subcategory of catalyzed photolysis has been termed *photoassistance* and C′ has been labeled the photoassistor or pseudocatalyst. Scheme 5.13 summarizes the proposed mechanism for the photoassisted oxidation of a cyclic alkene in the presence of oxygen and μ-oxobis(tetraphenylporphyrinato)iron(III), O[Fe(TPP)]$_2$ [18]. Upon photoexcitation, the dinuclear complex disproportionates to yield a ferryl species, OFe(TPP). Oxygen atom transfer from the latter to the alkene produces an epoxide. Reaction of Fe(TPP) with O$_2$ then regenerates the starting complex via a series of thermal reactions. The entire cycle can be repeated upon absorption of another photon as evidenced by the production of up to 300 moles of epoxide per mole of complex.

Catalysis by M may occur following the photoexcitation of O (eq. 5.7). One possible pathway involves a direct interaction between M and the electronically

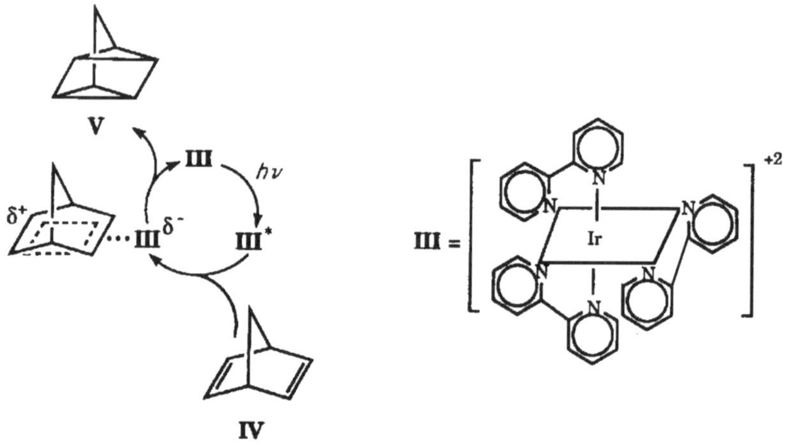

146   HOMOGENEOUS PHOTOCATALYSIS

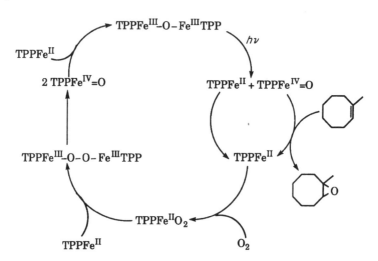

Scheme 5.11

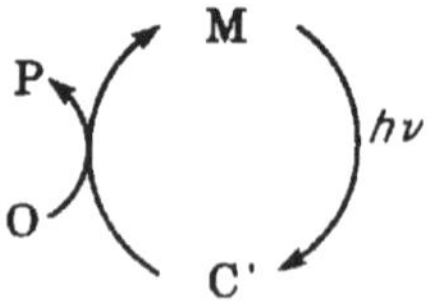

Scheme 5.12. Reproduced from reference 7. Copyright 1993 American Chemical Society

Scheme 5.13. TPP is the tetraphenylporphyrinato dianion. (From reference [6]. Reproduced by permission of Kluwer Academic Publishers)

TRANSITION METAL COMPLEXES 147

excited organic substrate (Scheme 5.14). Alternatively, M may catalyze the reaction of a photogenerated ground-state intermediate, R, to yield a final product that differs from the one obtained in the absence of the metal catalyst (Scheme 5.15). Great care must be exercised in assigning a mechanism of this type. The main uncertainty is the extent to which a ground-state M–O complex (eq. 5.8) plays a role in the overall transformation. While the absence of new ultraviolet (UV)-visible spectral features upon mixing M and O is usually regarded as *prima facie* evidence of the absence of complex formation, this criterion may be misleading if the features are weak or masked by other absorption bands.

Scheme 5.16 presents a plausible example of a system in which a metal catalyzes the photoreaction of an organic substrate [19]. Excitation of an

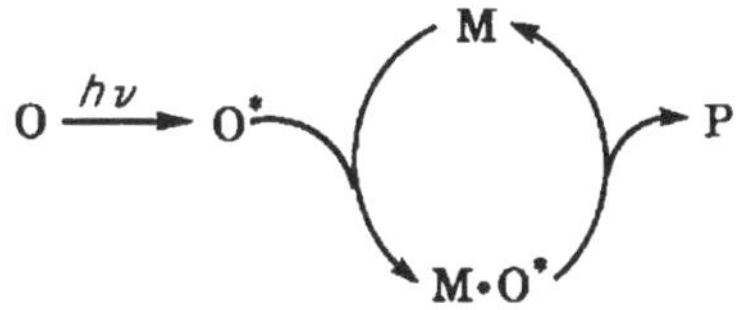

Scheme 5.14. Reproduced from reference [7]. Copyright 1993 American Chemical Society

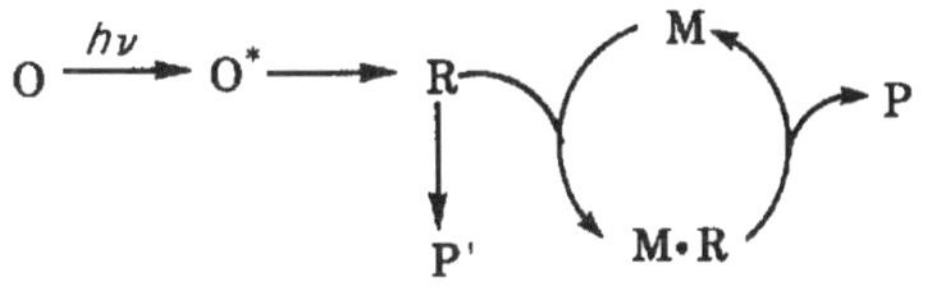

Scheme 5.15

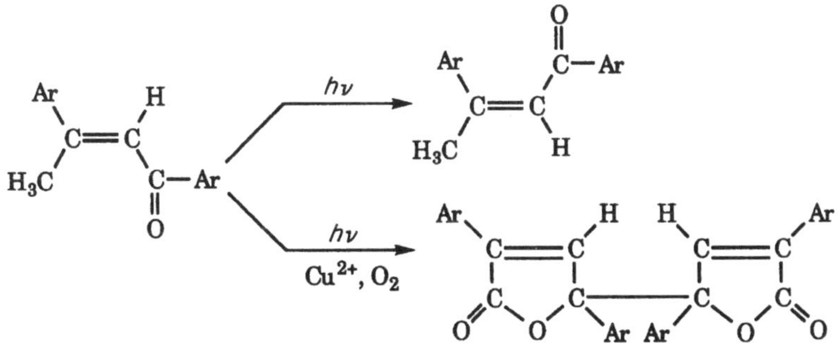

Scheme 5.16. Ar is phenyl, *p*-tolyl, or *p*-bromophenyl

α, β-unsaturated ketone in methanol results in exclusive *cis-trans*-isomerization about the carbon–carbon double bond. In the presence of Cu^{2+} and O_2, however, the ketone undergoes photooxidation to a dimeric lactone, **VI**. Both the absence of new UV spectral features upon mixing Cu^{2+} and the ketone, and the lower product yields at higher Cu^{2+} concentrations, presumably owing to competitive absorption by the metal ion, argue against the participation of a ground-state metal-substrate complex. Instead, the results support a mechanism in which Cu^{2+} interacts with an excited state or a primary photoproduct of the ketone.

3.3 A CAVEAT CONCERNING MECHANISTIC LABELS

Dividing photocatalysis into two mechanistic categories — reactions catalytic in photons and reactions not catalytic in photons — focuses attention upon the paramount role played by light. It also provides guidance in the design of practical applications that incorporate photocatalysis as an essential component. For example, applications that demand high photosensitivity should be best served by systems that undergo photogenerated catalysis with high chemical amplification. On the other hand, photochemical energy storage applications require endoergic transformations that only can be achieved via catalyzed photolysis.

The generality and complexity of the photocatalytic process have resulted in an unfortunate proliferation of mechanistic labels, with two or more labels sometimes referring to the same sequence of steps. While some consensus regarding the nomenclature in this area has begun to emerge, it appears unlikely that a universal set of labels will be adopted any time soon. To avoid confusion, it is imperative that any label be accompanied by a detailed mechanistic scheme describing the transformation under consideration. Thus, Scheme 5.4 precisely defines the photocatalytic behavior of the $Fe(CO)_5$/1-pentene/light system regardless of whether the mechanism is termed photogenerated catalysis or photoinduced catalytic reaction. In cases where mechanistic uncertainties exist, the general appellation photocatalysis should be used simply to denote that light and a catalyst are required for the transformation to occur.

4 PROPERTIES OF EXCITED STATES OF TRANSITION METAL COMPLEXES

The preceding section illustrates the wide variety of mechanisms by which transition-metal photocatalysis can occur. Common to each mechanism is the involvement at some point of an electronic excited state. A fundamental understanding of photocatalytic behavior thus requires a detailed knowledge of excited state properties. Transition metal complexes serve as the principal light-absorbing species in the majority of reported systems (Schemes 5.14, 5.15, and 5.16 are exceptions). Accordingly, in this section we review the key features of the electronic structures and reactivities of the different types of excited states that can

TRANSITION METAL COMPLEXES

arise in metal complexes. More comprehensive treatments of this topic can be found in a number of the references [8,9,20,21].

Consider the simple orbital energy diagram in Figure 5.1. Here are shown the molecular orbitals that result from the overlap of the valence orbitals of a d-block transition metal with the appropriate symmetry-adapted orbitals of six ligands situated at the vertices of an octahedron. Under the assumption of polar metal–ligand bonding (i.e. limited covalency), the molecular orbitals identified as σ_L, π_L, and π_L^a are localized largely on the ligands, whereas the orbitals labeled $d\pi$ and $d\sigma^a$ are predominantly metal in nature (the superscript, a, denotes antibonding character). Upon the absorption of light in the visible-ultraviolet region, metal complexes undergo electronic transitions from the ground state to various excited states. In the simple one-electron formalism, such transitions and the resulting excited states are labeled in terms of the orbitals that undergo a change in electron occupancy.

Ligand field (also termed d–d) excited states arise from transitions between valence d orbitals localized predominantly on the metal. The $d\pi \rightarrow d\sigma^a$ transition in Figure 5.1, for example, involves the promotion of an electron from a d orbital than can undergo a π-type interaction with the ligands to a higher energy d orbital that is strongly σ-antibonding with respect to the metal–ligand bonds. While this

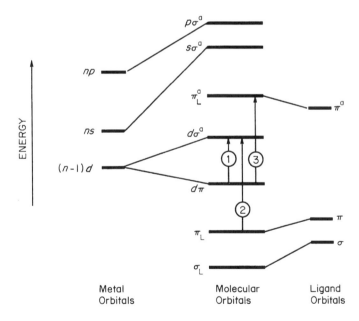

Figure 5.1. Qualitative energy level diagram of the molecular orbitals and electronic transitions in an octahedral coordination compound. Transition types are: (1) ligand field; (2) ligand-to-metal charge transfer (LMCT); (3) metal-to-ligand charge transfer (MLCT). For clarity, all orbitals of a given type are represented by a single energy level

angular redistribution of electron density leaves the formal oxidation state of the metal unchanged, it weakens the metal–ligand bonding in the complex and thereby increases the likelihood of ligand loss. Population of a ligand field excited state thus affords a convenient route to coordinatively unsaturated intermediates such as $Fe(CO)_3$(alkene) in Scheme 5.4.

Charge transfer excited states result from the redistribution of electron density between the constituents (metal(s) and ligands) of a complex or between the complex and the surrounding medium. The $\pi_L \longrightarrow d\sigma^a$ transition in Figure 5.1 occurs with the transfer of an electron from a ligand-based orbital to a metal d orbital and produces a ligand-to-metal charge transfer (LMCT) excited state. Electron transfer in the opposite sense, as occurs in the $d\pi \longrightarrow \pi_L^a$ transition, generates a metal-to-ligand charge transfer (MLCT) excited state. A transition that induces the transfer of an electron from one ligand to another in the same complex yields a ligand-to-ligand charge transfer (LLCT) excited state, while the photoinduced transfer of an electron between metal centers in a polynuclear complex gives rise to a metal-to-metal charge transfer (MMCT) excited state. A transition that involves the movement of electron density from the metal complex to the surrounding solvent affords a charge-transfer-to-solvent (CTTS) excited state. Lastly, a transition that results in electron transfer between the partners of an ion pair produces an ion-pair charge transfer (IPCT) excited state. Each of these transitions occurs with a radial redistribution of electron density that alters the oxidation states of the species involved (metal(s), ligands, solvent, or counterion). This property makes the resulting excited states susceptible to oxidation–reduction reactions. Ligand loss processes also may become important in systems where a change in oxidation state creates a substitutionally labile metal center. Moreover, changes in the charge distribution about a coordinated ligand may enhance its reactivity toward processes such as protonation and electrophilic substitution.

A few examples should suffice to illustrate the range of reactions that can originate from charge transfer excited states. Population of the low-energy LMCT states (X $\longrightarrow$ Co charge transfer) in $Co(NH_3)_5X^{2+}$ (X = Cl^-, Br^-, NCS^-) generates a substitutionally labile Co(II) center bound to a ligand radical [22]. In solution, rapid and irreversible ligand substitution by solvent leads to efficient redox decomposition of the complex (eq. 5.9). Population of a MLCT or CTTS excited state in $Ru(CN)_6^{4-}$ results in the production of a solvated electron [23]. In a chlorinated solvent such as chloroform, the net reaction is one-electron oxidation of the complex accompanied by irreversible reduction of the solvent (eq. 5.10). Under comparable conditions, complexes containing a metal center susceptible to a +2 change in oxidation state undergo photochemical oxidative addition (eq. 5.11; thpy is the ortho-C-deprotonated form of 2-(2-thienyl)pyridine) [24]. A highly distorted MMCT excited state in $O[Fe(TPP)]_2$ yields the active oxygenation catalyst, OFe(TPP), via a disproportionation reaction (Scheme 5.13) [25]. Population of a low-energy IPCT excited state in

TRANSITION METAL COMPLEXES

ion-paired $Ph_2I^+/Mo(CN)_8^{4-}$ (Ph is phenyl) affords the oxidized metal complex and the diphenyliodyl radical (eq. 5.12); the latter species then acts as a chain carrier in a cyclic process for the oxidation of alcohols (Scheme 5.17) [26].

$$(H_3N)_5Co^{III}X^{2+} \xrightarrow{h\nu} [(H_3N)_5Co^{II}, X^\bullet]^{2+} \xrightarrow{H_2O} Co_{(aq)}^{2+} + 5NH_3 + X^\bullet \quad \ldots(5.9)$$

$$Ru^{II}(CN)_6^{4-} + CHCl_3 \xrightarrow{h\nu} Ru^{III}(CN)_6^{3-} + Cl^- + {}^\bullet CHCl_2 \quad \ldots(5.10)$$

$$Pt^{II}(thpy)_2 + CHCl_3 \xrightarrow{h\nu} Pt^{IV}(thpy)_2Cl(CHCl_2) \quad \ldots(5.11)$$

$$Ph_2I^+/Mo^{IV}(CN)_8^{4-} \xrightarrow{h\nu} Ph_2I^\bullet + Mo^V(CN)_8^{3-} \quad \ldots(5.12)$$

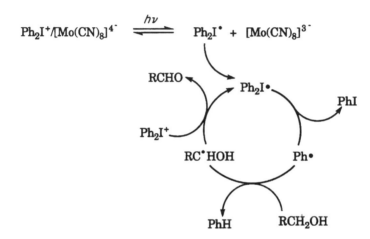

Scheme 5.17. From reference [6]. Reproduced by permission of Kluwer Academic Publishers

Complexes containing two or more metal atoms bonded directly to each other possess excited states that arise from transitions between molecular orbitals delocalized over the metal framework [27]. Transitions that result in the movement of electron density from a bonding or nonbonding orbital to an antibonding orbital (with respect to the metal framework) weaken the metal–metal bonding and thereby facilitate fragmentation of the cluster into smaller units. For $Mn_2(CO)_{10}$ and other dinuclear complexes containing a single, unsupported metal–metal bond, this type of transition causes homolytic cleavage to yield two metal-containing radicals (eq. 5.13):

$$Mn_2(CO)_{10} \xrightarrow{h\nu} 2\,^\bullet Mn(CO)_5 \quad \ldots(5.13)$$

These photogenerated species typically undergo secondary reactions such as ligand substitution, atom abstraction, and electron transfer. Photofragmentation decreases in importance for higher nuclearity metal clusters, and processes other than homolysis of a metal–metal bond may be involved.

Schemes 5.8 and 5.9 depict photocatalytic systems in which a photoexcited metal complex undergoes intermolecular electron transfer. The archetype of such behavior is Ru(bipy)$_3^{2+}$, whose lowest MLCT excited state is susceptible to both oxidative (eq. 5.14a; MV^{2+} is the methylviologen dication) and reductive (eq. 5.14b) quenching by suitable substrates.

$$\text{Ru(bipy)}_3^{2+} \xrightarrow{h\nu} \text{Ru(bipy)}_3^{2+*} \begin{array}{c} \xrightarrow{\text{MV}^{2+}} \text{Ru(bipy)}_3^{3+} + \text{MV}^{+\bullet} \quad \ldots(5.14a) \\ \xrightarrow{\text{N(C}_2\text{H}_5)_3} \text{Ru(bipy)}_3^{+} + {}^{\bullet +}\text{N(C}_2\text{H}_5)_3 \quad \ldots(5.14b) \end{array}$$

More generally, all types of electronic excited states are simultaneously stronger reductants and stronger oxidants than the corresponding ground state by the equivalent of the excitation energy, ΔE (neglecting structural reorganization effects). This photoenhancement of the driving force for intermolecular electron transfer reflects the change in the electron distribution within the complex caused by the absorption of a photon [28]. As illustrated in Figure 5.2, oxidation occurs more readily from an excited state since the electron that resides in the upper orbital is lost more easily than any electron in the ground state. Similarly, reduction proceeds more readily from an excited state because the vacancy in the lower

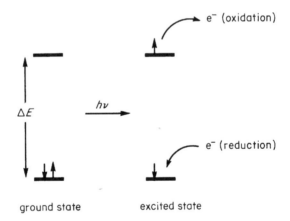

Figure 5.2. Alteration of the redox properties of a coordination compound upon photoexcitation. The excited state undergoes oxidation (loss of the electron from the higher-lying orbital) and reduction (addition of an electron to the partially filled lower orbital) more readily than the ground state. (Reproduced from reference 7. Copyright 1993 American Chemical Society.)

5 SELECTED EXAMPLES OF PHOTOCATALYSIS

5.1 HYDROGENATION AND HYDROSILATION

Photogenerated catalysis of the hydrogenation (eq. 5.15) and hydrosilation (eq. 5.16) of unsaturated organic substrates occurs for a wide range of transition metal complexes [29].

$$\mathrm{C{=}C} + H_2 \longrightarrow \mathrm{-C{-}C{-}} \quad \text{(H H)} \quad \ldots(5.15)$$

$$\mathrm{C{=}C} + R_3SiH \longrightarrow \mathrm{-C{-}C{-}} \quad \text{(H SiR}_3\text{)} \quad \ldots(5.16)$$

Quite commonly, the mechanism of these molecular addition reactions involves the initial photoinduced loss of one or more ligands from the metal. The resulting coordinatively unsaturated species then adds the substrate and H_2 or silane, and further reactions of these reagents within the coordination sphere of the metal yield the final product. These mechanistic features are illustrated in Scheme 5.18 for the $Fe(CO)_5$-photocatalyzed hydrogenation of 1-pentene [12,30]. The active catalyst in this system, $Fe(CO)_3$(1-pentene), results from the photochemical loss of two CO ligands. Following the addition of H_2, migratory insertion of the alkene into an Fe–H bond and reductive elimination of the resulting alkyl group and the remaining hydride afford the hydrogenated product. It should be noted that $Fe(CO)_3$(1-pentene) is also the species responsible for isomerization of the alkene (Scheme 5.4). Consequently, hydrogenation and isomerization to cis and trans-2-pentenes are competitive processes for 1-pentene under the reaction conditions.

Photocatalytic hydrogenation in the presence of $Fe(CO)_5$ occurs for a range of linear and branched alkenes, cycloalkenes, dienes, and acetylene. In contrast, $Cr(CO)_6$ and its Mo and W congeners are much more selective hydrogenation photocatalysts. Conjugated dienes, for example, undergo exclusive cis-1,4-addition of H_2 to yield the corresponding cis-monoalkene (eq. 5.17). Hydrogenation of norbornadiene affords two products, nortricyclene (**VII**) and norbornene (**VIII**) (eq. 5.18).

$$\text{diene} + H_2 \xrightarrow[Cr(CO)_6]{h\nu} \text{cis-monoalkene} \quad \ldots(5.17)$$

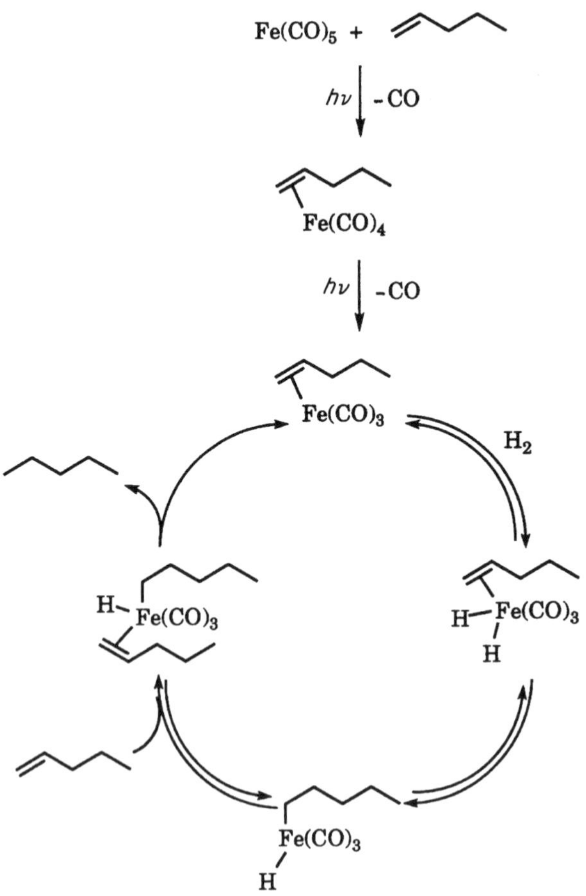

Scheme 5.18

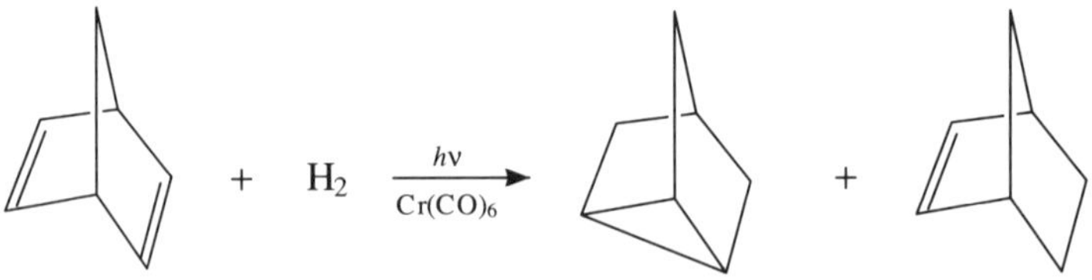

...(5.18)

VII   VIII

The mechanism of the latter reaction has been the subject of numerous studies, and only recently has a comprehensive picture emerged [31,32]. Photoexcitation of $M(CO)_6$ in the presence of norbornadiene (NBD) initially results in the loss of two CO ligands to form $M(CO)_4(\eta^4\text{-NBD})$. As depicted in Scheme 5.19, this product then undergoes photolabilization of another CO to yield the *fac* and *mer* isomers of $M(CO)_3(\eta^4\text{-NBD})(\eta^2\text{-NBD})$. These latter two species serve as

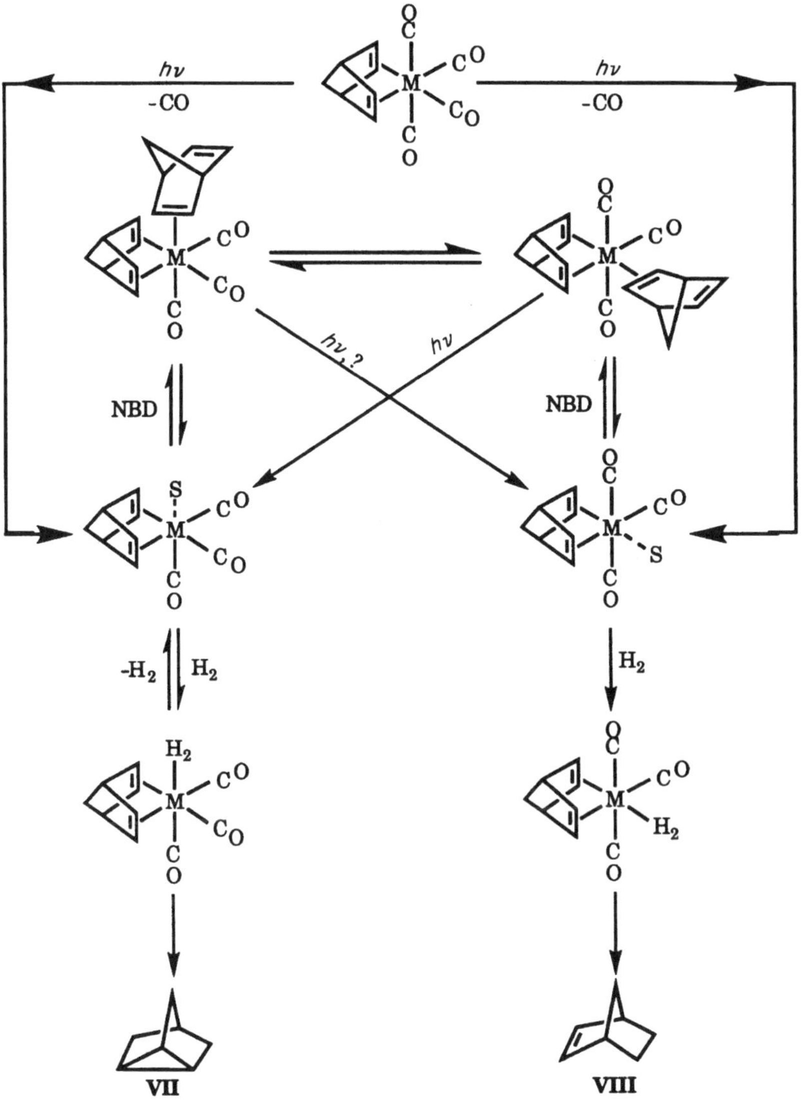

Scheme 5.19. Reproduced from reference [32]. Copyright 1990 American Chemical Soceity

reservoirs for the very reactive *fac* and *mer*-M(CO)$_3$(η^4-NBD)S (S is solvent), which add H$_2$ in a thermal process to form the corresponding isomeric dihydrogen complexes. Intramolecular transfer of H$_2$ to norbornadiene in the *mer* isomer yields norbornene, while the analogous transfer in the *fac* isomer gives nortricyclene.

A related mechanism (Scheme 5.20) has been proposed to explain the selective 1,6-addition of H_2 to 1,3,5-cycloheptatriene (CHT) in the presence of $Cr(CO)_3(\eta^6\text{-CHT})$ [33]. In this case, light causes a change in the hapticity of the cyclic ligand from η^6 to η^4, yielding a coordinatively unsaturated metal center that adds H_2 to form the *fac*-tricarbonyl complex. Intramolecular transfer of H_2 to the coordinated triene produces 1,3-cycloheptadiene (CHD).

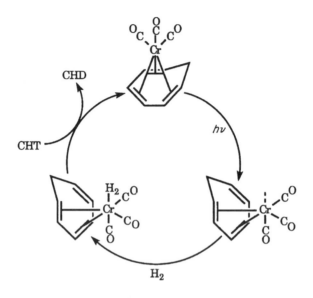

Scheme 5.20

Photoinduced loss of a phosphine ligand from Wilkinson's complex, $Rh(PPh_3)_3Cl$, yields an active catalyst for the hydrosilation of alkenes (Scheme 5.21) [34]. The rate of this process is enhanced in the presence of O_2 owing to the photochemical oxidation of free PPh_3 to the poorly coordinating $OPPh_3$. Since the phosphine oxide is less likely to recoordinate to the metal than the parent phosphine, higher concentrations of the catalytically active $Rh(PPh_3)_2Cl$ can be produced in aerated systems.

Recent studies of platinum complexes as photocatalysts for hydrosilation reactions offer an interesting divergence of mechanistic viewpoints. Boardman reported that irradiation of $CpPt(CH_3)_3$ (Cp is $\eta^5\text{-}C_5H_5$) in the presence of a silane initially yields a silyl methyl hydride complex (eq. 5.19) [35]:

$$CpPt(CH_3)_3 \xrightarrow[2HSiR_3]{h\nu} CpPt(SiR_3)(H)(CH_3) + CH_4 + R_3SiCH_3$$

$$\downarrow \text{decomp.}$$

$$Pt_X$$

...(5.19)

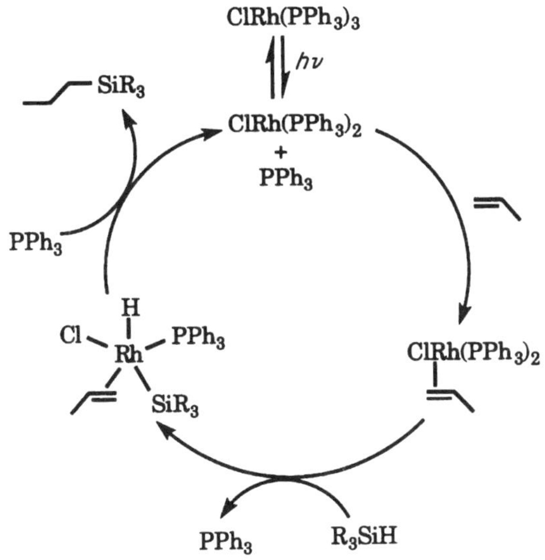

TRANSITION METAL COMPLEXES

Scheme 5.21

Subsequent thermal decomposition of this photoproduct produces platinum atoms which aggregate to a catalytically active colloid, Pt_x. Catalysis is thus viewed as a heterogeneous process occurring on the surface of the colloidal particles. In contrast, Lewis and Salvi proposed that photolysis of $Pt(acac)_2$ ($acac^-$ is the 2,4-pentanedionato anion) under typical hydrosilation conditions generates a highly active homogeneous catalyst [36]. As shown in Scheme 5.22, photolabilization of an $acac^-$ ligand yields a mononuclear complex containing coordinated alkene and silane. Hydrosilation occurs within the first coordination sphere of this species, which retains its catalytic activity for several hours at room temperature and is only slowly converted to a less active heterogeneous metal colloid. We can infer from these two studies that a range of platinum-containing species of different nuclearities and reactivities can exist throughout the course of photocatalyzed hydrosilation and, consequently, the assignment of mechanism as either homogeneous or heterogeneous will depend upon the precise experimental conditions employed and the point in the reaction at which the system is interrogated.

5.2 CYCLOADDITION

Transition-metal photocatalysis of alkene cycloaddition reactions provides an attractive route to novel and synthetically useful products. By far the most popular catalysts are copper(I) compounds, particularly simple salts such as CuCl, CuBr, and CuOTf (OTf^- is the triflate anion, $O_3SCF_3^-$). Copper(I) salts form stable ground-state complexes with a variety of cyclic and acyclic alkenes, and several mechanistic studies have found conclusive evidence that these complexes function as the principal photoactive species in the system (recall Scheme 5.6).

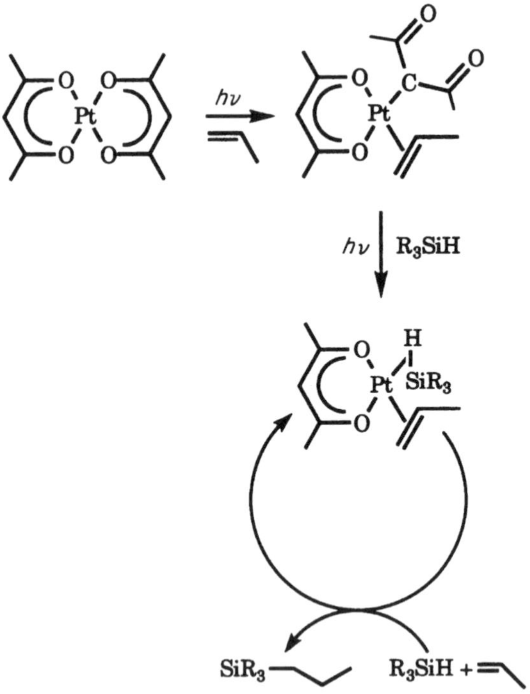

Scheme 5.22

Figure 5.3 shows the changes in the electronic absorption spectrum observed upon mixing CuCl with a large excess of *cis,cis*-1,5-cyclooctadiene (c,c-COD) [37]. The substantial increase in UV absorbance results from the formation of a complex of stoichiometry [CuCl(c,c-COD)]$_2$. Molecular orbital calculations performed on model copper(I)-alkene systems suggest that the maximum at 246 nm and shoulder at ~290 nm correspond to Cu $\longrightarrow$ (c,c-COD) MLCT transitions [38]. Photoexcitation of [CuCl(c,c-COD)]$_2$ in the wavelength region >290 nm causes *cis-trans* isomerization about one double bond to produce c,t-COD, while light of wavelengths >240 nm yields two additional products, t,t-COD and tricyclo[3.3.0.0^{2,6}]octane (TCO) [39]. Detailed quantum yield measurements establish that the intramolecular cycloadduct, TCO, is not a primary photoproduct; instead, it results from the secondary photolysis of complexed c,t-COD and both free and complexed t,t-COD [40]. Scheme 5.23 summarizes the 254 nm quantum yields for the complete set of transformations occurring in this complicated system. Several interesting aspects of these results should be noted. First, the involvement of c,t-COD and t,t-COD in the CuCl-photocatalyzed formation of TCO stands in sharp contrast to the Hg-photosensitized reaction of c,c-COD in the gas phase, which appears to

TRANSITION METAL COMPLEXES

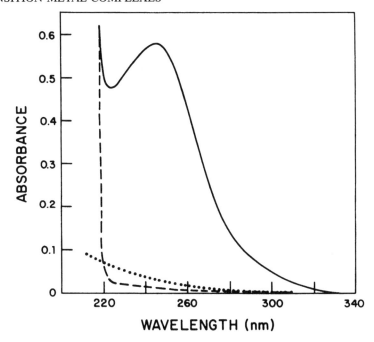

Figure 5.3. Electronic absorption spectra (. . . .) saturated solution of CuCl in ether, (– – – –) 0.0163 M cc-COD in ether, (———) saturated solution of CuCl in ether containing 0.016 M cc-COD. Cell path length is 1 cm. (Reproduced from reference 37. Copyright 1982 American Chemical Society.)

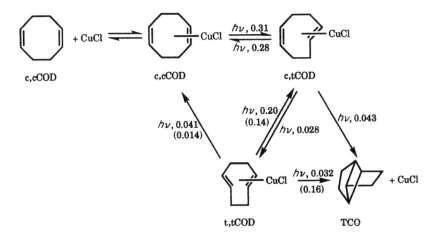

Scheme 5.23. Summary of 254 nm quantum yields for the CuCl-photocatalysed isomerization of cyclooctadiene; values in parentheses refer to isomerization of uncomplexed t,t-COD. For simplicity, complexes are shown as monomers

produce TCO without the intermediacy of the *trans*-bonded isomers. Second, the wavelength dependence of the product yields allows the convenient and high-yield synthesis of either c,t-COD or TCO by appropriate selection of the excitation wavelength. Finally, coordination of t,t-COD to CuCl enhances the photosensitivity of this highly strained alkene.

Salomon and Kochi popularized the use of CuOTf as a photocatalyst for alkene cycloaddition reactions. Unlike the copper(I) halides, this salt is soluble in a variety of organic solvents, and the ready displacement of the weakly coordinating triflate ion facilitates the formation of complexes containing up to four coordinated olefinic bonds. Irradiation of cyclohexene in the presence of CuOTf gives the two cycloaddition products, dimers **IX** and **X**, and 1-cyclohexyl cyclohexene (eq. 5.20) [41]. The preponderance of doubly *trans*-fused **IX** is remarkable, given the thermodynamic preference for the *cis*-fused isomers. The proposed mechanism (eq. 5.21) involves initial *cis-trans* isomerization about the double bond of the coordinated alkene to yield *trans*-cyclohexene.

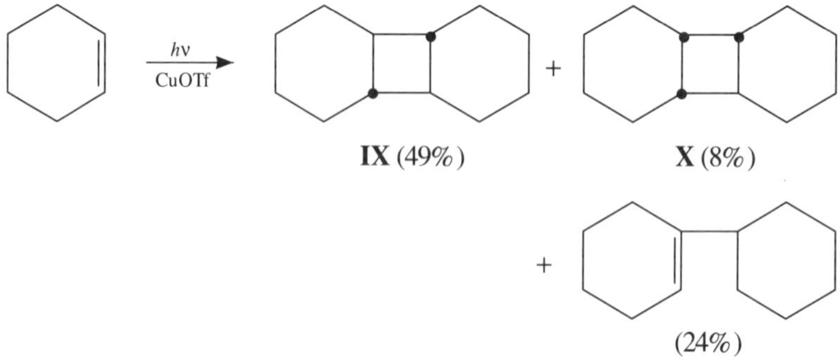

...(5.20)

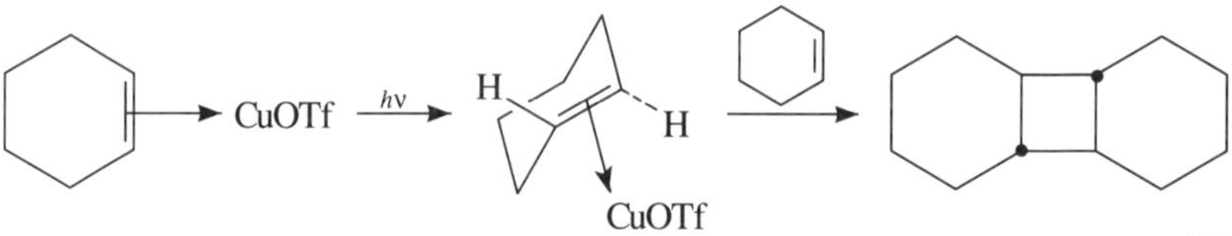

...(5.21)

Concerted thermal $[2\pi_a + 2\pi_s]$ cycloaddition of this reactive species with another (presumably coordinated) molecule of *cis*-cyclohexene produces **IX**. A similar pathway has been invoked to explain the novel cyclotrimerization of cycloheptene (eq. 5.22); thus CuOTf-photocatalyzed production of the *trans* isomer followed by the thermal cycloaddition of three *trans* alkene molecules at a single metal center yields a cyclic trimer, **XI**, with D_3 symmetry [42]. Surprisingly, irradiation of a mixture of cyclohexene and cycloheptene in the presence of CuOTf leads to the preferential formation of the C_{13} cycloadduct, **XII** (eq. 5.23) [43]. This result was attributed to the concerted photochemical $[2\pi_s + 2\pi_s]$ cycloaddition of the coordinated *trans* isomers of each cycloalkene.

TRANSITION METAL COMPLEXES

$$\text{cycloheptene} \xrightarrow[\text{CuOTf}]{h\nu} \text{XI} \quad \ldots(5.22)$$

$$\text{cyclohexene} + \text{cycloheptene} \xrightarrow[\text{CuOTf}]{h\nu} \text{XII} \quad \ldots(5.23)$$

Structural constraints preclude alkenes such as cyclopentene and norbornene from adopting a *trans* configuration about the carbon–carbon double bond. In these cases, CuOTf-photocatalyzed cycloaddition processes must occur via a mechanism that does not involve the intermediacy of a *trans* isomer [15]. It may be that the metal serves as a template for a concerted photochemical cycloaddition of the coordinated C=C bonds of a bis(alkene) complex as in path a of Scheme 5.24. Alternatively, excitation of the complex to an alkene ⟶ Cu LMCT excited state may favor the generation of a carbenium ion, **XIII**, as in path b of Scheme 5.24. This intermediate then proceeds to a metallocycle, **XIV**, which collapses to the final product by reductive elimination. Photogeneration of a carbenium ion in the CuOTf/cyclohexene system discussed above would account for the production of 1-cyclohexyl cyclohexene (eq. 5.24):

$$2 \text{ cyclohexene} \xrightarrow[\text{Cu}^+]{h\nu} \cdots \longrightarrow \cdots \xrightarrow{-\text{Cu}^+} \text{1-cyclohexyl cyclohexene}$$

$$\ldots(5.24)$$

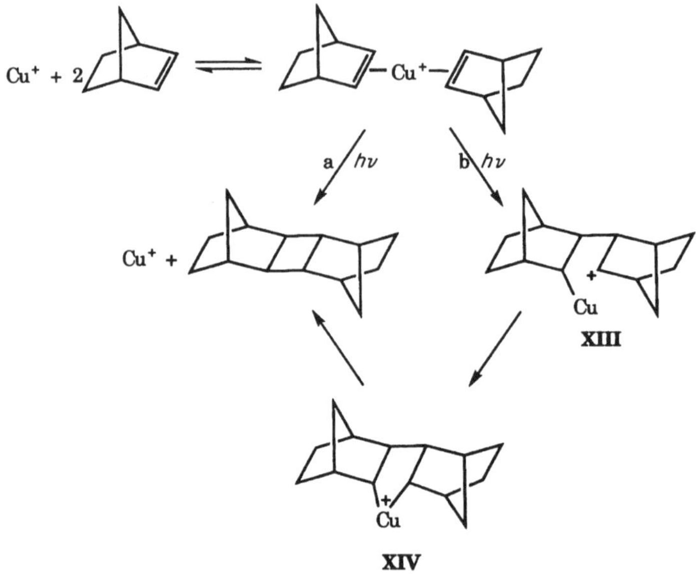

Scheme 5.24

Irradiation of the Cu^+ complex of diallyl ether yields **XV** (eq. 5.25), which contains the bicyclo[3.2.0] ring system [10]. This prototypical intramolecular cycloaddition process is useful for the construction of multicyclic carbon networks [44]. For example, the CuOTf-photocatalyzed conversion of **XVI** to **XVII** (eq. 5.26) is a key step in the total synthesis of the sesquiterpenes α and β-panisinsene [45].

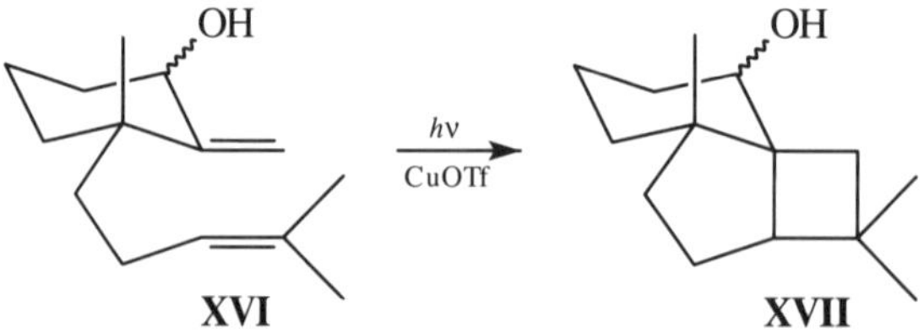

...(5.25)

...(5.26)

5.3 C–H BOND ACTIVATION

Selective functionalization of alkanes under mild conditions is of considerable technological interest, since it affords a means of converting a plentiful chemical feedstock into useful end-products. Early progress in this area was achieved by the groups of Bergman, Graham, and Jones, whose seminal studies demonstrated that the photochemical extrusion of ligands such as CO and H_2 from 18-electron rhodium and iridium complexes yields coordinatively unsaturated 16-electron species, which then undergo oxidative addition to the C–H bonds of alkanes (eq. 5.27; Cp' is η^5-$C_5(CH_3)_5$, RH is an alkane) [46].

$$Cp'Ir(CO)_2 \xrightarrow[-CO]{h\nu} Cp'Ir(CO) \xrightarrow[+RH]{} Cp'Ir(CO)(R)(H) \quad \ldots (5.27)$$

While representing a significant advance, these reactions are stoichiometric owing to the stability of the resulting alkyl hydride complex. The apparent key to catalytic behavior lies in the generation of a 14-electron metal complex that can oxidatively add the alkane to produce a 16-electron species which, by virtue of its residual coordinative unsaturation, can undergo further reaction at the metal center [47,48]. Scheme 5.25 depicts a photocatalytic cycle for the dehydrogenation of cyclooctane by $RhCl(CO)(PMe_3)_2$ (Me is methyl) [49]. Photoinduced loss of CO occurs with a quantum yield (λ_{ex} = 366 nm, 50 °C) of 0.1 to yield $RhCl(PMe_3)_2$, which undergoes oxidative addition of the alkane. Subsequent β-hydrogen elimination produces a dihydride species containing coordinated cyclooctene. Loss of this alkene followed by CO-assisted reductive elimination of H_2 regenerates the parent complex. Turnover numbers of up to 5000 have been measured in this system.

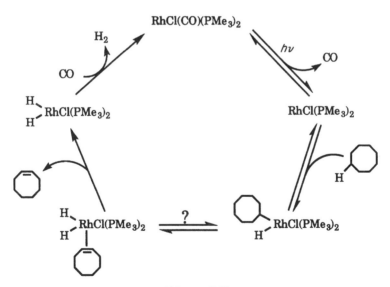

Scheme 5.25

Irradiation of $RhCl(CO)(PMe_3)_2$ in n-pentane under 1 atmosphere of CO results in the photocatalytic carbonylation of the alkane [47]. The proposed mechanism, summarized in Scheme 5.26, again features dissociation of CO as the primary photochemical step. Carbonylation of other acyclic alkanes, cyclohexane, and benzene occurs under similar conditions.

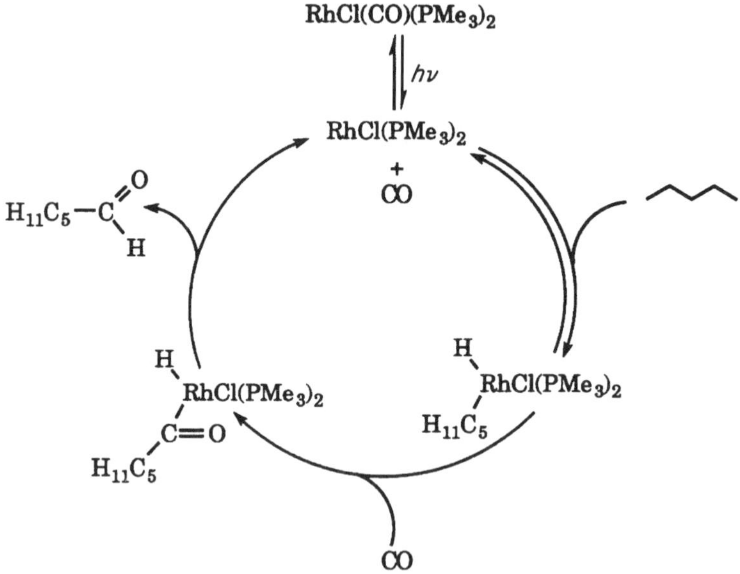

Scheme 5.26

Photocatalyzed functionalization of alkanes also has been achieved in the presence of soluble polyoxometallates such as $W_{10}O_{32}^{4-}$ and $PW_{12}O_{40}^{3-}$ [50,51]. These polynuclear metal clusters possess excellent oxidative and thermal stability, and their physical, spectral, and chemical properties can be tuned over a wide range. In the C–H activation systems considered above, photoexcitation of a metal complex serves to create a coordinatively unsaturated ground-state species which undergoes subsequent thermal reaction with an alkane. In contrast, irradiation of a polyoxometallate, P_{ox} (ox denotes the fully oxidized form), generates a strongly oxidizing charge transfer excited state that abstracts a hydrogen atom from the substrate. As shown in Scheme 5.27, the resulting radical pair, |PH • R|, diffuses apart to yield the reduced cluster, P_{red}, an alkyl radical, and a proton. Depending upon the experimental conditions, the radical may undergo any of several secondary processes, including oxidation to a carbenium ion, reduction to a carbanion, disproportionation to an alkene and alkane, or reaction with an added reagent such as CO (i.e. radical carbonylation [52]). Regeneration of the parent cluster results from the oxidation of P_{red} by H^+, O_2, or some other oxidizing agent in the system. Given the diversity of possible reaction paths, polyoxometallate photocatalysis represents an attractive strategy for the activation of alkanes.

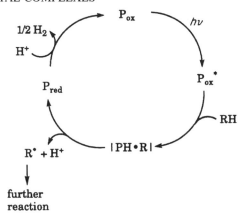

Scheme 5.27

6 FINAL REMARKS

Transition-metal photocatalysis of organic reactions provides a convenient, mild, and, in many cases, selective route to a diverse array of products. We have seen that light plays two key roles in this process. First, the absorption of a photon by the metal catalyst, organic substrate, or preformed metal–substrate complex creates an electronic excited state that either participates directly in the product-forming event (catalyzed photolysis) or yields a ground-state species which catalyzes product formation (photogenerated catalysis). Second, for endoergic transformations, light provides the external energy needed to overcome the unfavorable thermodynamics.

We also have learned that transition metal complexes can play several roles in the photocatalytic process. If serving as the principal chromophore, the metal complex may undergo an intramolecular excited-state reaction that produces the active catalyst. For example, photogeneration of a coordinatively unsaturated metal center via ligand dissociation, reductive elimination, or change in ligand hapticity creates one or more sites for the assembly and activation of substrates. Intermolecular excited-state processes such as electron transfer, energy transfer, and atom abstraction also result in substrate activation. In systems where the metal complex does not function as the principal light-absorbing species, it can influence photocatalytic behavior by interacting with the photoexcited organic substrate or binding to the substrate in the ground state prior to light absorption. Additional roles for transition metal complexes in photocatalysis undoubtedly will emerge as detailed mechanistic studies of a broader range of systems are undertaken.

In closing, it should be noted that promising applications of homogeneous transition-metal photocatalysis have been reported in areas such as organic synthesis, microimaging, coatings technology, biology and medicine, pollution remediation, and energy storage. More detailed accounts of these applications appear in two recent monographs [53,54].

ACKNOWLEDGMENTS

The author thanks the US National Science Foundation for financial support of his research program in the area of photocatalysis. He also is grateful to Dr Paul A. Grutsch for technical assistance in the preparation of this chapter.

REFERENCES

[1] C. Masters, *Homogeneous Transition-Metal Catalysis* (Chapman and Hall, London, 1981).
[2] F. A. Cotton and G. Wilkinson, *Advanced Inorganic Chemistry*, 5th edn. (Wiley-Interscience, New York, 1988) Ch. 28.
[3] J. A. Moulijn, P. W. N. M. van Leeuwen, and R. A. van Santen, (editors) *Catalysis*, Studies in Surface Science and Catalysis, Vol. 79 (Elsevier, Amsterdam, 1993).
[4] H. Kisch, in *Photocatalysis*, edited by N. Serpone and E. Pelizzetti, (Wiley-Interscience, New York, 1989) Ch. 1.
[5] F. Chanon and M. Chanon, in *Photocatalysis*, edited by N. Serpone and E. Pelizzetti, (Wiley-Interscience, New York, 1989) Ch. 15.
[6] H. Hennig, R. Billing, and H. Knoll, in *Photosensitization and Photocatalysis Using Inorganic and Organometallic Compounds*, edited by K. Kalyanasundaram and M. Grätzel (Kluwer, Dordrecht, 1993) pp. 51–69.
[7] C. Kutal, in *Photosensitive Metal-Organic Systems: Mechanistic Principles and Applications*, edited by C. Kutal and N. Serpone, ACS Advances in Chemistry Series, Vol. 238 Ch. 1 (American Chemical Society, Washington, DC, 1993).
[8] V. Balzani and V. Carassiti, *Photochemistry of Coordination Compounds* (Academic Press, New York, 1970) Ch. 5.
[9] C. Kutal and A. W. Adamson, in *Comprehensive Coordination Chemistry*, edited by G. Wilkinson, R. D. Gillard, and J. A. McCleverty, Vol. 1 (Pergamon Press, Elmsford, New York, 1987) pp. 385–414.
[10] R. G. Salomon, Tetrahedron, **39**, 485 (1983).
[11] C. Kutal, Coord. Chem. Rev., **64**, 191 (1985).
[12] M. A. Schroeder and M. S. Wrighton, J. Am. Chem. Soc., **98**, 551 (1976).
[13] R. L. Whetten, K. -J. Fu, and E. R. Grant, J. Am. Chem. Soc., **104**, 4270 (1982).
[14] C. Kutal, P. A. Grutsch, and D. B. Yang, Macromolecules, **24**, 6872 (1991).
[15] R. G. Salomon and J. K. Kochi, J. Am. Chem. Soc., **96**, 1137 (1974).
[16] H. Ikezawa, C. Kutal, K. Yasufuku, and H. Yamazaki, J. Am. Chem. Soc., **108**, 1589 (1986).
[17] P. A. Grutsch and C. Kutal, J. Am. Chem. Soc., **108**, 3108 (1986).
[18] L. Weber, G. Haufe, D. Rehorek, and H. Hennig, J. Chem. Soc., Chem. Comm., 502 (1991).
[19] T. Sato, K. Tamura, K. Maruyama, and O. Ogawa, Tetrahedron Lett., **43**, 4221 (1973).
[20] R. H. Hill, in *Encyclopedia of Inorganic Chemistry*, edited by R. B. King, Vol. 6 (Wiley, New York, 1994) pp. 3213–3227.
[21] A. Vogler and H. Kunkely, in *Photosensitization and Photocatalysis Using Inorganic and Organometallic Compounds*, edited by K. Kalyanasundaram and M. Grätzel (Kluwer, Dordrecht, 1993) pp. 71–111.
[22] J. F. Endicott, in *Concepts of Inorganic Photochemistry*, edited by A. W. Adamson and P. Fleischauer (Wiley-Interscience, New York, 1975) Ch. 3.
[23] A. Vogler, W. Losse, and H. Kunkely, J. Chem. Soc., Chem. Comm., 188 (1979).

[24] D. Sandrini, M. Maestri, V. Balzani, L. Chassot, and A. von Zelewsky, J. Am. Chem. Soc., **109**, 7720 (1987).
[25] M. W. Peterson, D. S. Rivers, and R. M. Richman, J. Am. Chem. Soc., **107**, 2907 (1985).
[26] R. Billing, D. Rehorek, and H. Hennig, Topics Curr. Chem., **158**, 151 (1990).
[27] T. J. Meyer and J. V. Caspar, Chem. Rev., **85**, 187 (1985).
[28] V. Balzani and M. Maestri, in *Photosensitization and Photocatalysis Using Inorganic and Organometallic Compounds*, edited by K. Kalyanasundaram and M. Grätzel (Kluwer, Dordrecht, 1993) pp. 15-49.
[29] G. L. Geoffroy and M. S. Wrighton, *Organometallic Photochemistry*, (Academic Press, New York, 1979) Ch. 3.
[30] M. E. Miller and E. R. Grant, J. Am. Chem. Soc., **109**, 7951 (1987).
[31] S. A. Jackson, P. M. Hodges, M. Poliakoff, J. J. Turner, and F. -W. Grevels, J. Am. Chem. Soc., **112**, 1221 (1990).
[32] P. M. Hodges, S. A. Jackson, J. Jacke, M. Poliakoff, J. J. Turner, and F. -W. Grevels, J. Am. Chem. Soc. **112**, 1234 (1990).
[33] I. Fischler, F. -W. Grevels, J. Leitich, and S. Ozkar, Chem. Ber., **124**, 2857 (1991).
[34] R. A. Faltynek, Inorg. Chem., **20**, 1357 (1981).
[35] L. D. Boardman, Organometallics, **11**, 4194 (1992).
[36] F. D. Lewis and G. D. Salvi, Inorg. Chem., **34**, 3182 (1995).
[37] E. Grobbelaar, C. Kutal, and S. W. Orchard, Inorg. Chem., **21**, 414 (1982).
[38] P. H. M. Budzelaar, P. J. J. A. Timmermans, A. Mackor, and E. J. Baerends, J. Organomet. Chem., **331**, 397 (1987).
[39] Y. L. Chow, G .E. Buono-Core, and Y. Shen, Organometallics, **3**, 702 (1984).
[40] I. Rencken and S. W. Orchard, J. Photochem. Photobiol, A: Chemistry, **45**, 39 (1988).
[41] R. G. Salomon, K. Folting, W. E. Streib, and J. K. Kochi, J. Am. Chem. Soc., **96**, 1145 (1974).
[42] T. Spee and A. Mackor, J. Am. Chem. Soc., **103**, 6901 (1981).
[43] P. J. J. A. Timmermans, G. M. J. de Ruiter, A. H. A. Tinnemans, and A. Mackor, Tetrahedron Lett., **24**, 1419 (1983).
[44] R. G. Salomon, S. Ghosh, and R. Raychaudhuri, in *Photosensitive Metal-Organic Systems: Mechanistic Principles and Applications*, edited by C. Kutal and N. Serpone, ACS Advances in Chemistry Series, Vol. 238 (American Chemical Society, Washington, DC, 1993) Ch. 16.
[45] J. E. McMurry and W. Choy, Tetrahedron Lett., **21**, 2477 (1980).
[46] R. G. Bergman, in *Homogeneous Transition Metal Catalyzed Reactions*, edited by W. R. Moser and D. W. Slocum, ACS Advances in Chemistry Series, Vol. 230 (American Chemical Society, Washington, DC, 1992) Ch. 14.
[47] M. Tanaka and T. Sakakura, in *Homogeneous Transition Metal Catalyzed Reactions*, edited by W. R. Moser and D. W. Slocum, ACS Advances in Chemistry Series, Vol. 230 (American Chemical Society, Washington, DC, 1992) Ch. 12.
[48] R. H. Crabtree, in *Photosensitization and Photocatalysis Using Inorganic and Organometallic Compounds*, edited by K. Kalyanasundaram and M. Grätzel, (Kluwer Dordrecht, 1993) pp. 391-405.
[49] J. A. Maguire, W. T. Boese, M. E. Goldman, and A. S. Goldman, Coord. Chem. Rev., **97**, 179 (1990).
[50] C. L. Hill, M. Kozik, J. Winkler, Y. Hou, and C. M. Prosser-McCartha, in *Photosensitive Metal-Organic Systems: Mechanistic Principles and Applications*, edited by C. Kutal and N. Serpone, ACS Advances in Chemistry Series, Vol. 238 (American Chemical Society, Washington, DC, 1993, Ch. 13).

[51] C. L. Hill and C. M. Prosser-McCartha, in *Photosensitization and Photocatalysis Using Inorganic and Organometallic Compounds*, edited by K. Kalyanasundaram and M. Grätzel, (Kluwer, Dordrecht, 1993) pp. 307-330.
[52] B. S. Jaynes and C. L. Hill, J. Am. Chem. Soc., **117**, 4704 (1995).
[53] D. B. Yang and C. Kutal, in *Radiation Curing: Science and Technology*, edited by S. P. Pappas (Plenum Press, New York, 1992) Ch. 2.
[54] *Photosensitive Metal-Organic Systems: Mechanistic Principles and Applications*, edited by C. Kutal and N. Serpone, ACS Advances in Chemistry Series, Vol. 238 (American Chemical Society, Washington, DC, 1993).

6 Photocatalytic Aspects of Silver Photography

J. BELLONI
Laboratoire de Physico-Chimie des Rayonnements associé au CNRS, Université Paris-Sud, Orsay, France

1	Introduction	170
2	Historical aspects	171
	2.1 Prehistory to photography	171
	2.2 Niepce's discoveries	172
	2.3 A decisive step: development	175
3	Principles of modern photography	175
4	Photosensitive material	176
	4.1 Photographic film	176
	4.2 Morphology of silver bromide crystals	177
	4.3 Chromatic sensitivity	177
5	Primary effects during exposure	180
	5.1 Photon effects	180
	5.2 Dye photosensitization	182
	5.3 Latent image	184
6	Development	186
	6.1 Principle of development	186
	6.2 Dry-silver processes	188
7	Fixing	188
8	Silver cluster properties	189
	8.1 Cluster thermodynamics	189
	8.2 Cluster deposition	191
	8.3 Solvated clusters	191
	8.4 Pulse radiolysis and laser photolysis	193
	8.5 Time-resolved observation of development	196
	8.6 Redox properties of solvated metal clusters	198
9	Model of the photographic development process	200
	9.1 Historical aspects of development theories	200
	9.2 Solvated cluster development model	201

Homogeneous Photocatalysis. Edited by M. Chanon
© 1997 John Wiley & Sons Ltd

	9.3 Development threshold	202
	9.4 Latent image regression	204
	9.5 Importance of development	205
10	Inversion-transfer processes	205
	10.1 Principle of inversion transfer	205
	10.2 Structure and development of the film	207
11	Color photography	207
	11.1 Interferential process	207
	11.2 Principles of trichromic color photography	208
	11.3 Structure of color emulsion	209
	11.4 Development of color image	210
12	Offset silver processes	213
	12.1 Principles of offset printing	213
	12.2 Mechanism of offset plate development	214
13	Cinematography	214
14	Holography	215
15	Conclusions	217
Acknowledgements		218
References		218

ABSTRACT

The silver photography is based on the photosensitivity of minute silver halide crystals. The image of the object is printed as an extremely weak latent image. The development consists of converting chemically into metal particles, the crystals containing a supercritical number of photoinduced silver atoms and transforming catalytically the latent image into a visible picture. All other photographic processes are derived from this basic principle of black-and-white photography.

1 INTRODUCTION

Nicéphore Niepce, the inventor of photography in 1824, gave this definition to his process: "The discovery I made, that I call heliography consists in reproducing spontaneously by the action of light, with the graduation of hue from black to white, the images received in the camera obscura" [1]. The definition still holds today even if silver salts have replaced the photosensitive bitumen layer used by Niepce.

In fact, he succeeded in materializing by an irreversible photochemical process, the weak insubstantial image projected on the rear of the simplest optical device. It was known from Antiquity that a small aperture through a wall of a dark room (or *camera obscura* from which our modern cameras are named) produced on the opposite wall the inverted image of the bright scene in front of the aperture.

Photography is now 170 years old, and ever since the first pictures were obtained, with fairly rudimentary equipment, the technique has been continuously advancing. Researchers and manufacturers vie with one another in developing more and more sophisticated equipment or processes, and ever more sensitive films. Photographic development, the chemical process that produces such an enormous amplification of the effects produced by light, was one of the most important advances. The chemical processes responsible for the required amplification were empirically discovered and then improved. They are based on catalytic reactions where a few silver atoms created by light act as nuclei to induce through a chemical reaction further production of a huge number of new atoms around them. This catalytic development is also exploited to generate the pigments of the colour image.

Niepce's invention, in demonstrating the feasibility of capturing "spontaneously" an image, initiated a new era. Now visual messages may diffuse everywhere and photography and derived technologies are present in everyday life. Some examples come to mind immediately, such as photography, black and white or colour, providing pictures for the press or for private use, for scientific research, for artistic purposes — art diffusion and creation through photography — but we should also include cinematography, reprography, telefacsimile, offset printing, microlithography.... This is why photography aroused an enthusiastic response that has never waned. It put the power of capturing ephemeral scenes within everyone's grasp. With a total of some 40 000 million photographs taken every year (not counting cine film), the photographic industry is one of the most flourishing in the world. It makes every effort to achieve two goals: a constantly increasing image quality, and greater and greater simplicity of operation that will nevertheless ensure successful photographs under any conditions. This simplicity is only apparent. It is in reality the result of extremely sophisticated technology, which is constantly improving.

2 HISTORICAL ASPECTS

2.1 PREHISTORY TO PHOTOGRAPHY

For thousands of years people were obsessed by the same dream: how to record permanently the fleeting image of scenes that had inspired them. Evidence for this is provided by the works of artists on the walls of caves and tombs, on pottery and, more recently, on canvas.

In fact, from the Renaissance onwards, painters, in their search for a more exact representation of nature, made wide use of the *camera obscura* or of a reduced version, a dark box equipped with a convergent lens instead of a simple aperture. When the rear wall was replaced by a screen like a ground glass today, it was named *camera ottica* and permitted the artist to observe directly the inverted image through the semi-transparent screen and to copy the main lines of the motif (Figure 6.1). Advances in the understanding of light properties and of optics

Figure 6.1. Camera obscura. In this example of dark box of eighteenth century, the image is reflected by the mirror and focussed on a ground glass

therefore had a strong influence on the art of painting through composition and the rules of perspective. Famous painters such as Leonardo da Vinci, Vermeer of Delft, and Canaletto, used currently the *camera obscura*, as it appears in their works. However, the projected image could not be made permanent without their art. Reciprocally, advances in photography were enhancing later research in optics.

Along the same long period of prehistory to photography, the irreversible effects of light on various materials, organic and inorganic, were commonly observed. Fading and bleaching of dyes, and the blackening of silver salts by the action of the light were known before the Middle Ages. In the beginning of the eighteenth century, Thomas Wedgwood had the idea to obtain the outlines of objects by placing them on leather impregated with a silver salt and exposing the whole to sunlight. But the parts previously unexposed to light also blackened when observed in daylight, as well as blackening the portions exposed in the camera. Thus the differentiation between exposed/unexposed areas could not be preserved and the silver image was lost. The chemist, Humphry Davy [2], also tried unsuccessfully to catch the image formed at the rear of the *camera obscura*.

2.2 NIEPCE'S DISCOVERIES

Nicéphore Niepce was the first to succeed in obtaining an image of a scene by combining the use of a *camera obscura* and of a chemical process capable of retaining the impression left by light. His early attempts in 1816 also depended on the blackening of photosensitive silver salts, but he was unable to prevent the image from further darkening. After having searched for a process where some irreversible change to light took place, he found around 1824 that bitumen of Judea or asphalt, containing heavy polyaromatic hydrocarbons, hardened under the action of light. Recently [3,4], the process was rediscovered. It is now possible to understand the process that took place which is basically a photopolymerization.

Although it is not completely elucidated, it can be described as follows. As a general mechanism of photopolymerization, primary effects of light cause excitation of hydrocarbon molecules which dissociate, generally at a carbon–hydrogen bond which is more labile (Figure 6.2 gives an example of photopolymerization of linear hydrocarbons). The molecule fragments, now having an unpaired electron at the break, constitute highly reactive free radicals. They are able to open one double bond present on a neighboring molecule and add to it, thus forming a dimer radical. This radical may again repeat a similar addition with a new molecule, so that the initial formation of a single radical pair by light results in the successive bonding of several molecules. The mechanism of photopolymerization is a chain reaction where the growing radical is propagating the reaction as long as it does not combine with another radical. In that case, the recombination step ends the process. The polymeric network obtained by cross-linking of several initial molecules gains new properties, in particular less solubility than the parent monomer.

In Niepce's heliography, the solubility of the exposed material was much reduced relative to the unexposed parts even if no visible effect was discernible before and after exposure of the brown bitumen layer. The unexposed bitumen was then dissolved when the plate was immersed in a mixture of lavender oil and light petroleum, and the image was fixed. However, this process suffered from a particular disadvantage: the exposure times required (controlling the total number of photons absorbed per surface unit of the sensitive layer) were very long, often a few days [3]. Nevertheless, discrimination between exposed and unexposed parts of the photosensitive layer was achieved for the first time.

Niepce derived several processes with bitumen coated on various supports such as glass, stone and polished metals: copper, tin or silver. By combining heliography with etching technics, he developed for example the photoengraving process [4] still used today. From his camera works, the only remaining piece is the famous picture on a tin plate, *Point de vue du Gras*, taken in 1827 from a window of the family home close to Chalon-sur-Saône.

The silver plate offered him the interesting property of being corrodible through attack by iodine vapors, restricted to the bare parts stripped after dissolution of the bitumen. Daylight then made the silver iodide turn into black divided silver metal. Finally, with dissolution of the hardened bitumen a positive image appeared, with black areas corresponding to the shadowed regions of the motif and polished silver corresponding to the bright ones [3]. Although silver iodide was not yet used as the sensitive layer in the camera, it was already employed to achieve inversion from the negative to the positive image on the same plate, while the bitumen mask ensured that the image contrast was preserved.

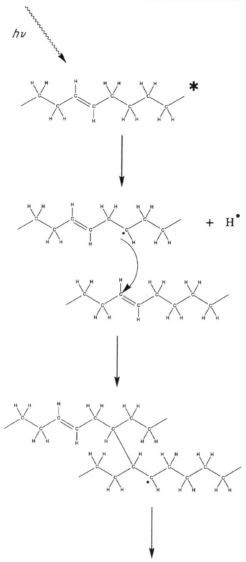

Figure 6.2. Mechanism of cross-linking through chain polymerization of hydrocarbons with unsaturated bonds

In 1829, Niepce entered into association with Louis Daguerre through a confidential 10-year contract. As a painter of theater scenery, Daguerre was familiar with the *camera obscura* and, when informed of Niepce's results, was interested in participating in the research, and then in overcoming the lengthy exposure times.

During this period, another new process was performed by Niepce and Daguerre [5]. Niepce died in 1833 without having obtained any recognition of his discovery.

2.3 A DECISIVE STEP: DEVELOPMENT

Research towards a faster process took a decisive step forward when Daguerre succeeded in obtaining images using silver iodide as the sensitive layer [6]. As previously, a polished silver plate was subjected to the action of iodine vapor and thus became covered in silver iodide. After exposure, the plate was immersed in a solution of sea salt. The silver iodide crystals in the unexposed areas were subsequently inactivated by complexation with chloride ions, thus fixing the image.

However, Daguerre eventually made a far greater contribution to photography. He introduced true photographic development, using mercury vapor which acted on iodized silver plates that had been exposed to very little light and which amplified its effect by reducing far more silver cations into silver atoms around those produced by light. The immense significance of this discovery, named daguerreotype, lay in the reduction of exposure times to light to just a few minutes. This advance indicated the real potential for photography, to capture not only still monuments or landscapes but also living subjects.

In 1839 François Arago announced at the Académie des Sciences in Paris the combined work of Niepce and Daguerre. The daguerreotype consisted of contrasted areas of polished silver and rough silver formed by tiny silver crystals. These crystals scatter light and appear black in ordinary lighting. The daguerreotype is therefore a negative image. However, when submitted to incident oblique light in a dark environment, the crystals appear bright in contrast with the polished areas appearing dark, so that the daguerreotype becomes a positive image.

At the same time, and completely independently, William Henry Fox Talbot, in Britain, and Hippolyte Bayard, in France, perfected photographic development with the use of gallic acid solutions to reproduce negative images on paper that had been impregnated with silver salts. W. H. F. Talbot opened the way to the production of multiple positive images from a single negative one. Dissolution of unexposed crystals by sodium hyposulfite, as proposed by Sir John Herschel, is still used in modern fixation techniques.

3 PRINCIPLES OF MODERN PHOTOGRAPHY

During the pioneering works in photography, the main concepts of photography such as sensitivity, development, contrast, inversion..., have been evoked. It will now be seen how these concepts evolved, leading to modern silver photography [7,8].

Before going into detail, we need to summarize, very broadly, the principles behind the modern photographic process. What is the actual basis of photography? Fundamentally, it utilizes the transformation caused in a substance, known as a photosensitive material, by the influence of light. The photosensitive material in present photography is still basically silver halide crystals and the aim is to replace silver ions by silver atoms through the photophysical effect. Then, the unexposed crystals are eliminated. The image obtained must reflect the different levels of illumination in the original scene as our eyes perceive it. Then, the unexposed crystals are eliminated. The light effect printed on the sensitive layer during a short period is much below the visibility threshold. An amplification of several orders of magnitude is therefore obtained through a catalytic chemical development. Eventually, the photographic image results from the contrast between the black of the divided silver metal grains and the white or transparency of the support.

4 PHOTOSENSITIVE MATERIAL

4.1 PHOTOGRAPHIC FILM

The first step is to take the picture with a camera. Photons of light reflected from the object hit the film during the short opening of the front aperture. The film consists of an inert plastic material covered with a layer of gelatine containing a suspension of minute crystals of silver bromide, which project through the surface (Figure 6.3). Mean sizes of the tiny crystals of this mosaic are of a few tenths of 1 μm. Each crystal contains about 10^9 Ag^+Br^- ion pairs arranged in a cubic structure. It constitutes the smallest element of the silver image (20–100 million elements in a 24×36 mm^2 film). The small size of this photodetector ensures a particularly high definition of the image (Figure 6.3).

The ranges of linear size differ by more than four orders of magnitude between the total image as seen by our eyes and the silver bromide crystal that controls the resolving power, and again by four orders of magnitude between the crystal and the ions ($r_{Ag+} = 0.126$ and $r_{Br-} = 0.196$ nm) which undergo interaction with photons at the atomic level. This initial interaction at the atomic scale produces, through the development, a visual effect in the macroscopic range.

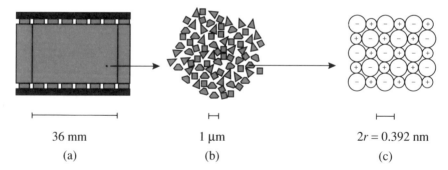

Figure 6.3. Structure of the photographic silver film: (a) film; (b) mosaic of AgBr crystals; (c) AgBr crystal structure

4.2 MORPHOLOGY OF SILVER BROMIDE CRYSTALS

The dimensions of the silver bromide crystals prepared by precipitation of silver nitrate with a bromide salt have to meet two specific requirements. On one hand, the desire for high definition in the image means that small crystals should be used. On the other hand, the *flux of photons*, or the number of photons available per second and per area unit, may be imposed by the environment and may be quite low. Therefore, under a given flux, a large surface area means that larger crystals are better at intercepting light, giving greater sensitivity at the expense obviously of the resolving power.

In early preparations of silver bromide in gelatine, precipitated crystals were of cubic shape and small. The resolution, which in principle should be high, actually suffered from the random orientation of the crystal faces relative to the light beam. The scattering of light indirectly led to unexpected exposure of neighboring crystals. It was discovered that maturation of the crystals in the precipitation solution slowly allowed them to increase their size with a gain in sensitivity. As shown later, image formation does indeed require a minimum number of atoms generated per crystal. Thus, for a given flux of photons per second and per surface unit, the higher the crystal area, the higher their sensitivity. The obviously correlated decrease in resolution is somewhat compensated for by less light scattering than is observed for minute crystals.

Sophisticated step-by-step controlled methods of silver bromide growth provide twinned flat crystals (tabular or T-grains). Many modern emulsions contain these crystals (Figure 6.4), which represent a good compromise between sensitivity and resolving power.

As has recently been emphasized [9], silver bromide possesses a range of properties (spectral sensitivity range, photoelectron lifetime, definition, visual aspects of the final image, etc.), which enable it to be a unique material for photographic applications.

4.3 CHROMATIC SENSITIVITY

The efficiency of the primary effect of light depends on different phenomena. The first element to be considered is the *absorption response* of the material

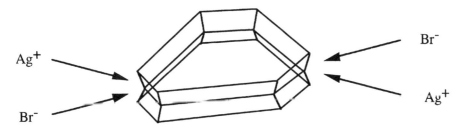

Figure 6.4. Structure of a twinned tabular crystal of AgBr

when excited by light and its *chromatic sensitivity*, possibly similar to that of human eyes.

The human eye is sensitive to a range of optical wavelengths, called the visible range, comprised between approximately 400 (violet) and 700 nm (red). In fact, the retina at the back of the eye includes two classes of photosensitive cells: the rods (125 million) are sensitive to the contrast between light and dark; the cones (6 million), are less sensitive, and three different types are capable of detecting specifically red, green or blue photons with optimum sensitivity at 560, 540 and 420 nm respectively [10] (Figure 6.5). They correspond to the so-called primary colours already distinguished by Thomas Young as early as 1802. The electric signals generated by photoexcitation of the photoreceptors are transmitted via the optical nerve to the brain, which compares the differences in intensities arising from rods and from blue, green and red cones. The brain is then able to extract from this processing a great deal of information on dimensions, distances, reliefs, movement and the color seen. Balanced excitation of the three kinds of cones yields the vision of white.

Ultrapure silver halide crystals are in fact poor absorbers of visible light. Pure silver chloride exhibits a sensitivity maximum at 340 nm in the ultraviolet, and the domain lies from 280 to 400 nm, below the visible range. Pure silver iodide is sensitive from 415 to 450 nm (maximum at 425 nm). Pure silver bromide is sensitive from 260 to 520 nm (maximum at 460 nm) (Figure 6.6a). However,

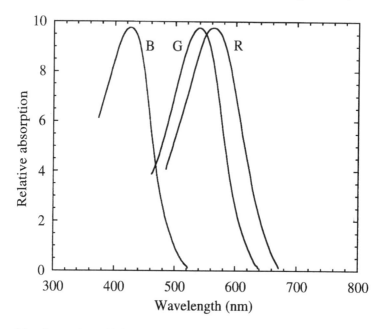

Figure 6.5. Spectral sensitivity to blue (B), green (G), and red (R) of the three types of cones in the human retina [10]

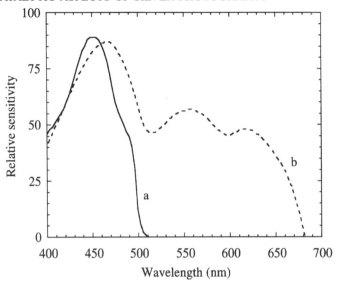

Figure 6.6. Spectral sensitivity of AgBr crystal [7]: (a) emulsion of pure AgBr; (b) sensitized panchromatic emulsion

addition of small amounts of AgI to the bromide extends its sensitivity to 560 nm. Thus silver halides are transparent to photons in the yellow and red regions of the spectrum that are seen by human eyes.

As for semiconductors in general, the photon absorption and photoconductivity properties of halides strongly depend on slight changes in their structure or their composition. Structure defects such as anion vacancies occupied by excess electrons (F centers) may exist (Figure 6.7). Other defects are produced by the displacement of a silver cation, leaving an interstitial vacancy in the regular

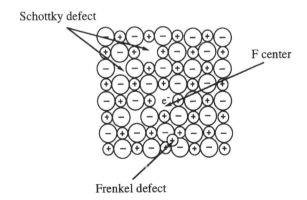

Figure 6.7. Structure defects in AgBr crystal

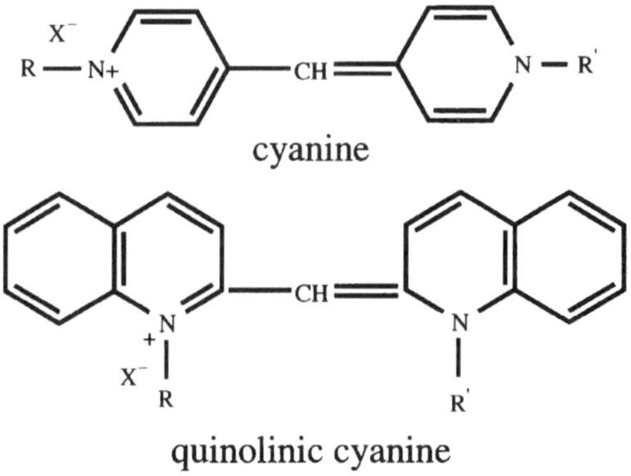

Figure 6.8. Examples of cyanine sensitizers

arrangement of ions (Frenkel defect), or by a double vacancy of a cation and an anion (Schottky defect). The defects generate new energy states located between the valence (VB) and the conduction bands (CB) of the energy diagram. The spectrum of possible phototransitions in the semiconductor is thus extended. For instance, inclusion of I$^-$ ions in AgBr crystals contributes, due to the higher I$^-$ radius relative to that of Br$^-$, to the generation of defects in the structure. The presence of defects also enhances the generation yield of silver atoms by light as shown in the next paragraph.

In order to confer more sensitivity to silver halides for photons of low energy, following H. Vogel's discovery (1873) [7,8], organic dyes are added as *photosensitizers* which enhance the photon absorption in this range (Figure 6.6). The energy of a photon is related to the wavelength λ of the spectrum by

$$E = hc/\lambda \qquad \ldots (6.1)$$

with $c = 3 \times 10^{10}$ cm s^{-1} being the speed of light and $h = 6.624 \times 10^{-27}$ erg.s Planck's constant. Since $E(\text{erg.}) = 6.2 \times 10^{11} E$ (eV), the photon energy in eV is related to the wavelength in nm by $E(\text{eV}) = 1240/\lambda$ (nm).

A classical example of photosensitizers are the series of cyanines, characterized by heteroatoms in aromatic cycles bound by an aliphatic chain (Figure 6.8). Conjugation of double bonds and resonance between the two nitrogen sites cause transition in the visible. Substitution of carbon–hydrogen bonds in the cycles by various chromophore groups allows one to adjust the spectral range of their absorption.

5 PRIMARY EFFECTS DURING EXPOSURE

5.1 PHOTON EFFECTS

When absorbed by the semiconducting crystal of silver halide, AgBr, the photon creates a pair of opposite charges: electron–hole. The *primary quantum yield* is unity, that is, one charge pair is produced per photon absorbed. In fact, the *effective quantum yield* may be much higher if one photon initiates a chain reaction involving several molecules. Photopolymerization described above illustrates this case (Figure 6.2 p. 174). If the propagation of the chain is particularly efficient, this amplification suffices to transform the molecular action of the light into a visible effect.

However, the photophysical interaction on silver halides is not enhanced by a chain reaction. On the contrary, the effective quantum yield is much less than unity due to fast charge recombination that reversibly suppresses most of the initial pairs, particularly when the crystal structure is almost perfect.

The mechanism by which the initial photoinduced electron–hole pair interacts with the crystal was given by the model of R. W. Gurney and N. F. Mott [11] and by that of J. W. Mitchell [12].

$$AgBr + h\nu \longrightarrow AgBr^* \qquad \ldots (6.2)$$

$$AgBr^* \longrightarrow e_{AgBr}^- + h_{AgBr}^+ \qquad \ldots (6.3)$$

$$h_{AgBr}^+ + Br^- \longrightarrow Br^\circ \qquad \ldots (6.4)$$

$$e_{AgBr}^- + Ag^+ \longrightarrow Ag^\circ \qquad \ldots (6.5)$$

$$e_{AgBr}^- + h_{AgBr}^+ \longrightarrow AgBr \qquad \ldots (6.6)$$

$$Ag^\circ + Br^\circ \longrightarrow AgBr \qquad \ldots (6.7)$$

$$Ag^\circ + h_{AgBr}^+ \longrightarrow AgBr \qquad \ldots (6.8)$$

$$e_{AgBr}^- + Br^\circ \longrightarrow AgBr \qquad \ldots (6.9)$$

Once produced, the electron and the hole equivalent to a bromine radical Br° (reaction 6.4) diffuse by a cascade of charge transfers between adjacent sites. The hole diffusion results from an electron transfer through the Br^- network toward the initial Br° (Figure 6.9). At the surface, a hole trap donates irreversibly an electron.

Simultaneously with the hole diffusion, photoelectrons created in the crystal move randomly with high mobility toward positive potential traps around the structural defects (Figure 6.7) and are finally scavenged by interstitial silver cations (reaction 6.5 and Figure 6.9). Some molecules such as silver sulfide Ag_2S deposited at the surface favor this reaction.

Nevertheless, electron and hole may encounter again and recombine (reactions 6.6–6.9). The effective quantum yield Φ_{eff} is therefore equal to the difference between the primary quantum yield Φ equal to unity (reaction 6.2) and the recombined fraction R resulting from the last four reactions (6.6–6.9).

$$\Phi_{eff} = 1 - R \qquad \ldots (6.10)$$

In order to improve the effective charge separation, and thus silver atom formation, a competitive process to the recombination reactions (6.6–6.9) should be

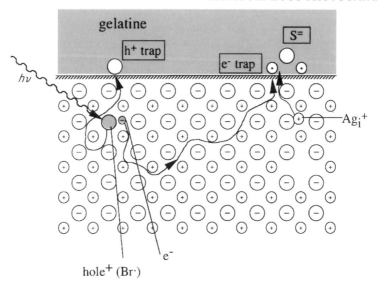

Figure 6.9. Diffusion and trapping of the photoinduced electron–hole pair

offered to the positive hole h^+ or to its product, the radical bromine atom $Br°$.

$$h_{AgBr}^+ + \text{hole trap} \longrightarrow \text{trap}^+ \qquad \ldots (6.11)$$

$$Br° + \text{hole trap} \longrightarrow \text{trap}^+ \qquad \ldots (6.12)$$

In the past, a huge increase in photographic sensitivity was achieved as soon as the silver halide crystals were no longer isolated on the support, such as in the daguerreotype, but were embedded in chemical systems such as wet collodion (cellulose nitrate) (F. S. Archer, 1860) [7,8], and later gelatine. It is now well known that the amino acid compounds of gelatine contain efficient hole traps (reactions 6.11, 6.12 and Figure 6.9). Thus, instead of being just slightly lower than unity, the recombination ratio R due to reactions (6.6)–(6.9), which in early processes let escape only a very small fraction of initial pairs, dropped to a much smaller value and therefore the effective quantum yield increased by several orders of magnitude.

5.2 DYE PHOTOSENSITIZATION

The principle is to induce an electron transfer to Ag^+ through photons which cannot excite AgBr directly. The role of the added dye is to extend the range of absorbed energy below the band gap of pure AgBr.

The photosensitizer present in the emulsion thus extends its chromatic sensitivity. A photon of energy sufficient to be absorbed by the dye yields first an excited state of the dye, having an electron donor character (Figure 6.10). The

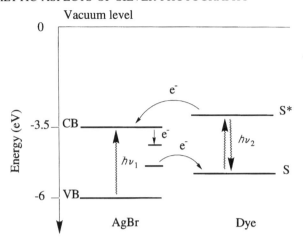

Figure 6.10. Diagram of energy levels in AgBr sensitized by a dye

energy levels are such that the dye in its fundamental state does not react with silver ions. The emulsion is therefore stable in the dark. The electron transfer must be photoinduced and occurs only under exposure. The lifetime of the dye excited state, $t_{1/2}$, the time necessary for desactivation of half of the states, must be long enough to allow electron transfer to the AgBr crystal. Otherwise, this competitive process of desactivation reduces much of the dye efficiency (reaction 6.15). Another condition necessary for the electron transfer to occur is that the energy level of the excited state must be higher than that of the AgBr conduction band.

$$\text{Dye} + h\nu \longrightarrow \text{Dye}^* \quad \ldots (6.13)$$

$$\text{Dye}^* \longrightarrow \text{Dye}^+ + e^- \quad \ldots (6.14)$$

$$\text{Dye}^* \longrightarrow \text{Dye } t_{1/2} \quad \ldots (6.15)$$

The free electron is then transferred from the AgBr conduction band to electron traps consisting of interstitial silver cations. The dye cation is able to transfer the positive charge to a hole trap at the AgBr surface, provided the energy levels are also appropriately located.

Intensive research continues to synthesize new photosensitizer molecules fulfilling, in addition to chromatic specificity, the above conditions of photophysical and redox properties [13].

$$e^- + \text{trap} \longrightarrow e^-(\text{trap}) \quad \ldots (6.16)$$

$$e^-(\text{trap}) + \text{Ag}^+ \longrightarrow \text{Ag}^\circ \quad \ldots (6.17)$$

$$\text{Dye}^+ + \text{hole trap} \longrightarrow \text{Dye} + \text{trap}^+ \quad \ldots (6.18)$$

The overall effect of the mechanism is to produce, as in unsensitized material, a silver atom and the oxidation of the dye molecule acting as a hole trap, although the photon energy is much less than the threshold of the band gap energy required to directly excite a pure AgBr crystal. This paradoxical situation in fact results from the new energy states created by the strong interaction between the dye and the crystal, which must now be considered as a whole.

If regeneration of the dye is possible (reaction 6.18), the dye is again able to absorb another photon, and behaves as a photocatalytic relay to transfer the photon energy to the silver bromide crystal.

The net effect of light is that photons cause the bromide anions Br^- to yield electrons, which neutralize neighbouring silver cations Ag^+. The gelatine does not act just as a cement between the mosaic of crystals. It also indirectly helps to maintain the efficiency of the incident light by reacting with the bromine atoms or positive holes and by preventing them from recapturing the electrons trapped by the silver ions. However, the highest effective quantum yield obtained in modern emulsions is no greater than $\Phi_{eff} = 0.3$, which means that at least 70% of pairs are still lost by recombination [14].

5.3 LATENT IMAGE

Photon impacts are randomly distributed over the crystal surface. Initially, the charge pairs are thus created in different sites. However, since all photoelectrons generated in the same crystal tend to diffuse toward a single site having the highest positive potential, the atoms of silver thus produced cluster together at this site. Only one single silver cluster or speck of nuclearity n is formed per crystal.

Depending on the intensity of the photon flux and on the duration of the illumination, the crystals receive greater or lesser numbers of photons. After the exposure, each crystal therefore contains one speck including several silver atoms (in the brightly illuminated areas), no silver atoms (in the very darkest areas), or varying numbers of atoms (in the intermediate areas of the image). The overall distribution constitutes *the latent image*, which is far too weak to be visible to the naked eye (Figure 6.11). In fact, the selection of the brevity of the exposure time is made so as to just reach the threshold of cluster nuclearity which will be

---

Figure 6.11. Principle of silver photography. During the exposure (A) the photons of visible light excite an isolated crystal of silver bromide held in the gelatine. The atoms of silver thus formed cluster together. Depending on the intensity of the exposure, the crystals receive a greater or lesser number of photons. After the exposure, they therefore contain a single cluster of several silver atoms (in bright areas); no atoms (in the darkest areas); or varying numbers (gray areas). The overall distribution forms the latent image. Development (B) consists of completing via a chemical electron donor the conversion of all the Ag^+ ions into metallic silver within any crystal where a sufficiently large cluster of atoms has been created by the light. The whole mass of minute particles of silver appears black to the eye. The fixer (C), dissolves and washes away the unexposed crystals that are otherwise still photosensitive. The image thus obtained is a negative of the original scene: the black silver crystals correspond to bright areas

PHOTOCATALYTIC ASPECTS OF SILVER PHOTOGRAPHY 185

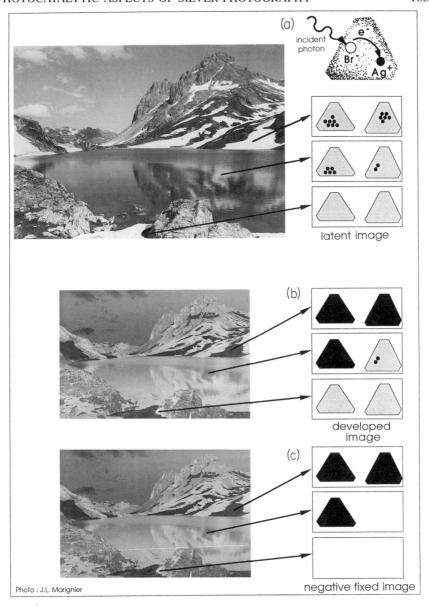

Photo: J.L. Marignier

developable. In this way, a given emulsion is used under optimized conditions of sensitivity. Usually the nuclearity of clusters in the latent image ranges from 0 to 10.

6 DEVELOPMENT

6.1 PRINCIPLE OF DEVELOPMENT

Without the discovery of the development process, the silver image had to be obtained by direct blackening produced by the action of light alone. This demanded a prolonged exposure, which had to last until a sufficient number of photons had accumulated to convert the silver halogen crystals in the illuminated areas of the photosensitive layer into black crystals of silver. If a mean photon flux with a given effective quantum yield is able to produce about 10 silver atoms per crystal after 0.001 s exposure, it is clear that the total reduction of about 10^9 ions in a crystal, by light instead of by chemical development, will require a day-long exposure. The success of photography lies in the replacement of this single long process by two distinct stages:

1. The exposure, which is often very short (a fraction of a second) but suffices to record the imprint left by the light as a latent image.
2. The development, which, in a second separate step, changes the invisible latent image into a visible one. This conversion is effected by a chemical agent that is an electron donor, the developer.

In fact, development consists of converting all the Ag^+ ions into metal atoms in those crystals where clusters of sufficient or *critical size* n_c have been formed by the action of light during the exposure. Crystals that are too little exposed, or unexposed, remain inert toward the developer and thus intact (Figure 6.11b). The developer has the task of multiplying through a catalytic reaction the few atoms, that are the result of the exposure, into several billion, which is the number of Ag^+ ions in a small crystal in an emulsion. The gain in sensitivity ($\approx 10^8$) offered by development was crucial in making possible very short exposures, provided Φ_{eff} is also close to Φ (eq. 6.10).

$$Ag^+ + e_{trap}^- \longrightarrow Ag^\circ \qquad \ldots (6.17)$$

$$n Ag^\circ \longrightarrow Ag_n \qquad 0 \leq n \leq 10 \qquad \ldots (6.19)$$

The number of atoms per cluster ranges from 0 to about 10 units.

To be reduced by the developer (Dev) Ag^+ ions must be associated with a *supercritical* cluster ($n \geq n_c$). When in contact with the developer, a supercritical cluster Ag_n of the latent image acts as a catalytic germ inducing further electron transfers from the developer to the silver ions in the vicinity. While the photoreduction of silver ions into atoms was achieved during exposure in the solid crystal, the development occurs at the interface of the crystal with a solution (developing bath) of the electron donor. Various classes of developers are used, such as hydroquinones.

$$Ag_n + Ag^+ \longrightarrow Ag_{n+1}^+ \quad n \geq n_c \quad \ldots(6.20)$$

$$Ag_{n+1}^+ + Dev \longrightarrow Ag_{n+1} + Dev^+ \quad \ldots(6.21)$$

$$Ag_{n+1} + Ag^+ \longrightarrow Ag_{n+2}^+ \quad \ldots(6.22)$$

$$Ag_{n+2}^+ + Dev \longrightarrow Ag_{n+2} + Dev^+ \quad \ldots(6.23)$$

$$Ag_{n+i} + Ag^+ \longrightarrow Ag_{n+i+1}^+ \quad \ldots(6.24)$$

$$Ag_{n+i+1}^+ + Dev \longrightarrow Ag_{n+i+1} + Dev^+ \quad \ldots(6.25)$$

When the developer is a dielectronic reducing agent such as a hydroquinone, Dev^+ may itself transfer a second electron to clusters. The redox potentials of the two successive couples are at their lowest value above pH 12 where the unprotonated forms predominate [15].

$$Ag_{n+1}^+ + HO\text{-}\bigcirc\text{-}OH \longrightarrow Ag_{n+1} + HO\text{-}\bigcirc\text{-}O^\bullet + H^+ \quad \ldots(6.26)$$

hydroquinone → semiquinone

$$Ag_{n+2}^+ + HO\text{-}\bigcirc\text{-}O^\bullet \longrightarrow Ag_{n+2} + O\text{=}\bigcirc\text{=}O + H^+ \quad \ldots(6.27)$$

semiquinone → quinone

The oxidized forms of the developer stay in the developing bath while the emulsion is removed after development and washed. The catalytic character of the series of reactions (6.20–6.25) lies in that the successive electron transfers require the existence of the primary supercritical cluster. Then the process continues to add to the growing cluster one additional atom per monoelectronic developer molecule oxidized and ends when the crystal containing the germ is completely transformed into a reduced silver crystal. After development, the crystals are therefore either reduced as a whole or preserved, as a consequence of the discrimination by the developer. The black and white image reproduces the distribution of supercritical and subcritical clusters of the latent image. The density of grays results from the local concentration of developed crystals.

The specificity of the development mechanism is that it starts and ends within the boundaries of a single silver bromide crystal, independently of others in the image. For that reason, even after development, the new silver particle replacing the exposed AgBr crystal with an equal number of Ag atoms still constitutes the truly smallest element of the image. The development process, despite the high level of amplification, therefore preserves all the minute details of the latent image.

The gain is fixed by the ratio of total atoms in the crystal to the critical nuclearity of the germ: it can be improved either by increasing the crystal size or by decreasing the minimum nuclearity of developable germs. It is known that the strongest developers are able to reduce crystals containing a germ of just two atoms, which is considered as the extreme limit. In fact, the risk in that case exists of developing unexposed crystals that contain silver atoms as an impurity. Although extreme precautions are taken to avoid light exposure and microsparks during the automatic industrial production of emulsions, atoms may still be accidentally present in crystals or be generated during long storage (particularly by cosmic rays). They give rise to extra random development or *fogging*. Fogging is obviously less probable if the threshold of developability is of higher nuclearity.

6.2 DRY-SILVER PROCESSES

Silver development as described above is a wet process. For some applications, dry processes have also been derived that are based on catalytic reduction of silver ions around photoinduced germs. In the *dry-silver* process, silver ions are present in two forms. A small fraction is actually photosensitive silver bromide, the rest is in the form of a long-chain aliphatic acid salt, such as silver behenate $C_{21}H_{43}COOAg$, with the specific property of being insensitive to light. Moreover, the reduction of the salt by the hydroquinone developer already included in the same layer does not occur at room temperature even in daylight and requires heating to 110–140 °C to overcome an activation barrier.

The dry layer is a structureless intimate mixture of the photosensitive silver salt, the silver behenate and the developer. Thus the resolution is not controlled by crystal boundaries but by the heating time which governs the extent of development around the germ. Indeed, after exposure, the film coated with the dry emulsion is submitted to brief heating and the germs catalyze the reduction of silver bromide and behenate in the vicinity. When heating stops the reaction ends and the emulsion again becomes insensitive or fixed. This thermal process, which does not require diffusion of reactants nor developing and fixing baths, is immediate (a few seconds). It is particularly useful in instant photography and for other fast recording processes, an example being the dry-silver paper for black-and-white telefacsimile systems. Even instant reprography of colored images has been recently achieved by combining the advantages of the dry-silver process with those of the color inversion-transfer principle (§11).

7 FIXING

Development is the first process that the film undergoes in the laboratory, in the dark. The second process is fixing, which eliminates the unexposed crystals, that are still photosensitive, by dissolving and washing them away.

The fixing agent is a complexing ion such as thiosulfate, that substitutes for the bromide, making the silver ions soluble:

$$x\text{AgBr} + y\text{Na}_2\text{S}_2\text{O}_3 \longrightarrow (\text{S}_2\text{O}_3)_y\text{Ag}_x^{(2y-x)-} + x\text{Br}^- + 2y\text{Na}^+ \quad \ldots (6.28)$$

This insoluble complex of silver sodium thiosulfate yields, with excess thiosulfate, the soluble $\text{Na}_3(\text{Ag}(\text{S}_2\text{O}_3)_2):2\text{H}_2\text{O}$ which is removed upon rinsing. Note that a small amount of thiosulfate is sometimes added to the developing bath, which favors the release of silver ions at the moment of the reducing development around supercritical clusters (*physical development*).

The image, now insensitive to light, thus obtained after elimination of unexposed or crystals containing subcritical clusters, is a negative version of the initial imprint left by the light: the minute crystals of silver, which appear black to the eye, correspond to areas that were brightly illuminated, in contrast with the transparency of the film now free of silver which corresponds to shadowed areas (Figure 6.11c p. 185).

A positive image results from negative–positive inversion. For this purpose, a second photograph is made either by direct contact or by projection enlargement of the negative film on to sensitive paper, which is then treated through development and fixing as just described.

8 SILVER CLUSTER PROPERTIES

The answer as to how the developer is capable of discriminating supercritical nuclearities among the population of clusters lies in the properties of the latent image. However, its direct observation is not easy. The clusters escape observation due to their extremely small size and to the intrinsic photosensitivity of the substrate AgBr that makes most high-resolution methods inappropriate. Moreover, the silver atoms constituting a cluster differ from the surrounding silver ions of the crystal just by their neutral charge.

In fact, our knowledge of the latent image comes indirectly from its behavior under development or from theory. An example is arrested development, in which the developer concentration is low so as to only partially develop the exposed crystals. Electron micrographs of these partially developed crystals clearly show that no more than one single site of development exists per crystal, generally located at the crystal edge.

8.1 CLUSTER THERMODYNAMICS

Our understanding of clusters in the gas phase is increasing much more rapidly than for clusters in emulsions due to the use of supersonic molecular beams: the sudden expansion of a population of metal atoms in a vapor by transfer to a very low pressure chamber causes the coalescence of atoms into clusters. After a given time of flight, they are detected through ionization and transfer to a mass spectrometer and their size distribution is determined. Associated with a variety of probe systems, this powerful technique provides plentiful information on the size dependence of various cluster properties and confirms a concept that first appeared in gases in the early 1980s [16–18]:

The thermodynamic properties of a metallic cluster vary with the number of atoms n that it contains. The theory is that an isolated atom, or a few atoms linked together as in a molecule, possess distinct electron levels. This is the opposite of what occurs in a macroscopic metallic crystal, where the quasi-free electron levels are structured into bands. This is known as a *quantum-size effect*, which is one of the significant concepts in modern research into materials in an ultrafine state. It enables us to explain, for example, how the transition from a nonmetallic to a metallic state occurs at a specific number of atoms in gas phase clusters [16–18].

Under these conditions, the number of atoms contained within a cluster has a strong influence for example on the ionization potential of silver [19] or copper [20] clusters as found in the gas phase through molecular beams experiments (Figure 6.12). The coulomb attraction between the electron and the positively charged cluster exhibits discontinuities due to the layered electron structures as well as fluctuations corresponding to changes in the numerical parity of the electrons: odd-numbered clusters are more stable against the loss of an electron than even-numbered ones. The properties of this ultrafine material — somewhere between atoms and crystals — and the quantum effects that depend on size [21],

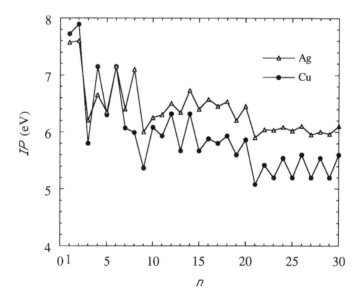

Figure 6.12. Dependence of the ionization potential of silver [19] and copper clusters [20] on their nuclearity

PHOTOCATALYTIC ASPECTS OF SILVER PHOTOGRAPHY   191

have aroused increasing theoretical and experimental interest, particularly as it concerns the gas phase, over the last two decades.

8.2 CLUSTER DEPOSITION

An interesting approach to photography is to prepare mass-selected clusters and to deposit them through *soft landing* onto the surface of cubic or tabular-shaped AgBr microcrystals, without excess energy in order to avoid further dissociation or other side reactions (6.23) [22,23]. The cluster properties are then studied, in particular regarding their developability under conditions comparable with photographic development. For a given nuclearity of selected clusters, the developability is evaluated by counting the fraction of effectively developed crystals in the electron micrograph of the substrate (Figure 6.13). It was confirmed by direct observation that only clusters above a critical size are indeed developed. The critical nuclearity is four atoms for conventional developers and is lower if the developer redox potential is more negative (Figure 6.14).

8.3 SOLVATED CLUSTERS

For a long time, chemists were often faced with puzzling phenomena, difficult to explain, when experiments involved ultradivided matter. Formation of a new phase, a precipitate from a solution or a deposit, is classically known to obey empirical rules, not yet rationalized, which diverge from the behavior of macroscopic matter. The behavior of the latent image in photography was just one example among many others.

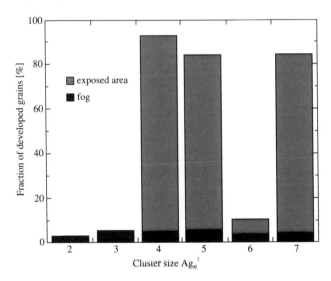

Figure 6.13. Developability of AgBr crystals after deposition of variable nuclearity Ag_n^+ clusters [24]

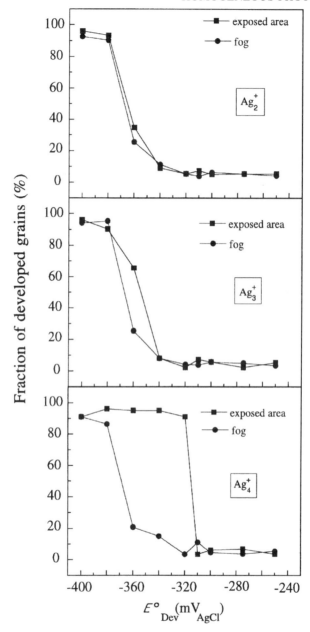

Figure 6.14. Dependence on the developer redox potential of the developability of AgBr crystals after deposition of selected nuclearity Ag_n^+ clusters [24]

A particular example of these unexplained results is given by the formation yield of a metal precipitate when metallic salt solutions are irradiated with high-energy radiation such as X or γ photons or particle beams. There is a similarity between the interaction of such radiation and that of light. Radiographic silver films are able to "see" this kind of invisible radiation whose energy is much higher than the band gap of AgBr. We may recall that Henri Becquerel discovered radioactivity in 1896 because his photographic plates were fogged by radiation emitted from a piece of pitchblende (uranium ore) that was placed in the same drawer. Nuclear silver emulsions were used for a long time to materialize the track of a fast nuclear particle, revealed by the succession of silver clusters induced along the track.

In the past, radiation chemists often used metal ions to scavenge the primary reducing species induced by ionizing radiation [24–26]. However, metal precipitation was lower than expected, even for noble metals. An extreme case was presented by copper clusters induced by irradiation of cuprous ions. The clusters were unstable and ultimately undetectable. Also, an excess yield of molecular hydrogen was measured [27].

To explain these results, a new concept was proposed in 1973, assigning to native metal clusters, in a quasi-atomic state, the ability to be oxidized in solution [27], whereas the massive metal is noble and not oxidized under the same conditions. This size dependence of redox properties of metal particles could answer many earlier problems encountered in radiolytic or chemical metal precipitation.

8.4 PULSE RADIOLYSIS AND LASER PHOTOLYSIS

Direct observation of metal clusters solvated in solution requires specific techniques, since they are short-lived and coalesce through free diffusion when not immediately oxidized by the medium. Moreover, detection by mass spectrometry cannot be used as in the gas phase. The short lifetime of solvated atoms or aggregates of a few metal atoms makes their observation possible only through pulse techniques. These techniques allow the individual steps in a wide range of chemical mechanisms to be determined. The principle is to generate, within a time much shorter than further reactions (typically a few nanoseconds), species to be studied with a sufficient concentration to allow their direct observation (10^{-5}–10^{-4} mol l^{-1}). This obviously implies short powerful pulses. A synchronized time-resolved detection apparatus, such as one based on optical absorption, is capable of measuring the time dependence of the optical absorption spectrum and therefore the kinetics of the transient species.

The generation of silver atoms may be realized through a light pulse provided by a *pulsed laser* through a mechanism similar to flash photography. This equipment is indeed quite useful for studying some intermediate steps of the primary effects of light such as the conductivity of photoelectrons, the lifetime and reactions of excited photosensitizers, etc. However, in the particular case where optical absorption detection is used to observe the transient species, the system under study must itself be insensitive to light. This is why an alternative method must be used to generate the primary silver atoms.

In *pulse radiolysis* [28–30] (Figure 6.15), an accelerator delivers a short pulse of electrons able to generate silver atoms even in a solution of silver ions insensitive to light. In fact, the high energy of the MeV electron beam exceeds by far the bond energy of any molecule. In a pure AgBr crystal, it would create several hole–electron pairs. In water solutions the more abundant solvent molecules undergo most of the ionization and excitation events, yielding first H_2O^* and the H_2O^+/e^- pair. These species readily give rise to molecular and radical products, the latter highly reactive toward the solute, in this case silver cations.

$$H_2O \xrightarrow{e^- \text{ beam}} e_{aq}^-, H_3O^+, H^\bullet, OH^\bullet, H_2, H_2O_2 \qquad \ldots (6.29)$$

The solvated electron e_{aq}^- is the strongest reducing agent. This property is measured by the standard redox potential with reference of the normal hydrogen electrode (NHE). The oxidized form of e_{aq}^- is H_2O and the standard potential of the redox couple is $E°(H_2O/e_{aq}^-) = -2.7 \, V_{NHE}$ [28]. The hydrated electron quickly reacts with silver cations at a diffusion-controlled rate and yields silver atoms in a reaction that does not interfere with subsequent reactions.

$$e_{aq}^- + Ag^+ \longrightarrow Ag° \qquad \ldots (6.30)$$

The bimolecular rate constant of reaction (6.30) is $k_{30} = 3.6 \times 10^{10} \, l\,mol^{-1}\,s^{-1}$ [31] which means that silver atoms are produced within ≈ 100 ns. The $OH^\bullet$ radicals are scavenged by additives such as alcohols to avoid reverse oxidation of $Ag°$. This method of time-resolved spectrophotometry reveals the very rapid steps occurring in the process: formation of the first atom of silver, association reactions of the atoms with excess Ag^+ ions, and dimerization and condensation of the

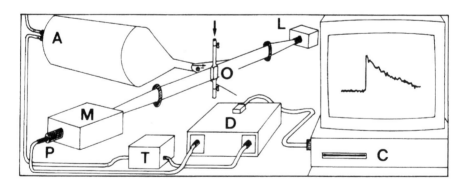

Figure 6.15. Pulse radiolysis set-up: (A) electron accelerator, (O) optical cell, (L) flash lamp, (M) monochromator, (P) photomultiplier, (T) trigger for synchronisation, (D) transient digitizer, (C) computer

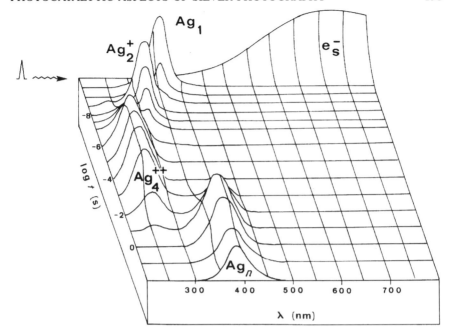

Figure 6.16. Time evolution of the optical absorption spectrum in an Ag_2SO_4 solution imbedded in a polymeric ion-exchange membrane after an electron pulse of 3 ns. Under strong acidic conditions, absorbance of clusters Ag_n with $n \leq 10$ may decrease due to corrosion [37]

small clusters until they coalesce into stable aggregates [31–33] (Figure 6.16).

$$Ag° + Ag^+ \longrightarrow Ag_2^+ \quad \ldots(6.31)$$

$$Ag_2^+ + Ag_2 \longrightarrow Ag_4^{2+} \quad \ldots(6.32)$$

$$Ag_2^+ + Ag^+ \longrightarrow Ag_3^{2+} \quad \ldots(6.33)$$

$$Ag_4^{2+} + Ag_4^{2+} \longrightarrow Ag_8^{4+} \quad \ldots(6.34)$$

$$Ag_{i+x}^{x+} = Ag_{j+y}^{y+} \longrightarrow Ag_n^{z+} \quad \ldots(6.35)$$

Elementary steps of the mechanism, the nature of the reactions, together with their absolute rate constants, may be determined with this method from examination of the kinetics in the evolution of the optical absorption spectrum [31–33]. Depending on the time that has elapsed since the pulse, the solution contains isolated atoms, atoms combined with silver ions, or larger and larger clusters also combined with ions, causing them to have a positive charge.

This part of the mechanism is analogous to the exposure stage in photography, except for the fact that the atoms and electrons are free to diffuse through the solution, rather than being confined to a single silver bromide crystal suspended in the gelatine. As the method relies on the absorption of light, it avoids employing a system that is itself photosensitive.

8.5 TIME-RESOLVED OBSERVATION OF DEVELOPMENT

It is possible to study the reactivity of these different species, despite their ephemeral nature, by adding the desired reactant beforehand to the solution to be irradiated. The method also allows one to choose the conditions so as to intervene at the desired stage of clustering. The different sizes of clusters that coexist in the latent photographic image are, in the pulse radiolysis simulation experiments, separated in time, and are in fact identified by the elapsed time and by their spectra (Figure 6.16).

Since the clusters are transients, their properties — particularly their developability — must be probed by kinetic methods. Considering that the redox properties of clusters may change as their nuclearity increases, addition of a molecule with a known redox potential allows one to monitor the cluster sizes for which the electron transfer between the monitor and the clusters is effective. The probe molecule is also generated by the pulse from a suitable precursor simultaneously with the silver atoms generated from Ag^+ ions. In order to permit the direct observation of the reaction of the monitor with the silver clusters, the reduced or oxidized form of the probe must exhibit an intense absorption spectrum [34]. When possible, it is also preferable that the redox couple be monoelectronic. For this reason, sulphonatopropylviologen in its reduced form (SPV^-) was chosen as one of the probes used to act as a developing reactant. When the pulse occurs, transforming the Ag^+ ion into an atom of silver, the molecule SPV is transformed into the electron–donor ion SPV^-, which is blue in colour and exhibits a redox potential of -0.41 V_{NHE}. The transfer of electrons from SPV^- to a silver cluster Ag_n^+ is thermodynamically possible only if the potential of the cluster is greater than that of the donor. Therefore, the donor decay is expected to start only after a certain time has elapsed after the pulse, a delay required to allow the growing cluster to reach a potential more positive than the donor.

The experiment indeed shows that the SPV^- species does not undergo an immediate reaction (Figure 6.17). A certain lapse of time occurs during which the charged clusters are still very small and thus have a lower tendency to capture electrons than the developer itself. After this critical time (the induction time), which corresponds to the growth of the clusters above a minimum size n_c, the difference between the potential of the clusters and that of the developer is reversed and transfer can begin.

$$Ag_{n \geq n_c}^+ + SPV^- \longrightarrow Ag_n + SPV \qquad \ldots (6.36)$$

$$Ag_n + Ag^+ \longrightarrow Ag_{n+1}^+ \qquad \ldots (6.37)$$

$$Ag_{n+1}^+ + SPV^- \longrightarrow Ag_{n+1} + SPV \qquad \ldots (6.38)$$

$$Ag_{n+1} + Ag^+ \longrightarrow Ag_{n+2}^+ \qquad \ldots (6.39)$$

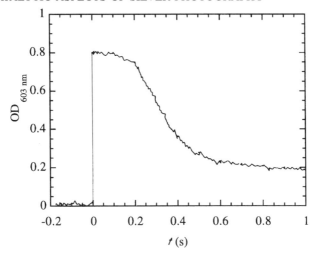

Figure 6.17. Example of electron donor decay in the presence of silver clusters. The decay starts after an induction delay corresponding to the time necessary to reach supercritical sizes through coalescence [34]

Coalescence (reactions 6.31–6.35) is now in competition with the cascade of electron-transfer reactions (6.36–6.39) restricted to $n > n_c$. Due to this complexity, the quantitative analysis of kinetics must be achieved through computer numerical simulation.

Detailed kinetic study of the induction time, including the known coalescence kinetics and analysis of the donor decay, shows that SPV^- becomes an electron donor as soon as a cluster contains five atoms of silver having one positive charge, Ag_5^+ [34]. Subsequently, repetitive capture of electrons onto this nucleus, (reaction 6.38) alternating with the accretion of new Ag^+ ions (reactions 6.39), causes the cluster to grow provided the Ag^+ or SPV^- species are not exhausted. The system behaves exactly like a collection of nuclei that grows with a velocity proportional to the abundance of germs and to that of the SPV^- donor. This is known as autocatalyzed growth.

It is remarkable that the turnover rate does not depend on the free silver cations Ag^+ present in the system and acting as a reservoir for the growth. Reactions (6.37) or (6.39) are fast and are not controlling steps of the mechanism. More important is the fact that no direct reaction occurs between the donor and free Ag^+. This confirms the strong reducing character of the single silver atom $Ag°$ [33,35].

Incidentally, pulse radiolysis provides also a clear explanation of why a silver salt solution when reduced through a reducing agent does not yield silver metal particles in the bulk but a silver mirror on the walls of the vessel. The usual reducing agents are not strong enough to donate an electron to isolated Ag^+ and therefore to start the reaction in the bulk. On the contrary, silver ions adsorbed on the walls benefit of a more positive redox potential: their reduction is achieved as soon as the electron donor has diffused to them. Autocatalytic growth on the preexisting radioinduced germs in the presence of an electron donor is very similar.

8.6 REDOX PROPERTIES OF SOLVATED METAL CLUSTERS

From the above results it was concluded that the potential of the Ag_5^+/Ag_5 couple solvated in water is $E°(Ag_5^+/Ag_5) = -0.4 V_{NHE}$.

By changing the nature of the electron donor, and thus its redox potential, it was found that the critical size of the nucleus also changed. The potential of the cluster of this size was determined as previously described. Table 6.1 presents some cluster potentials obtained by this kinetics method. For bielectronic donors, such as hydroquinones, the actual potential of each monoelectronic transfer must be considered [15]. Note that with H_3O^+, the electron transfer was observed from the cluster acting as a donor toward H_3O^+, similar to corrosion.

Figure 6.18 presents the nuclearity dependence of the silver cluster redox potential in water [34]. In order to allow comparison with the ionization potential (IP in eV) of clusters of the corresponding nuclearity in the gas phase [19], the ordinate scale of $E°$ is translated relative to the vacuum level by 4.5 eV (the Fermi level of the normal hydrogen electrode).

$$-IP(Ag_n)_{solv} = e \times E°(Ag_n^+/Ag_n) - 4.5 \text{ eV} \qquad \ldots (6.40)$$

The results available in water are not sufficiently numerous to assume a possible parity effect as in the gas phase. The important feature of the results in Figure 6.18 is that the redox potential and the ionization potential of silver clusters, respectively in water and in vapor, show opposing dependences on nuclearity. This can be explained by considering that electron loss in water is followed by cation solvation. For example, $\Delta E_{solv}(Ag^+) = 5.6$ eV. This important energy

Table 6.1. Size-dependent redox potential of silver clusters in water

n	Cluster system	Redox monitor		Ref.
		Couple	$E°$ (V_{NHE})	
1	Ag^+/Ag_1	From $E°$ (Ag^+/Ag_∞) and ΔH_{sub}.	−1.8	[35]
2	Ag_2^+/Ag_2	Calc.	−0.62	[33]
5	Ag_5^+/Ag_5	SPV/SPV$^-$	−0.41	[34]
10	Ag_{10}^+/Ag_{10}	H_3O^+/H_2	0	[37]
11	Ag_{11}^+/Ag_{11}	Cu^{2+}/Cu^+	+0.15	[36]
85	Ag_{85}^+/Ag_{85}	Q^-/QH_2 pH 3.9[a]	+0.22	[38]
550	Ag_{550}^+/Ag_{550}	Q^-/QH_2 pH 4.8[a]	+0.33	[38]

[a]QH_2: hydroquinone of naphtazarin.

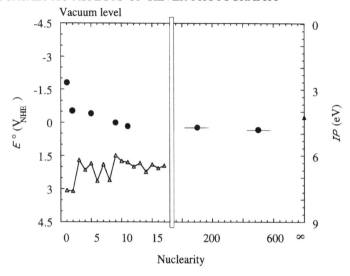

Figure 6.18. Nuclearity dependence of silver cluster redox potential [38] (Table 6.1) as compared with the ionization potential in the gas [19]

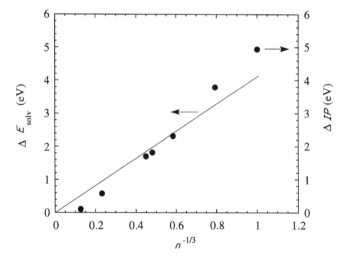

Figure 6.19. Size dependence of the difference in ionization potential between free and solvated clusters of same nuclearity [38]. Experimental data from Fig. 18, — Calculated data from Born equation

represents the difference between IP and IP_{solv} for the atom. The same difference for larger sizes may also correspond to the solvation energy of the clusters $\Delta E_{solv}(Ag_n^+)$. When plotted vs. $n^{-1/3}$, the difference $\Delta IP = IP - IP_{solv}$ can be compared with the Born model of solvation ΔE_{solv} [38] (Figure 6.19):

$$\Delta E_{\text{solv}} = \frac{e^2}{2 \times 4\pi\varepsilon_0 r_n} \left(1 - \frac{1}{\varepsilon_s}\right) \qquad \ldots (6.41)$$

with ε_0 and ε_s being the permitivity of the vacuum and of the solvent, respectively.

Assuming that Ag_n^+ is spherical and using the same radius for Ag^+ or an atom, the radius r_n may be expressed as $r_n = r_0 \times n^{1/3}$. The straight line corresponds to the Born equation (6.41). From the correspondence between ΔIP and ΔE_{solv}, it can be seen that this very simple model nevertheless accounts fairly well for the results [34].

9 MODEL OF THE PHOTOGRAPHIC DEVELOPMENT PROCESS

9.1 HISTORICAL ASPECTS OF DEVELOPMENT THEORIES

The chemical basis for the considerable gain obtained by development lies in the way in which the developer acts as an electron donor, and the way in which the Ag^+ cations act as receptors. But not all the silver bromide crystals in the emulsion are equally blackened. A fortunate distinction protects the crystals that have not been exposed to light, or have been only slightly exposed, from any alteration, thus preserving the information content of the image. How can this be explained? Why is it that only crystals containing clusters that are larger than a certain critical number of atoms are able to accept electrons from the developer and thus darken? Despite the extensive use and technological improvements of photographic development from its origin, the explanation of the specificity of the process has defied our understanding for a long time.

For over a century, numerous theoretical models [8,39,40] have been proposed to explain how the supercritical clusters of silver atoms created by the light act as nuclei to catalyze development. In general, the transfer of electrons effected by the light, and that of electrons donated by the developer, have been assumed to obey the same laws. In fact, as has been shown in §8, the mechanisms are governed by distinct parameters.

The oldest models suggested that the primary silver atoms created by light remained dispersed within the silver bromide as in a supersaturated metastable phase. The next phase, produced by the developer, could grow only when concentration was higher than a certain threshold value of supersaturation. Above this threshold, the nucleus that acted as a catalyst to development was able to form. In this type of *phase model* [39] the cluster of silver atoms was assumed, as soon as formed, to have the same properties as the metal in bulk form, particularly regarding the electron acceptor character of Ag_n^+. Thus, any silver cation was presumed to accept an electron, but the discrimination by the developer against certain clusters was attributed in that case to an increasing development speed with the cluster size. This contradicted chemical knowledge: massive silver is a noble metal and resists attack by weak oxidizing agents. Yet in an exposed, but undeveloped film, the clusters of a few silver atoms are very sensitive to the action of various compounds, even weak oxidizing agents, and behave on the contrary as electron donors. We must therefore assume that when silver is in an ultrafine form — that is, when it consists of clusters of just a few tens of atoms — it possesses properties different from those of the massive metal. Moreover, the smallest size of developable cluster is not determined by the contact time with the developer but essentially by its redox properties.

The subsequent group of models known as *atomistic models* [12,40] does not extrapolate the macroscopic properties of the solid to derive the behavior of a cluster containing just a few atoms. The *ab initio* models were based on a the size-dependent properties of clusters.

Since the mechanism of photographic development essentially rests on the transfer of electrons to a nucleus that contains a supercritical number of silver atoms, and which is positively charged by the surrounding silver cations, it is, therefore, important to know the tendency that a cluster has to capture electrons as a function of n. As the theoretician J. F. Hamilton emphasized in 1977 [8,41], the transfer of electrons from the developer should become more and more difficult as n increases, judging by the properties measured in the gas phase. In other words, the behavior of clusters in the gas phase cannot account for the existence of a minimum critical size, despite this being well known to every practicing photographer. In addition, in the gas phase, an isolated atom is more stable relative to ionization than any cluster. The transfer of electrons from the developer to an Ag^+ ion to produce an atom of silver ought therefore to be spontaneous, but this certainly does not occur in photographic development.

This ignores the fact that photographic development brings an emulsion containing solid crystals into contact with an aqueous solution of an electron donor, the developing agent. These atomistic theories were thus unable to explain the growth of the nucleus and hence the development mechanism. The attempts to extend the concept of size-dependent properties to the photographic development process failed in so far as:

1. Electrons produced by the action of light and electrons donated by the developer were considered similarly;
2. The size-dependent potential was taken from data on gas phase clusters.

9.2 SOLVATED CLUSTER DEVELOPMENT MODEL

The important role of the solvent in the development mechanism and the specificity of the electron transfer relative to photoelectron generation were demonstrated in the cluster study through pulse techniques.

The various aspects of development revealed by the kinetic studies of solutions [34] correspond with a number of characteristics known empirically to photographers. These include growth by an autocatalytic chain reaction, critical size required for a nucleus (or speck) to grow, no direct reaction with Ag^+, the susceptibility of subcritical clusters to oxidation and the variation in the critical size increasing with the redox potential of the developer.

Nonetheless, certain differences exist. In the pulsed experiment, the creation of developable silver nuclei and their development occur successively in the same system, so that the very fast initiation of development can be directly observed by the time-resolved method. In photography, on the other hand, these two stages are separated in time. The other difference is that the silver atoms created by the pulse are able to diffuse freely throughout the solution. Development takes place on a free silver nucleus, which is positively charged by surrounding Ag^+ cations and is accompanied by sulfate anions. In photography, however, the silver atoms remain trapped within a crystal of silver bromide that is surrounded by gelatine. Development therefore takes place at the interface between the silver-bromide crystal and the developing solution. Overall the two situations are very similar. The resemblance is therefore sufficiently close for one to be able to assume that the conclusions about the catalytic growth of nuclei in solution can be applied to photographic development (Figure 6.20).

9.3 DEVELOPMENT THRESHOLD

The explanation of the development mechanism is therefore as follows [34]. The latent image, produced by the action of light, consists of clusters of silver atoms, each within a crystal of silver bromide. The numbers of atoms in these clusters vary (being zero, subcritical, or supercritical). When in contact with a solution, the redox potential of these clusters together with the neighbouring Ag^+ ions is assumed directly to increase with the number of atoms. If the solution contains a developer (an electron donor of given redox potential), transfer can occur only toward clusters that contain more atoms than a certain threshold number. This threshold is governed by the redox potential of the developer exclusively, as emphasized by Malinowski [42]. It occurs when the potentials of the developer and of the cluster are equal (Figure 6.20). In effect, therefore, the whole emulsion consists of a myriad of microscopic electrical cells, of which one pole (the developer solution) is fixed, and the other (the population of crystals of silver bromide that have been more or less exposed to light) is variable. The passage of electrical current requires a positive difference in potential. The discrimination shown by the developer is therefore the consequence of a quantum effect caused by size, whereby the redox potential of the silver nucleus (or its ionization potential) increases with the cluster number. It is noteworthy that a similar trend (increase of cluster ionization potential with the nuclearity) had already been suggested by Trautweiler [43] in a speculative model designed for explaining the photographic development process. The author stressed the major point, i.e. the potential of supercritical sizes should lie *above* that of the developer. However, the only data then available (IP values of vapour phase clusters which exhibit a size dependence with an opposite trend) did not confirm this view.

PHOTOCATALYTIC ASPECTS OF SILVER PHOTOGRAPHY

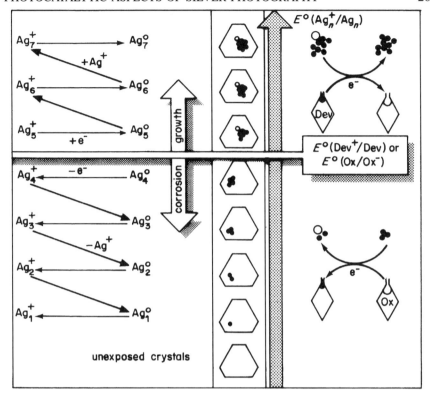

Figure 6.20. Model of the photographic development [34]. The redox potential of photoinduced clusters increases with their nuclearity n On the other hand, the potential of the developer, an electron donor, has a constant value. It is only thermodynamically possible to transfer electrons from the developer to clusters whose potential is greater than that of the developer. Transfer, therefore, occurs only above a certain critical size and it subsequently becomes more and more favored. Conversely, if the developer were to be replaced by an electron acceptor, the smallest clusters would behave as electron donors. The transfer of electrons from the clusters to the oxidant would cause the progressive corrosion of the silver nucleus until the latter was completely dissolved, as happens when the latent image is destroyed

Concerning the actual role of the silver bromide substrate in shifting the redox potential values relative to free clusters in solution (Figure 6.18), it is known that semiquinones/hydroquinones couples with $E°(QH°/QH_2)$ close to $+0.025$ V_{NHE} in basic medium [15] develop photographic emulsions at threshold close to $n = 4$, while this size is reached in free water with a donor potential of -0.41 V_{NHE}. The difference of 0.43 V, at least for this nuclearity could correspond to the effect of the AgBr support, making the reduction easier [44].

With each new electron transfer, the cluster surrounded by Ag^+ ions increases by one unit and the potential becomes more and more favourable for transfer.

The developer being present in excess, the silver ions in a crystal are therefore successively neutralized by a catalytic electron transfer to exhaustion, provided that the crystal originally contained a nucleus that was of supercritical size. The developed image is formed of crystals totally reduced into silver. After fixing, the effects of black, white or gray are given by the local density of black microcrystals (§6.1). It is important to note that there are no grounds for believing that from this point onwards the rate of development increases with the size of the nucleus: according to the above results the rate is zero below the critical size, and fast but constant above it. In photography, the diffusion rate of the developer through gelatine is certainly lower than the autocatalytic reduction process.

It was also seen that for a given emulsion, it is possible to vary the development threshold and thus the contrast by altering the choice of developer and its redox potential. The greater the (negative) value of this potential the smaller will be the critical size of the clusters. This will accentuate on the positive image the bright areas at the expense of the gray tones.

Although a given developer imposes the critical nuclearity, the exposure time of an emulsion of given nominal sensitivity may be yet lowered (and therefore the sensitivity somewhat enhanced). Actually, when the object is only weakly illuminated, it is possible to resort to a two-stage process: the actual exposure, which may be preceded or followed by a second, weak, even (i.e. neutral gray) exposure of the whole negative. This second exposure creates a few atoms of silver in all the crystals, but these are insufficient in themselves to initiate development. When these atoms are added to the atoms created by the actual exposure, however, the size of certain clusters crosses the critical threshold. A short contact of the emulsion with molecular hydrogen before exposure (chemical hypersensitization) similarly enhances the sensitivity.

In conclusion, the creation during the exposure of electrons, precursors of silver atoms, results from the excitation by light of a sensitized solid, while, during development, the electron transfer from the donor occurs under solvated conditions and is controlled by the size-dependent redox potential of the clusters and that of the developer.

9.4 LATENT IMAGE REGRESSION

If an oxidant is added to the solution instead of a developer, the inverse transfer process can take place: clusters whose redox potential is less than the potential of the oxidant lose an electron (Figure 6.20). Such a loss lowers the potential of the cluster, causing it to become even more susceptible to reaction. Nothing can stop this process, which continues until all the clusters with strongly negative potentials are completely dissolved. This explains why the latent image is so prone to degradation by even mild oxidants. Such fragility (regression of the latent image) is known to occur easily in emulsions which are not developed soon.

Unlike the latent image, a true image consists of a dense clump of silver crystals, which are naturally minute but which contain sufficient atoms for them to behave as a metal that cannot be oxidized, just like massive silver. This is why a developed silver photograph resists aging so well.

9.5 IMPORTANCE OF DEVELOPMENT

Ever since the first results on development were achieved, this key stage in the photographic process, like all the others, has benefited from innumerable empirical improvements. Various additives have been combined with the emulsion or the developer to achieve better control of the process. It is quite possible that some of these additives influence the interrelationship between the potentials that we have discussed above.

This enormous gain in sensitivity makes possible, without any great sacrifice in exposure performance, the use and the manufacture of emulsions with finer and finer crystal. The result has been a definition which still amazes our eyes, which have perhaps become somewhat jaded by the positive bombardment by images that they undergo. It is not commonly realized, for example, that an ordinary photographic emulsion in everyday use contains such a large number of microscopic, juxtaposed crystals of silver bromide (3 million mm^{-2}) that the entire information content of a television image can be stored, without loss, in just 5 mm^2 of silver emulsion (Figure 6.3). This resolving power is applied in photographic microfilms to store an enormous quantity of information in a tiny space. Conversely, such high definition enables large images to be projected onto enormous screens, and also the enlargement of scientific images (telescope imaging for astronomy, electron microscopy, etc.) so that minute details become visible.

As will be shown, development is a key step which also governs other processes derived from silver cluster growth (§§10,12). Via the oxidized state that arises in the developer after it has reacted with exposed silver crystals, colored molecules can be synthesized, thus producing the basis of still or cine photography in colour (§11).

10 INVERSION-TRANSFER PROCESSES

10.1 PRINCIPLE OF INVERSION TRANSFER

The aim of the inversion-transfer processes is to obtain a positive image of the object directly and thus to avoid the double series of operations of exposure–development–fixation imposed by the passage through a negative. This principle opens the possibility of one-step processing and thus of instant photography like the polaroid system. The process allows the image to migrate close to the surface of the emulsion (receiving layer), where also other physicochemical treatments are feasible and confer specific properties to the image (see §12).

In fact, the effect of light on a silver emulsion discriminates two latent images (black-and-white photography). One is constituted by the ensemble of developable crystals with supercritical nuclei: it will generate a black image, therefore negative, after development. The second one, complementary and also made up of silver bromide crystals, but normally undevelopable, is usually eliminated during the fixing step. Instead the inversion-transfer process takes advantage of this shadow latent image to produce directly the black areas corresponding to the shadows, that is, the positive image. In addition to the reducing properties of the processing bath, two conditions need to be fulfilled. The first is to provide the silver cations of the unexposed crystals with the nuclei required to catalyze their chemical reduction into divided metal, which looks black (Figure 6.21). The second is to ensure contact between the ions and the nuclei through diffusion.

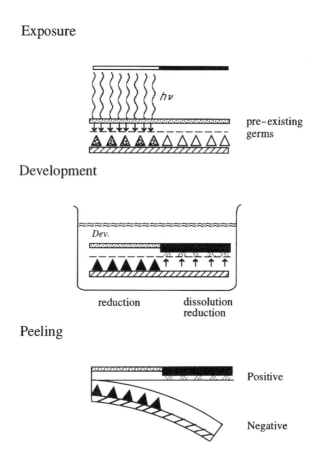

Figure 6.21. Inversion-transfer process. Preexisting germs are included in the upper layer of the emulsion. They catalyze the development of the complementary image after diffusion of silver ions. The positive image is finally separated from the negative by peeling

10.2 STRUCTURE AND DEVELOPMENT OF THE FILM

For this purpose, the unexposed crystals are dissolved, as in fixing (reaction 6.28), by a complexing agent, added to the developing agent. The ions are thus permitted to diffuse toward the layer of exogen nuclei. These germs are included in a thin receiving layer which must be placed in contact with the silver ions of the emulsion at the moment of development–transfer (Figure 6.21). The rapid recycling of the silver halide complexing agent that follows the reduction of Ag^+ at the nuclei accelerates the dissolution and the diffusion of new Ag^+ ions. Simultaneously, silver ions contained in exposed crystals of AgBr are in the immediate vicinity of the germs photoinduced and, even dissolved by the complexing agent, are immediately developed *in situ* without migration. The success of the process depends therefore on the competition between the fast development rate of the exposed crystals and the transversal diffusion of the ions of the complementary image toward the germs of the receiving layer. Finally, all silver ions are reduced (fixing is no longer needed), but the divided silver metal is now distributed in two separate layers. The quality of the initial image must obviously be preserved by ensuring almost no diffusion parallel to the layer plane.

A physical separation between the negative and the complementary positive images in their respective layers is then made, provided each of them follows the corresponding appropriate support in the peeling (Figure 6.21). Instant photography must control carefully the relative rates of diffusion between the layers and of catalytic development inside the layers. The advantage is to provide a quite compact system of exposure and immediate development in the camera itself to produce the final image.

11 COLOR PHOTOGRAPHY

11.1 INTERFERENTIAL PROCESS

The principle of the interferential process lies in the wave property of light. When two beams of light issued from the same source are crossing, and the difference of length between the respective paths is a multiple of the wavelength, the intensities correspond to the same phase and are added, while, if the difference is one-half of a wavelength longer, vibrations in opposite phases compensate and the intensity is zero. This phenomenon yields specific patterns known as fringes. The physicist Gabriel Lippmann [45] first proposed in 1891 to use the interferential fringes so created to reproduce the color of the beam.

The silver emulsion was deposited on a mirror-like support which reflected the light. The silver bromide crystals had to be very small (0.01 μm) in order to resolve the fringes separated by one-half of the wavelength. Incident light penetrating the emulsion and the second beam reflected from the support produce, at a given wavelength, interference sites where the exposure is maximized and which are materialized by exposed crystals regularly spaced in the depth of the emulsion. After development and fixing the image is exclusively constituted of silver particles corresponding to bright parts of the object. However, their specific stratified distribution in depth according to the fringes of the different colors of the exciting light is capable of also selectively reflecting the light and restores directly the original colors. Such a process implies maximum control of the photosensitive material structure to avoid deformation of local fringe periodicity and therefore faithfully reproduce the colored image.

The interferential process is no longer used for color photography. Nevertheless, the production of fringe patterns in the emulsion is exploited today in three-dimensional photography to create holographs (§14).

11.2 PRINCIPLES OF TRICHROMIC COLOR PHOTOGRAPHY

Color photography is founded on the trichrome theory proposed in 1802 by Thomas Young. According to this theory, three fundamental or primary colors — red, green, and blue — suffice for the eye to reproduce all other colors (Figure 6.5). The action of light in these three colors is combined in the retina, and the process is therefore known as *additive*. In 1859, the physicist James Clerk Maxwell demonstrated the physical principle of trichromatic vision by successfully obtaining the first additive color synthesis. The method of applying trichrome theory to photography was proposed in 1869, simultaneously, but independently, by Charles Cros, who merely established the principles, and by Louis Ducos du Hauron. In addition, the latter was the first to produce actual color photographs. The visible radiation coming from the object to be reproduced was separated into three bands: red, green, and blue, and the color picture resulted from the superimposition of three distinct images realized separately in each of the colors.

Nowadays, the colors in photography are produced according to a *substractive* color synthesis. Every color of an object is defined by the hue, corresponding to a spectral composition, the saturation (an unsaturated color seems to contain a certain amount of white) and the brightness, which is measured by the intensity of the transmitted or reflected light per wavelength. A white light source (or a white support) W emits (or reflects) a sum of B, G and R radiation ($W = B + G + R$). A yellow filter Y, for example, has the property to absorb selectively the blue light: the eye is thus excited by the sum $G + R$ which feeds the brain the sensation of yellow ($Y = W - B = G + R$). Similarly, a magenta (purple) filter corresponds to $M = W - G = B + R$ and a cyan (blue-green) filter to $C = W - R = B + G$. Yellow, magenta and cyan are complementary colors of blue, green and red respectively. The higher the absorbance of the complementary filter, the lower the fraction of the intensity of the primary color transmitted.

Yellow, magenta and cyan dyes play precisely the role of local filters in photography. Their superficial concentration controls the transmitted intensity of each primary color and therefore the excitation of the retina and perception by our brain. Thus, the substractive color synthesis offers a powerful means of printing the hue and the brightness of an image. It is applied to the different color systems in negative–positive or inversion processes.

11.3 STRUCTURE OF COLOR EMULSION

In practice, the color separation is achieved today by the use of three superimposed photosensitive layers coated onto the same transparent base, separated by filters and specifically sensitized to B, G and R (Figure 6.22a):

In black-and-white photography, the greater the brightness of an object, the greater the fraction of exposed crystals developed into black silver particles in

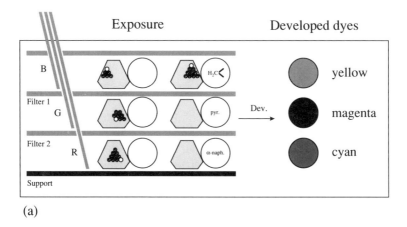

(a)

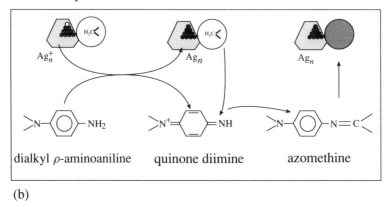

(b)

Figure 6.22. Structure and development of a negative color emulsion (a) Multilayer emulsion with the three layers sensitized to blue (B), green (G) and red (R) separated by intermediate filters and containing AgBr crystals associated with a coupler which is a precursor of the required dye (b) At the development stage, the chromogenic developer selects the supercritical clusters as for black-and-white photography. Its oxidized form reacts with the associated coupler to generate the appropriate dye. During fixing, the undeveloped crystals are eliminated, together with the metallic silver that would otherwise form a gray veil

the image. Similarly, the negative color image must be obtained in synthesizing in each layer the complementary color or absorber of the incident excitation light. The local concentration of the dye formed must also be correlated with the number of photons absorbed during the exposures. In the negative–positive color process, the complex problem of reproducing colors and brightness has been achieved by the specific synthesis of appropriate dyes indirectly governed by the excitation of silver bromide.

Each layer of the three-layer emulsion for color photography contains silver bromide associated with a coupler which is a precursor of the dye (Figure 6.22a). The silver bromide crystals in these three layers are thus capable of being excited selectively by photons of the three basic colors. The first layer does not contain a chromatic sensitizer and absorbs only blue. The second and third layers are respectively sensitized to green and red selected by the intermediate filters.

The coupler is in intimate association with each crystal of silver bromide in this multilayer emulsion. It serves as a reservoir of specific molecules that are destined to form a given color. For example, in a negative color film (Figure 6.22a) an acetyl–acetone derivative is used as a coupler in order to produce azomethine, which yields a yellow image in the upper layer sensitive to blue, pyrazolone is used as a coupler to yield the magenta (purple) image in the layer sensitized to green and α-naphthol as a coupler to yield the cyan (blue-green) in the layer sensitized to red. The action of light during exposure is to produce a certain number of silver atoms in each AgBr crystal in the bright part of the blue, green and red images.

11.4 DEVELOPMENT OF COLOR IMAGE

At the development stage (Figure 6.22b) the multilayer emulsion is developed by a unique chromogenic developer. The oxidized form of the chromogenic developer is destined to react with the associated coupler to generate the appropriate dye.

In the first step, the developer selects the supercritical silver clusters as for black-and-white photography (§9.3). The electron donor catalytically reduces the silver ions around these nuclei according to reactions (6.20–6.25) (with two equivalents if the developer is dielectronic as in Figure 6.22b: the number of oxidized molecules of developer is reduced to that of half the silver atoms). Classical chromogenic developers are the dialkyl-p-amino-anilines. When oxidized by Ag^+ ions, they yield quinone-dimine molecules that are formed in close vicinity to the crystals of the latent image and the organic coupler.

$$Ag_{n+2}^{2+} + R_2N-C_6H_4-NH_2 \longrightarrow Ag_{n+2} + R_2N^+=C_6H_4=NH + H^+$$
$$\ldots (6.42)$$

Quinone-diimine is therefore spatially distributed as the negative silver image. Its conjugated double bonds make it reactive toward the couplers. The next step after formation of the developed silver image is an addition of the coupler on the diimine double bond, yielding a *leuco*-derivative. In the case of the coupler cyanomethylphenylketone addition occurs on the methylene carbon:

$$R_2N^+=C_6H_4=NH + \underset{CN}{\underset{|}{\overset{O=C-C_6H_5}{\overset{|}{CH_2}}}} \longrightarrow R_2N-C_6H_4-NH-\underset{CN}{\underset{|}{\overset{OC-C_6H_5}{\overset{|}{CH}}}} + H^+ \quad (6.43)$$

leuco

The *leuco*-derivative is itself oxidized by a second quinone-diimine molecule giving rise to the insoluble yellow azomethinic dye, thus fixed close to the negative silver image.

$$\underset{\text{leuco}}{R_2N-C_6H_4-NH-\underset{CN}{\overset{OC-C_6H_5}{CH}}} + R_2N^+=C_6H_4=NH \longrightarrow \underset{\text{azomethine}}{R_2N-C_6H_4-N=\underset{CN}{\overset{OC-C_6H_5}{C}}} + 2H^+$$

$$+ R_2N-C_6H_4-NH_2 + H^+ \qquad \dots(6.44)$$

The intensity of the dye image also reproduces that of the negative silver image as four silver ions have been used *in situ* to synthesize stoichiometrically one dye molecule. Similarly, couplers such as pyrazolone and α-naphthol are used to generate molecules with conjugated double bonds, respectively the magenta (purple) and the cyan (blue-green) dyes. The overall reactions are written as follows:

$$2\; \underset{R}{\overset{R}{N}}{+}{=}\!\!\!\left\langle\!\!\!\begin{array}{c}\\ \end{array}\!\!\!\right\rangle\!\!\!={NH} + H{-}\underset{CN}{\overset{O=C-C_6H_5}{\underset{|}{C}}}{-}H \longrightarrow \underset{R}{\overset{R}{N}}{-}\!\!\!\left\langle\!\!\!\begin{array}{c}\\ \end{array}\!\!\!\right\rangle\!\!{-}N{-}\underset{CN}{\overset{O=C-C_6H_5}{\underset{|}{CH}}}$$

substituted methylene   azomethine *(yellow)*   ...(6.45)

$$+ 2H^+ + \underset{R}{\overset{R}{N}}{-}\!\!\!\left\langle\!\!\!\begin{array}{c}\\ \end{array}\!\!\!\right\rangle\!\!{-}NH_2$$

$$2\; \underset{R}{\overset{R}{N}}{+}{=}\!\!\!\left\langle\!\!\!\begin{array}{c}\\ \end{array}\!\!\!\right\rangle\!\!\!={NH} \;+\; \underset{\underset{O}{\overset{\|}{C}}}{H_2C}\!\!\overset{\overset{R}{\underset{|}{N}}}{\underset{}{\diagdown}}\!\!\underset{}{N}{-}C_6H_5$$

pyrazolone   ...(6.46)

$$\underset{R}{\overset{R}{N}}{-}\!\!\!\left\langle\!\!\!\begin{array}{c}\\ \end{array}\!\!\!\right\rangle\!\!{-}N{=}\underset{\underset{O}{\overset{\|}{C}}}{C}\!\!\overset{\overset{R}{\underset{|}{N}}}{\underset{}{\diagdown}}\!\!N{-}C_6H_5 + 2H^+ + \underset{R}{\overset{R}{N}}{-}\!\!\!\left\langle\!\!\!\begin{array}{c}\\ \end{array}\!\!\!\right\rangle\!\!{-}NH_2$$

indazolone *(magenta)*

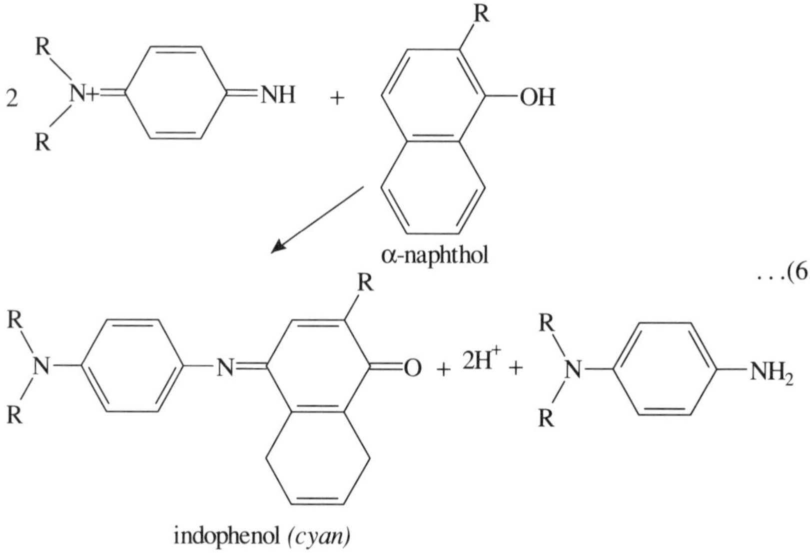

$$2 \underset{R}{\overset{R}{N}}-\hspace{-2pt}\langle=\rangle=NH + \text{[α-naphthol with R, OH]} \longrightarrow$$

α-naphthol

$$\text{[indophenol structure]} + 2H^+ + \underset{R}{\overset{R}{N}}-\langle\rangle-NH_2 \qquad \ldots(6.47)$$

indophenol *(cyan)*

The oxidized form of the developer, instead of being washed away as in the case of black-and-white photography, is thus used to create the cyan, magenta and yellow dyes, which are the complementary colors to red, green and blue. Their absorbance is correlated to the brightness of the object. The negative color image is produced in this manner and the process naturally has to be repeated to obtain a positive image.

In the positive emulsion, the order of organic couplers may be different from that in the negative film. However, the dyes Y, M and C are still synthesized in the layers excited by B, G and R photons respectively. Now, yellow areas of the negative image excite the R and G layers of the positive emulsion where M and C dyes are produced, giving the eye the sensation of blue. The double inversion blue object ⟶ yellow negative image ⟶ blue positive image is effectively achieved.

As a final step, non-reduced silver bromide crystals are eliminated as well as the crystals of metallic silver that would otherwise form a gray veil (bleaching). The silver image has thus been used twice as a transient step only to control the development of the three latent images and to produce catalytically the dyes via the oxidized developer. In the end silver is totally removed and recyclable. However, color photography must be considered indeed as a silver process.

The same principle of dye synthesis is applied with an appropriate order of couplers for *direct positive* color processes, such as in slides for projection. But the inversion on a single film is now obtained in synthesizing the complementary dyes in the *unexposed* part of the emulsion. For this purpose, the exposed silver crystals of the latent image are developed in a first step by an electron donor which consumes catalytically all silver ions (reactions 6.20–6.25). The negative image becomes unable to catalyze other redox reactions. Then, the emulsion is exposed to a white light source to create silver clusters in the rest of the image. The last step consists in synthesizing *in situ* the dyes through the chromogenic developer in this complementary image (reactions 6.42–6.47), followed by bleaching as above to eliminate silver. The dyeing of the complementary image by a complementary color yields a direct positive.

Instant color photography on paper combines the inversion transfer (§10) with the substractive color synthesis. In this process, the dyes are preexisting in the emulsion. The specific molecules constituting the chromophores are linked to the developer by a chemical chain and the couple dye-developer is included in the layer just below the AgBr layer sensitized to the corresponding radiation. The solubility, and thus the ability of the couple dye-developer for diffusing, depends on the oxidation state of the developer. Therefore, when the developer is allowed to react with the AgBr latent image by pressing the multilayer film in the camera itself immediately after exposure, the insoluble couple dye-oxidized developer remains fixed around the crystals of the developed latent image. Conversely, the non-oxidized developer-dye couples present in the rest of the emulsion diffuse to the final receiving layer containing a mordant. Both developed and unexposed silver are removed by peeling the original sensitive layer.

In conclusion, instant color photography constitutes a process of extreme sophistication, consisting of a multilayer emulsion integrating all chemical ingredients required to achieve in the camera successively formation of the trichromic latent image, silver development, diffusion of the positive image colors to the final support and elimination of silver. This system realizes the most advanced synthesis of the different processes introduced in photography.

12 OFFSET SILVER PROCESSES

12.1 PRINCIPLES OF OFFSET PRINTING

The principle of offset printing is to produce a plate able to adsorb ink in contrasted areas corresponding to the image, in order to then repeatedly transfer the ink onto a support. The process is used to print at low cost thousands of paper copies. The image integrates any kind of picture or written text alike, in black and white, or in color, provided distinct plates print successively the three colored inks plus one black ink for shadow. Offset processes represent an important application of photographic technology.

The image of the original composition to be reproduced is projected onto a photosensitive material. The film may be an organic substrate able to photopolymerize, or a silver halide emulsion. In the case of a photopolymer, the contrasted areas, exposed/unexposed, may be used either to engrave selectively the metal support or to coat it with a second metal of a different affinity for ink. Printing may be also based on contrasted absorption of the ink by the polymer and by the unmasked support.

214   HOMOGENEOUS PHOTOCATALYSIS

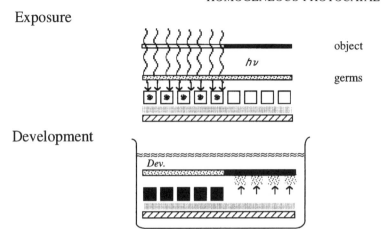

Figure 6.23. Structure and development of a silver offset plate. The positive image is obtained by inversion transfer. Unexposed silver ions diffuse to the top layer of the emulsion where they are reduced catalytically by the germs into an oleophilic silver image

In the case of silver emulsion for offset application, a metallic silver image with hydrophobic properties is formed at the top of upper surface of the plate. During the printing stage the contrast with the hydrophilicity of the support ensures the distribution of ink with a resolution comparable to that of photography.

12.2 MECHANISM OF OFFSET PLATE DEVELOPMENT

The silver image in normal emulsions is embedded in the transparent gelatine and is unable to provide a contrast, except visual, for some superficial physical property such as ink receptivity. Offset emulsions (Figure 6.23) are therefore coated with a specific layer containing catalytic germs for development as in the inversion-transfer process (§10). The aim of these added germs is to reduce the silver ions of the unexposed crystals, after complexation and diffusion to the top layer of the emulsion that will be later in contact with the ink. This oleophilic image will absorb ink and is a direct positive. The development process is very similar to that in inversion transfer (Figure 6.23). The negative image is also developed *in situ* inside the gelatine, but it does not need to be removed since it does not interfere with the printing.

13 CINEMATOGRAPHY

With its extraordinary amplification factor, development has enabled considerable gain in sensitivity to be achieved. It opened the way for the great technological advances in photography that have been obtained. Exposure times have been reduced by several orders of magnitude, an important advance in itself, but also one that gave rise to derived technologies, among them cinematography. After all, cine photography is merely a flood of images, each of which is static, but which were all obtained by a succession of extremely short exposures at very short intervals. The quick replacement of an image by the next one during the projection must be achieved within the response time of our retina (J. Plateau, 1820) which is close to 0.04 s. This principle, well known from earlier devices to obtain motion effects by quick superimposition of static drawings, allowed pioneers such as E. Muybridge to take a series of photographs with minimum exposure times which decomposed the movement with a series of still cameras, and J. E. Marey had already used a single camera for his chronophotograph.

Recomposition after decomposition of motion was made possible 100 years ago by Thomas Edison with the kinetoscope and by Auguste and Louis Lumière with the cinematograph which allowed projection. The total period of time between two successive images being less than 0.04 s, the cinema required emulsions for short exposure (of high sensitivity) in order to exist. At the same time, the emulsions must also be of high resolution since the cinematographic image has to be projected onto very large screens where otherwise the enormous enlargement (about 200 times in one dimension) would reveal the detailed structure of the image.

Cinematography obviously continues to receive the benefit of all improvements in optics and photographic emulsions technology. The structure and development of black-and-white or color emulsions after exposure derive from those used in still photography.

14 HOLOGRAPHY

Holography, invented by D. Garbor in 1948, is capable of recording the image of an object with all the features of photography, and in addition information in three dimensions. The principle is that the emulsion is exposed to a double beam of light: one is issued directly from a laser as a coherent light source, the second comes from reflection of a part of the same laser beam by the object (Figure 6.24a). The wavelength being fixed, interference between the beams in the plane of the emulsion gives rise to fringes with maximum intensity when the beams are in phase (or when the path difference between direct and reflected beams is equal to a multiple of half the laser wavelength). As in the Lippmann process (§11), fringes are printed in the depth of the emulsion in the form of thin stratas of tiny exposed AgBr crystals (0.01 μm). However, the phase difference corresponding to a single point of the object now depends on the part of the holographic emulsion from which it is seen. Each part of the hologram contains the complete information on the object seen under a specific angle.

For this reason, when the hologram is observed under the same laser light, after the usual development and fixing, our eye sees the virtual three-dimensional

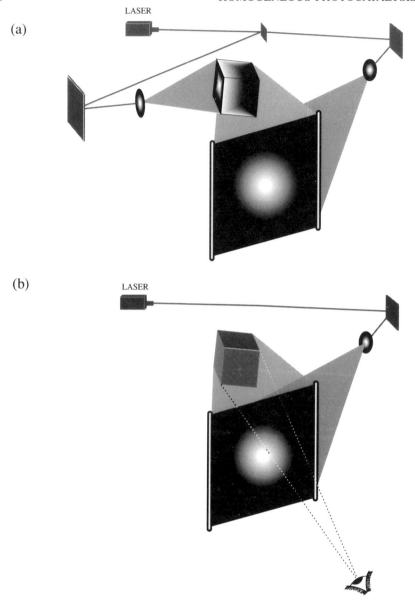

Figure 6.24. Formation and observation of a silver holographic image. a) Interference between the direct beam issued from the laser and the beam reflected by the object gives rise to fringes printed in the depth of the emulsion. b) The virtual three-dimension image is recomposed from the fringes of the hologram observed under the same laser light

image recomposed from the fringes of the emulsion (Figure 6.24b). The laser light reflected by each fringe interferes with the light of the other fringes and recomposes the spatial brightness of the object with a perspective depending on the position of the eye relative to the hologram. The larger the hologram, the larger the angle of possible perspectives for the observer.

The low sensitivity of the very small crystals of holographic emulsions requires long exposures concomitantly with a total absence of movement or vibration. With exposure from three different lasers delivering the basic colors, thus combining the effects observed in the Lippmann process with holography, a color hologram may be obtained in principle.

15 CONCLUSIONS

Niepce's discovery, photography, has been considerably improved by advances in different fields of science: organic and inorganic chemistry, optics, photophysics, photochemistry, catalysis, chemical kinetics. It has become the basis of an important and dynamic industry. Silver salts are still the unique photosensitive material and find their main application in photographic technologies. Their resolving power reaches an unequaled sharpness compared to electronic devices.

Photography has given rise to various derived processes with a high degree of sophistication, for which the above descriptions fall far short of covering all aspects. The emulsions, particularly for color photography, are of great complexity and contain up to 200 components in order to control optimized conditions of image formation. The mechanisms, involving innumerable chemical reactions, are not yet totally understood.

Autocatalytic development is a key step in the enormous amplification which preserves the finest details of the image. It confers to photography an extreme sensitivity and is present as a transient step in all types of photographic processes. The specificity of development results from the size-dependent redox potential of silver clusters. In fact, photography was the first application of quantum-size effects.

Modern techniques of physical chemistry have provided a deeper understanding of the mechanisms at the atomic level, even for fast electron transfer from sensitizers to silver crystals or for the silver autocatalytic reduction. This knowledge promises improvements in technique for the future. Systems allowing greater simplicity in use and industrial processes that preserve the quality of our environment are also of interest to photography researchers.

Moreover, the limits of this powerful technology have not been reached. Certainly, the sensitivity may be still increased through an increase of the effective quantum yield (by inhibiting the electron–hole recombination) and also a decrease in the critical size of the developable speck. Our new understanding of the development mechanism will undoubtedly be applied to practical photographic processes. In addition, it will also cast light on the phenomena of nucleation and growth of crystals, as well as on the mechanisms underlying the transfer of electrons that are catalyzed by metals in the ultrafine state.

In the course of 170 years, taking photographs has become an extremely commonplace process, but it has still not lost its magic.

ACKNOWLEDGEMENTS

The author is gratefully indebted to Dr Jean-Louis Marignier and Dr Denise Parent for their valuable comments and helpful suggestions about the manuscript, and to Dr Jamal Khatouri for his collaboration in the illustrations.

REFERENCES

[1] N. Niepce, 'Notice sur l'Héliographie', 1829, in L. Daguerre, *Histoire et description des procédés du daguerréotype*, edited by A. Giroux, 1839 or Rumeur des Ages, Paris reactive 1982.
[2] T. Wedgwood and H. Davy, J. Royal Inst. Great Brit., **1**, 170 (1802).
[3] J. L. Marignier, Nature, Lond., **376**, 115, (1990).
[4] J. L. Marignier, in *Processes in Photoreactive Polymers*, edited by V. V. Kronghaus and A. D. Trifunac (Chapman & Hall, New York, 1995) p. 3.
[5] J. L. Marignier, *Proc. IS& T'S 48th Annual Conference, Washington*, 1995, p. 222.
[6] L. Daguerre, *Historique et description des procédés du daguerréotype*, edited by A. Giroux, 1839 or Rumer des Ages, Paris 1982.
[7] P. Glafkides, *Chimie et physique photographiques*, 5th edn (Editions de l'Usine Nouvelle, Paris 1987).
[8] J. F. Hamilton, *Theory of the Photographic Process*, 4th edn., edited by T. H. James (Macmillan, New York, 1977).
[9] Physics Today, **42**, (9), (Sept 1989); Mat. Res. Soc. Bul., **14** (5) (1989).
[10] G. Dartnall, J. Bowmaker, and J. Mollon, Proc. Roy. Soc. Lond. **B220**, 115 (1983).
[11] R. W. Gurney and N. F. Mott, Proc. Roy. Soc. Lond., **A 164**, 485 (1938).
[12] J. W. Mitchell, J. Photogr. Sci., **31**, 148 (1982); Photogr. Sc. Eng., **25**, 170 (1981). *Proc. IS& T 48th Annual Conference, Washington*, (1995) p. 136.
[13] P. B. Gilman, *Proc. 47th IS& T Conf.*, Vol. I, (1994) p. 119.
[14] T. Tani, Photogr. Sc. Eng., **26**, 111, (1982); **27**, 75 (1983); Phys. Today, **42**, 36 (1989); *The Present Status and Future Prospect of Silver Halide Photography* (Oxford Univ. Press, Oxford 1995).
[15] P. Wardman, J. Phys. Chem. Ref. Data, **18**, 1637 (1989).
[16] H. Haberland, (editor), *Clusters of Atoms and Molecules* (Springer-Verlag, Berlin 1994).
[17] M. D. Morse, Chem. Rev., **86**, 1046 (1986).
[18] E. Schumacher, Chimia, **42**, 357 (1988).
[19] C. Jackschath, I. Rabin and W. Schulze, Z. Phys. D — Atoms, Molecules and Clusters, **22**, 517 (1992); G. Alameddin, J. Hunter, D. Cameron, and M. M. Kappes, Chem. Phys. Letters, **192**, 122 (1992).
[20] D. E. Powers, S. G. Hansen, M. E. Gensic, D. L. Michalopoulos and R. E. Smalley, J. Chem. Phys., **78**, 2866 (1983). M. B. Knickelbein, Chem. Phys. Letters **192**, 129 (1992).

[21] F. Cyrot-Lackmann (editor), *Entre l'atome et le cristal: les agrégats* (Editions de Physique, 1981).
[22] P. Fayet, F. Granzer, G. Hegenbart, E. Moisar, B. Pischel and L. Wöste, Phys. Rev. Letters, **55**, 3002 (1985).
[23] Ch. Rosche, S. Wolf, T. Leisner, F. Granzer, L. Wöste, *Proc. 47th IS& T Conf.*, Vol. I, (1994) p. 54.
[24] M. Haïssinsky, *La chimie nucléaire et ses applications*, (Masson, Paris, 1957); M. Haïssinsky, in *Radiation Chemistry*, edited by J. Dobo and P. Hedvig, (Akad. Kiado, Budapest 1972) Vol. II, p. 1353.
[25] J. H. Baxendale, in *Pulse Radiolysis* (Academic Press, London, 1965) p. 207.
[26] G. V. Buxton, J. Phys. Chem. Ref. Data, **7**, 513 (1988).
[27] M. O. Delcourt and J. Belloni, Radiochem. Radioanal. Lett., **13**, 329 (1973).
[28] E. J. Hart and M. Anbar, *The Hydrated Electron* (Wiley, New York 1970).
[29] J. H. Baxendale and F. Busi (editors), *The Study of Fast Processes and Transient Species by Electron Pulse Radiolysis*, NATO ASI Vol. 86 (Reidel, Dordrecht, 1981).
[30] C. D. Jonah, in *Chemical reactivity in Liquids. Fundamental Aspects*, edited by M. Moreau, P. H. Turq, C. Troyanowsky and J. Belloni, (Plenum Press, New York, 1988).
[31] J. Von Pukies, W. Roebke and A. Henglein, Ber. Bunsenges. Phys. Chem., **72**, 842 (1968).
[32] E. Janata, A. Henglein and B. G. Zrshiw, J. Phys. Chem., **98**, 10888, (1994).
[33] R. Tausch-Treml, J. Lilie and A. Henglein, Ber. Bunsenges. Phys. Chem., **82**, 1335 (1978).
[34] M. Mostafavi, J. L. Marignier, J. Amblard and J. Belloni, Z. Phys. D-Atoms, Molecules and Clusters, **12**, 31 (1989); J., Radiat. Phys. Chem., **34**, 605 (1989); J. Imag. Sc., **35**, 68 (1991).
[35] A. Henglein, Ber. Bunsenges. Phys. Chem., **81**, 556 (1977).
[36] A. Henglein and R. Tausch-Treml, J. Còll. Interf. Sci., **80**, 84 (1981); A. Henglein, Current Chem., **143**, 113 (1987).
[37] J. Amblard, O. Platzer, J. Ridard and J. Belloni, J. Phys. Chem., **92**, 2341 (1992).
[38] J. Belloni, J. Khatouri, M. Mostafavi and J. Amblard, in *Ultrafast Reaction Dynamics and Solvent Effects*, edited by Y. Gauduel and P. Rossky (AIP, 1994), p. 541.
[39] E. Moisar and F. Granzer, Photogr. Sci. Eng., **26**, 1 (1982); E. Moisar, in *Contribution of Clusters Physics to Material Science and Technology*, edited by J. Davenas and P. M. Rabette, NATO ASI Series E, Applied Sciences No. 104, (Nijhoff, 1986) p. 311.
[40] R. C. Baetzold, in *Contribution of Clusters Physics to Material Science and Technology*, edited by J. Davenas and P. M. Rabette, NATO ASI Series E, Applied Sciences No. 104, (Nijhoff, 1986), p. 195.
[41] J. F. Hamilton, in *Growth and Properties of Metal Clusters*, edited by J. Bourdon, (Elsevier, Amsterdam 1980) p. 289; J. F. Hamilton Adv. Physics, **37**, 359 (1988).
[42] J. Malinowski, in *Growth and Properties of Metal Clusters*, edited by J. Bourdon, (Elsevier, Amsterdam, 1980) p. 303.
[43] F. Trautweiler, Photogr. Sc. Eng., **12**, 138 (1968).
[44] J. Belloni, *Proc. IS& T' 48th Annual Conférence, Washington*, (1995) p. 315.
[45] G. Lippmann, Compt. Rendus. Ac. Sc., **112**, 274 (1891); J. M. Fournier, J. Optics, **22**, 259 (1991).

7 Immobilized Photosensitizers and Photocatalysis

MICHEL JULLIARD
Faculté des Sciences St-Jérôme Marseille, France

1	Introduction	222
2	Immobilized photosensitizers: types, syntheses and behaviour	223
	2.1 Silica and alumina adsorbed photosensitizers	223
	2.1.1 Organic medium	223
	2.1.2 Aqueous medium	224
	2.2 Ion-exchange resin complexed photosensitizers	224
	2.3 Glass beads supported photosensitizers	226
	2.4 Photosensitizing polymers	226
	2.4.1 Energy transfer photosensitizers	226
	2.4.2 Electron transfer photosensitizers	229
	2.4.3 Multifunctional polymeric photosensitizers	230
	2.5 Photosensitizers incorporated into polymeric films	231
	2.6 Photosensitizers incorporated into membranes	232
	2.7 Photosensitizers incorporated into zeolites	233
	2.8 Polymer bound photosensitizers	233
	2.8.1 Rose Bengal	234
	2.8.2 Benzophenones	236
	2.8.3 Electron transfer photosensitizers	237
	2.8.4 Porphyrines	237
	2.9 Silica-bound photosensitizers	239
	2.9.1 Rose Bengal	239
	2.9.2 Electron-transfer photosensitizers	239
3	Advantages and drawbacks of immobilized photosensitizers	245
	3.1 Silica-adsorbed photosensitizers	245
	3.2 Ion-exchange resins supported photosensitizers	245
	3.2.1 The two phases model	245
	3.2.2 Stereoselective system	246
	3.2.3 Light absorption in a heterogeneous system	246
	3.3 Glass beads supported photosensitizers	247

Homogeneous Photocatalysis. Edited by M. Chanon
© 1997 John Wiley & Sons Ltd

         3.4   Polymeric Photosensitizers 247
               3.4.1   Energy migration along the chain 247
               3.4.2   Photosensitizing efficiency 248
               3.4.3   Electron-transfer photosensitizers 249
               3.4.4   Multifunctional polymeric photosensitizers 249
               3.4.5   Degradation of polymeric photosensitizers 250
               3.4.6   Photophysics of polymeric photosensitizers 251
         3.5   Photosensitizers incorporated into polymeric films 251
         3.6   Photosensitizers incorporated into membranes 252
         3.7   Photosensitizers incorporated into zeolites 252
         3.8   Polymer-bound photosensitizers 253
               3.8.1   Rose Bengal 253
               3.8.2   Photosensitizing efficiency 253
               3.8.3   Electron-transfer polymer-bound photosensitizers 254
               3.8.4   Bleaching of polymer-bound photosensitizers 255
               3.8.5   Degradation of polymer-bound photosensitizers 256
               3.8.6   Photophysics of polymer-bound photosensitizers 256
         3.9   Silica-bound photosensitizers 256
4   Concluding remarks 257
References 258

1 INTRODUCTION

Mechanistic and synthetic studies on photosensitized reactions sometimes encounter difficulties or problems. Depending upon the temperature, the concentration to be used and the nature of the chromophore or the solvent, the photosensitizer may form dimers or higher aggregates of reduced efficiency [1,2]. Problems also arise from photosensitizer and products separation which may be difficult when the photoproducts do not precipitate. In the field of photobiology, the sensitized photoxidation of enzymes is often complicated by binding of the photosensitizing dye to the protein.

These and other problems can be partially or totally overcome through the use of an appropriate heterogenized system for carrying out the photosensitized reaction. In photochemical syntheses, what better way to separate the products from the photocatalyst than by anchoring the photosensitizer to an insoluble support so that the products can be washed away? Moreover, the heterogenized system would limit possible side reactions between the sensitizer and the substrate or the products.

Immobilized photosensitizers may also be useful in mechanistic investigations: they can be employed in solvent systems in which the homogeneous photosensitizer would be insoluble.

First proposed by Kautsky [3], this idea was then continued by Merrifield [4] who used the knowledge developed in polypeptides synthesis. Kautsky's work [3] was carried out on photosensitized oxidations using systems in which both the photosensitizing dye and the substrate were linked to a solid support.

Because immobilized Rose Bengal (RB) is largely used for singlet oxygen photogeneration, we would make a short comment on this commercial name. It derives from the similarity in its color to the cosmetic dye "sincur" used as a dot in the center of the forehead of Bengali women to symbolize marriage. The original red Indian dye may be a root extract of "cinnabar" [5-6].

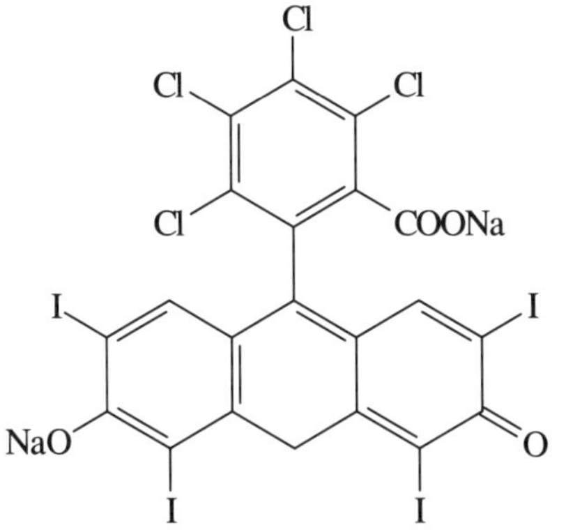

(1)

Rose Bengal

2 IMMOBILIZED PHOTOSENSITIZERS: TYPES, SYNTHESES AND BEHAVIOUR

2.1 SILICA AND ALUMINA ADSORBED PHOTOSENSITIZERS

2.1.1 Organic Medium

Photosensitizers can be immobilized by adsorption on a solid inorganic support. This technique is generally applied with silica gel as the support [7-8]. As an example, the adsorption of Ir(bipy)$_3$OH^{2+} on the surface of silica gel affords a heterogeneous photosensitizer usable for the endoergic valence isomerization of norbornadiene to quadricyclene. Thus ionic transition metal photosensitizers which might be incompatible with organic systems, could function when immobilized on a support [9].

Rose Bengal photosensitizer is described as carrying out singlet oxygen oxidations when deposited on the surface of silica gel with a degree of coating representing less than a monomolecular layer of dye. The $^1O_2$ generation was demonstrated by oxygenating tetramethyl ethylene, 1,3-diphenylisobenzofuran and tryptophan. The efficiency is of the same order as that obtained with the homogeneous sensitizer [7]. In the case of methylene blue, a uniform, durable and resistant coating can be obtained by neutralizing a slurry of silica gel and $AlCl_3$ to a weakly basic pH [7].

$$\text{Methylene blue} \quad (2)$$

Other solid supports such as alumina, celite or powered cellulose were also explored with RB. All are much less effective than silica gel and this is probably due to their much smaller surface area [10].

A special design is to immobilize a $^1O_2$ photosensitizer on silica gel particles which are attached on one face of a glass plate. The plate is irradiated sensitizer-side down, above the surface of the reaction medium leaving a small air space between the photosensitizer and the solution [11].

2.1.2 Aqueous Medium

Some attempts were made to find supports suitable for use in aqueous solutions. Silica gel particles cause problems by their swelling and by disintegration. By contrast, alumina does not swell and can be coated with pyrene. Its efficiency as oxidizing sensitizer was tested for oxidation of tryptophan. Owing to the smaller surface area of alumina the rate of the reaction was low (1–3% conversion for a 2×10^{-5} M solution in 2 h) compared to the rate of the reaction photosensitized by RB adsorbed on silica gel [10].

$$\text{Pyrene} \quad (3)$$

The adsorption method is simple, convenient, but usually leads to weak binding so that the photosensitizer is easily eluted by polar solvents. Moreover, in any part of the support where the percentage of adsorption is high, stacking and aggregation of the sensitizing chromophore may occur.

2.2 ION-EXCHANGE RESIN COMPLEXED PHOTOSENSITIZERS

When photosensitizing molecules are charged, they can form strong complexes with appropriate ion exchange resins [12]. The anionic photosensitizers RB and eosin can be attached to the basic anion exchange resins Amberlite IRA-400 [13–14] or Amberlite XAD 7 [15].

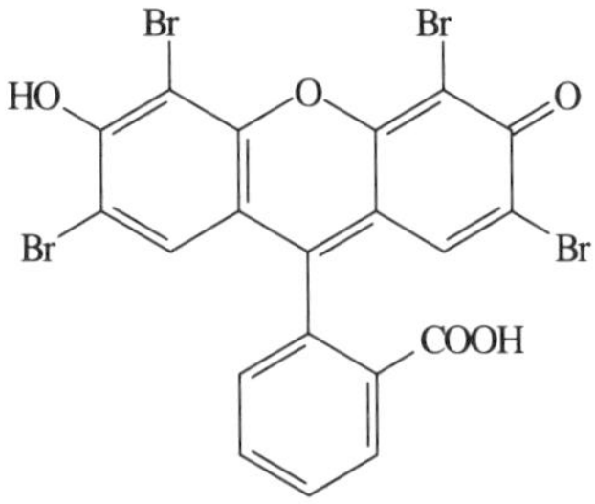

$$\text{(4)}$$

Eosin

A hydrophilic supported photosensitizer linked by ionic bond can be prepared by exchanging Rose Bengal anions with chloride anions from a suitable exchange resin as Dowex [14]. The physical properties of this ionic RB were compared with those of covalent RB synthesized by Schaap's method [16]. The quantities of RB linked to the support in both cases are roughly equal. The photosensitizing ability of the ionic bound RB is slightly higher than that of the covalent bound RB and increases with the polarity of the solvent. Moreover the ionic RB still retains its good sensitizing efficiency after repeated uses [14].

The cationic photosensitizing dye methylene blue can be complexed to acidic cation exchange resins such as Amberlite IRC 200 [13]. The generation of $^1O_2$ by this photosensitizing resin was demonstrated by oxidizing 2-methyl-2-butene. As expected if $^1O_2$ was formed, this substrate undergoes the characteristic "ene" reaction, leading to a mixture of hydroperoxides [13]. In the same way, anthracene reacts via the characteristic Diels–Alder reaction to yield the expected endoperoxide.

Similarly, Ru(bpy)3^{2+} (bpy = 2,2'-bipyridine) binds to the Dowex 50 W-X1 cation-exchange resin [17]. It then exhibits a limiting quantum yield of 0.90 for the $^1O_2$ oxidation of tetramethylethylene in CH_3OH [17].

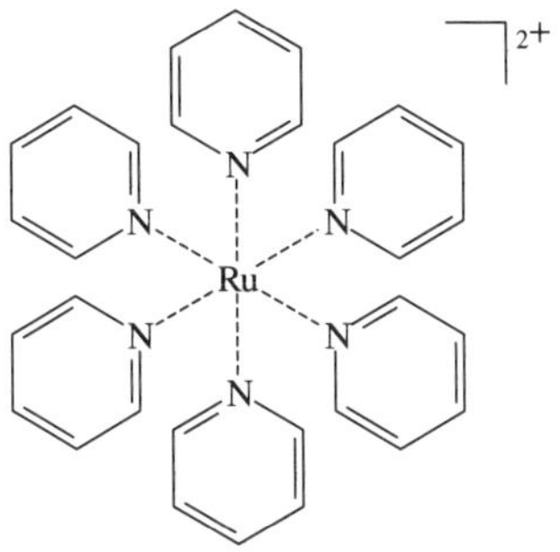

$$\text{(5)}$$

Ru(bpy)3^{2+}

An ion exchange adduct of optically active Ru(bpy)3^{2+} and montmorillonite has also been used to produce optically active sulfoxides from phenyl alkyl sulfides [18].

2.3 GLASS BEADS SUPPORTED PHOTOSENSITIZERS

Glass beads grafted photosensitizers are usable in a variety of solvents, in acidic or basic aqueous media. They can be stored and reused [19]. As an example, methylene blue can be immobilized on activated glass beads (6) and is stable for at least a year when stored in 10% acetic acid [19].

$$\text{glass}-O-\underset{\underset{OCH_3}{|}}{\overset{\overset{OCH_3}{|}}{Si}}-CH_2-CH_2-O-CH_2-\underset{\underset{OH}{|}}{CH}-CH_2-O-\text{[methylene blue structure]}$$

(6)

The oxidation of methionine residues in polypeptides with light and methylene blue beads is effective and the selective photooxidation of the methionyl residues of lysosyme was successfully carried out without adsorption of the enzyme onto the beads [19].

2.4 PHOTOSENSITIZING POLYMERS

The potential use of polymer-bound photosensitizers in photochemical synthesis and in chemical-based solar energy systems has promoted the study and development of these polymers [20–22] (see Chapter 8).

The properties of an effective polymeric photosensitizer are:

1. Low singlet excited state energy but rather high triplet energy;
2. High intersystem crossing yield;
3. Long lifetime in the triplet state;
4. Favorable structure in which triplet energy migrates effectively [23].

Several benzoylated or naphthoylated polymers and copolymers are expected to satisfy the above properties. As a rule, they display less efficient photosensitizing ability than monomers.

2.4.1 Energy-transfer Photosensitizers

A difficulty to overcome in building UV-absorbing photosensitizers is the prevention of polymer degradation resulting from secondary reactions of the excited chromophore. The most common process is hydrogen abstraction from the polymer which leads to cross-linking and loss of the chromophore.

The degradation can be prevented by hindering abstraction. One way is to replace the hydrogen atoms with fluorine atoms. Another way is to start from a monomer as *p*-divinylbenzophenone (8) and to build highly rigid polymer structures as poly(*p*-divinylbenzophenone) (7) which tend to render hydrogen abstraction from the polymer backbone less likely [24].

In rigid polymers, the benzophenone carbonyl chromophore is more inert toward photochemical degradation than in polyvinyl or polystyrene based benzophenones. Thus, whereas benzophenone is completely inactive after 15 h irradiation, polymeric benzophenone remains active after 50 h [24].

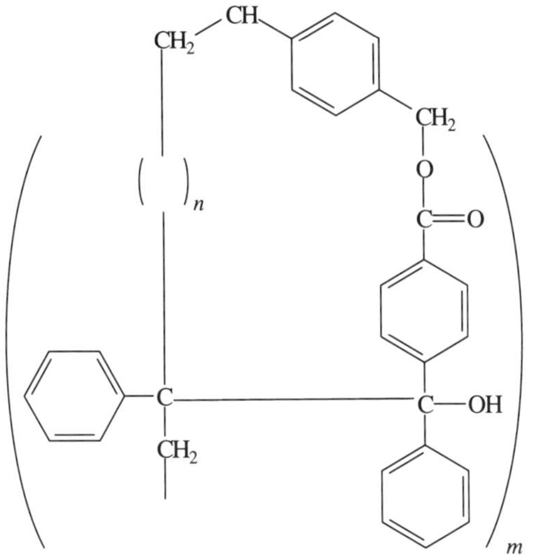

(7)

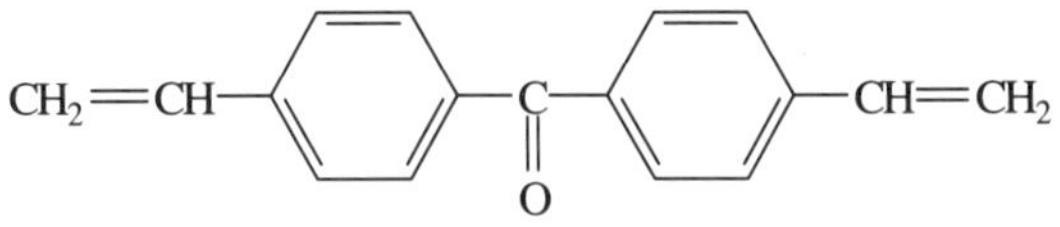

$$CH_2=CH-\underset{O}{C_6H_4-\overset{\|}{C}-C_6H_4}-CH=CH_2 \quad (8)$$

Benzoylated polystyrene-divinylbenzene can be prepared from chloromethyl poly(styrene-divinylbenzene) copolymer beads with ethyl acetate as a solvent and triethylamine as a catalyst [25].

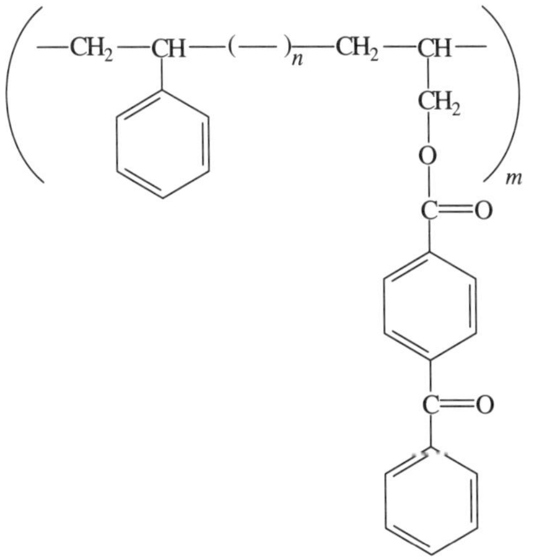

(9)

This polymeric photosensitizer was used successfully to promote the conversion of coumarin to its dimer, the cycloaddition of tetrachloroethylene to cyclopentadiene and the dimerization of indene [26]. This same polymeric benzophenone can be also prepared by a Friedel and Crafts' process [27]. It was tested as photosensitizer for the 2 + 2 photocycloaddition of cyclohexene and maleic anhydride [28].

Copolymers of poly(divinylbenzophenone) and methyl methacrylate, acrylonitrile or styrene-divinylbenzene are effective photosensitizers. They were compared with free benzophenone as energy donors in two sensitized photochemical reactions [24]: the photocycloaddition of benzo (b) thiophene and dichloroethylene and the photosensitized decomposition of benzoyl peroxide. The efficiency of all these polymers as photosensitizer is much lower than free benzophenone. By contrast, the rate constant for hydrogen abstraction from tetrahydrofuran is roughly twice as large for poly(4-vinylbenzophenone) than for benzophenone itself [29].

Polystyrene can be naphthoylated with α- and β-naphthoyl chloride. Approximately two-thirds of the polystyrene phenyl rings are naphthoylated [30].

(10)

The resultant polymers (10) were used to sensitize the photoisomerization of cis- and trans-stilbene. They show the same quantum efficiencies as their corresponding model compounds except for the isomerization of trans-stilbene photosensitized by the α-naphthoylated polymer [30].

Polymers having photosensitive nitroaryl groups were synthesized by the nitration of polyacenaphthylene or by the ionic polymerization of 5-nitroacenaphthylene. The photoisomerization of trans-cinnamic acid in the presence of these photosensitizers shows that the partially nitrated polyacenaphthylene displays a better photosensitizing activity than the polymer having 100% photosensitizer unit [31].

Several models of monomers, homopolymers and copolymers containing the photosensitizing group 4-nitro-1-naphthyl-carbamoyl (11) were tested as triplet energy donors with trans stilbene as a quencher. Their efficiency is closer to that of benzophenone and higher than that of the acetophenone [32].

(11)

Water-soluble copolymers of poly(sodium styrenesulfonate-styrene-vinylbenzyl chloride) and poly(sodium styrenesulfonate-2-vinylnaphthalene-vinylbenzylchloride) containing various amounts of RB chromophore attached to the polymer chain have been synthesized. They can solubilize large hydrophobic compounds and are efficient generators of $^1O_2$ toward hydrophilic oxygen acceptors solvated in the hydrophobic polymeric microdomains [33].

Finally, polymeric benzaldehyde can be formed photochemically [34] and used as a triplet energy donor toward classical quenchers such as 1–3 pentadiene or 1–3 cyclohexadiene [35].

2.4.2 Electron Transfer Photosensitizers

Despite the increasing importance of this class of reactions, few photosensitizing polymers working through electron transfer have been reported. Photoexcited aromatic esters are good electron acceptors. Then the simplest choice is to use an aromatic polyester. Some experiments involving polyethylene terephthalate were performed: the observed cyclodimerization of phenyl vinyl ether is the proof of the photochemical generation of radical ions [36]. Other examples can be found with the phenyl-vinyl ketone-2-vinylnaphthalene copolymer which can also act as an electron-transfer photosensitizer [23].

As the monomer, the polymeric 2,4,6-triphenylpyrylium salts (12) are electron-transfer photosensitizers in both the Diels–Alder dimerization of 1,3-cyclohexadiene and the dimerization of phenyl vinyl ether. They behave as efficient electron acceptors and secondary processes are avoided if one uses only the long-wavelength excitation part of the pyrylium salts absorption [37].

2.4.3 Multifunctional Polymeric Photosensitizers

Comparing the complementarity of functional groups and the effect of reaction media in enzymes which are typical polymeric catalysts, several authors synthesized multifunctional polymeric photosensitizers containing a photosensitizing chromophore and a substrate-attracting group.

Such polymeric photosensitizers containing both pendant benzophenone [38] or nitroaryl group [39] as a photosensitizing moiety and pendant carboxylate anion, quaternary ammonium 13b or phosphonium 13a salts [38,40,41] as a substrate-attracting group were synthesized by radical copolymerization of p-(4-nitrophenoxymethyl) styrene or p-(4-nitro-1-naphthoxymethyl) styrene with methacrylic acid followed by neutralization of the resulting copolymers with various bases.

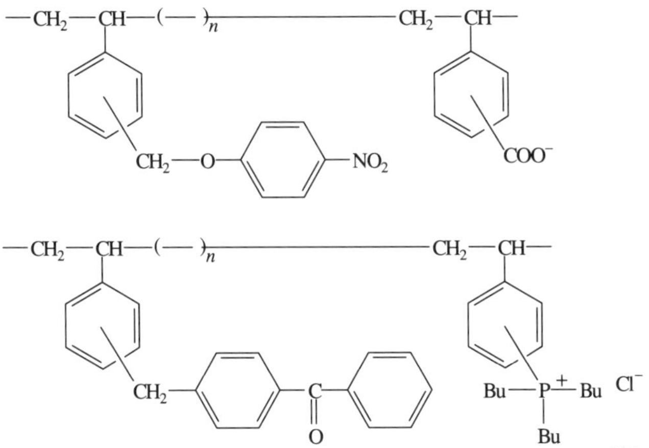

[Structural formulas of polymer photosensitizers]

(**13** *continued*)

These photosensitizers are soluble in methanol or water and were tested on the photoisomerization of (2-cinnamoyloxyethyl)trimethylammonium bromide [39].

Another way is the partial substitution reaction of poly(*p*-chloromethyl)styrene or insoluble chloromethylated polystyrene bead with potassium 4-benzoylphenoxide, 4-nitro-1-naphthoxide or 4-nitrophenoxide followed by reaction of the remaining chloromethyl groups in the polymers with tertiary phosphines or amines [40,41].

2.5 PHOTOSENSITIZERS INCORPORATED INTO POLYMERIC FILMS

Twenty-six commercially available dyes have been immobilized in cellulose acetate films [42]. All were prepared by the solvent-casting technique: the polymer is dissolved in a solvent mixture and cast onto a glass plate with a device that can be set to lay the solution at any desired thickness. The relative abilities of the dyes to sensitize the $^1O_2$ photooxidation of dimethylanthracene were tested. Rose Bengal, Safranine O and Ru(bpy)$_3^{2+}$ were found to be the most efficient [42].

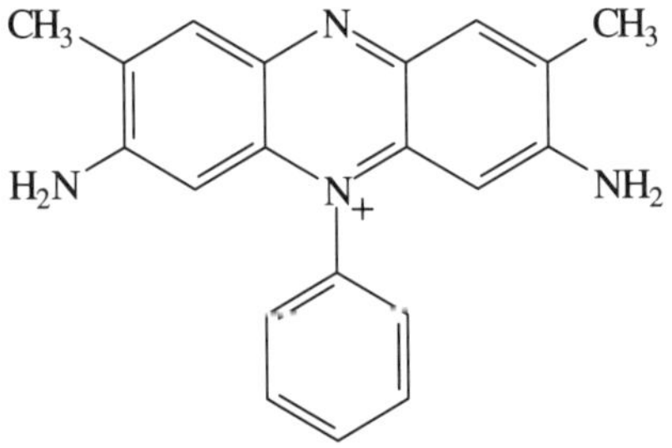

(**14**)

Safranine O

Rose Bengal, Safranine O and Ru(bpy)$_3^{2+}$ tetrafluoroborate can be also immobilized in poly(vinyl chloride) PVC, poly(butyl methacrylate) and polystyrene films [43]. PVC was the only material investigated that could be used to make satisfactory films containing all sensitizers. Leaching of the dyes from PVC films is negligible except in the case of long reaction times. Di-n-butylsulfide is photooxidized to di-n-butyl sulfoxide [43]. The better quantum yield is observed with RB.

2.6 PHOTOSENSITIZERS INCORPORATED INTO MEMBRANES

Commercially available cellulose acetate membranes can be modified to incorporate zinc tetraphenylporphyrin (ZnTPP) (15). The procedure involves swelling the fibers in tetrahydrofuran solution saturated with ZnTPP followed by an aqueous rinse which contracts the fibers and entraps the ZnTPP. The membrane contains up to 3% ZnTPP by weight [44]. The membrane remains sufficiently stable to sustain a flow system. It is impermeable to ionic solutes [44]. It exhibits fluorescence with a short lifetime <10 ns and a spectrum consistent with that of ZnTPP. The membrane is effective as an heterogeneous photosensitizer for the $^1O_2$ generation attested by the oxygenation of aqueous 2-furylmethanol.

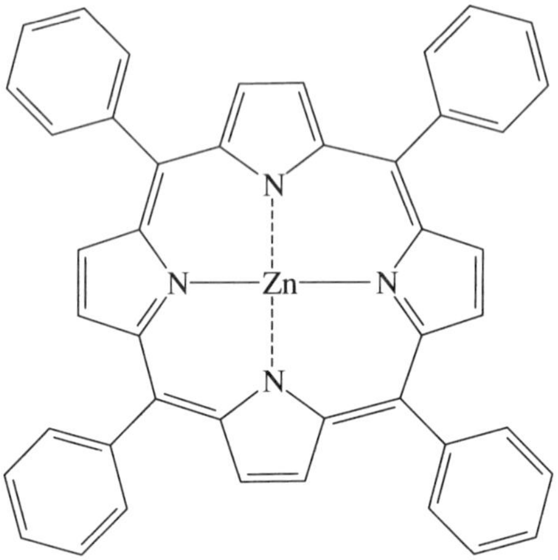

(15)

Zinc tetraphenylporphyrin

The ultimate goal of the ZnTPP modified fibers is the accomplishment of photosensitized redox reactions in solution. A variety of acceptors and donors have been tested. The donors are EDTA, triethanolamine or tertiary amines. The acceptors are viologens. The irradiations with microheterogeneous ZnTPP indicate a lower efficiency caused by reduced access of the solution-phase donors and acceptors to the sensitizer or to shortening of the excited state lifetime of the photosensitizer [44].

2.7 PHOTOSENSITIZERS INCORPORATED INTO ZEOLITES

Ru(bpy)$_3^{2+}$ can be incorporated into zeolite X by a simple impregnation [45]. The same heterogeneous photosensitizer can be prepared in an alternative way by inclusion within the cavity of zeolite Y [46]. This is obtained by allowing 2,2'bipyridine to react with Ru(NH3)$_6^{3+}$ in the cavity of zeolite Y. The complex stays intact and can undergo one-electron transfer with N,N,N',N'-tetramethyl-p-phenylenediamine, 10-phenylphenothiazine and tetrabromo-o-benzoquinone. The exchanged zeolite is able to sensitize $^1O_2$ formation. For the photooxygenation of 1-methyl-1-cyclohexene, the product distribution is similar to the one observed with other $^1O_2$ photosensitizers [46]. This result shows that $^1O_2$ can migrate freely from the cavity and reacts in normal fashion with its target. Ru(bpy)$_3^{2+}$, trapped in the dehydrated zeolite Y supercage, can photoreduce dimethylviologen as evidenced by the formation of the MV$^+$ radical cation [47]; 2,4,6-triphenylpyrylium cation (16) has also been encapsulated into zeolite Y. It shows moderate activity as an electron-transfer photosensitizer, but is able to promote the isomerization of *cis*-stilbene to *trans*-stilbene via the corresponding radical cation [48].

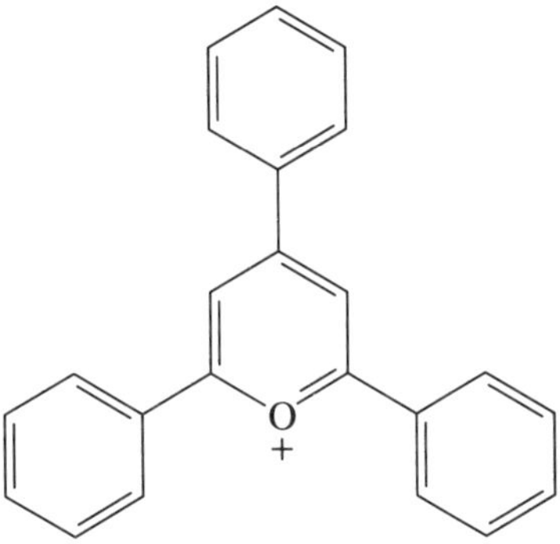

(16)

2,4,6-triphenylpyrylium cation

2.8 POLYMER-BOUND PHOTOSENSITIZERS

Several preparations of polymer-bound sensitizers are carried out using the experimental technique first developed by Merrifield [4,49] in the field of polypeptide synthesis.

In this procedure, chloromethylated styrene-divinylbenzene copolymer beads are converted to carboxylate esters in a nucleophilic displacement reaction. The result is a reagent which is grafted on a polymer matrix which is compatible with most organic solvents but is insoluble in them.

$$P + CH_3-O-CH_2-Cl \xrightarrow[CHCl_3]{SnCl_4} P-CH_2-Cl + CH_3OH$$

$$P-CH_2-Cl + R-COONa \longrightarrow P-CH_2O-CO-R + CH_3OH$$

(17)

2.8.1 Rose Bengal

Rose Bengal contains both carboxylate and phenoxide functional groups each of them being capable of reaction with a chloromethyl side chain of the polymer. One in five chloromethyl residues on the surface of the polymer bead can be converted to RB derivative [6,50–51]. In fact, the reaction of RB with chloromethylated poly(styrene-divinylbenzene) beads involves nucleophilic displacement of chloride by one or both of the anionic centers C-2 or C-6 of the dye. Lamberts and Neckers [52–53] established that the reaction of the chloromethylated Merrifield polymer with RB produces a covalent bond to the polymer bead only by reaction at the C-2 carboxylate.

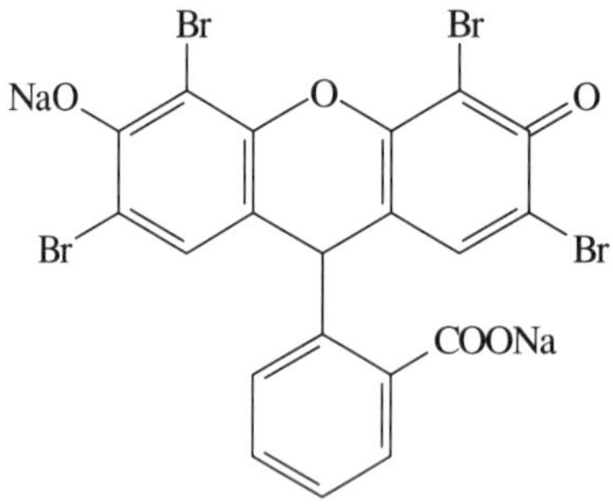

a) Eosin Y

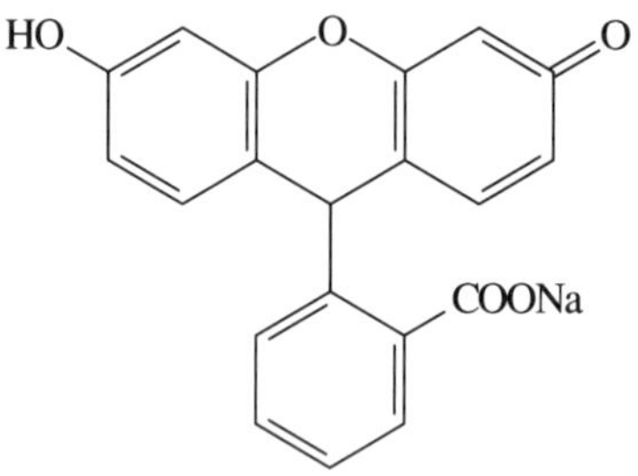

b) Fluorescein

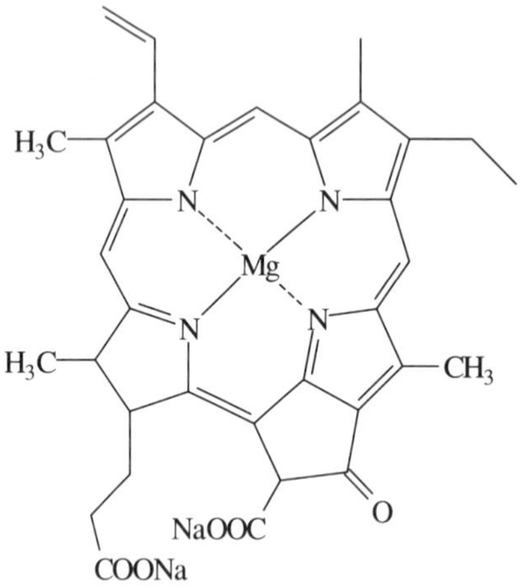

c) Chlorophillin

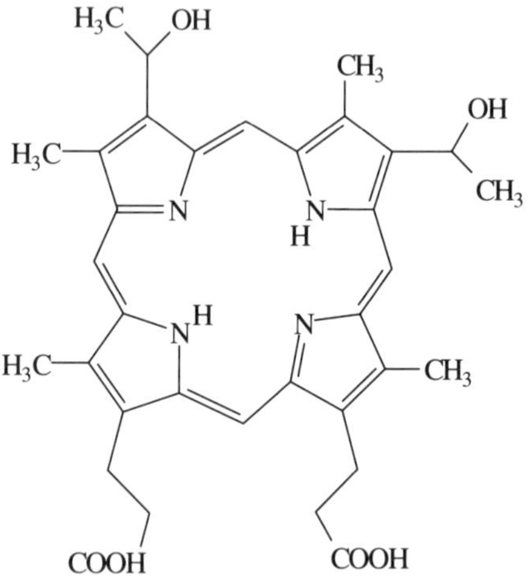

d) Hematoporphyrin (**18**)

This RB P–RB bonded to poly(styrene-divinylbenzene) support [54] was used for the photosensitized oxidation of 2,3 diphenyl-1,4 dioxene to peroxides in organic solvents [54]. Since the conversion of this substrate is a $^1O_2$ mediated reaction, the experiment indicates that $^1O_2$ is efficiently formed by energy transfer from P–RB to oxygen. Now marketed as Sensitox, P–RB has been used for a number of other photosensitized reactions [55]. Its advantages are: (1) its stability compared to RB in solution; (2) the polystyrene renders the RB, an ionic dye, compatible with non-polar solvents. If copolymers of poly(sodium styrene sulfonate-styrene-vinylbenzylchloride) are used as a support, RB plays the same role as an antenna into a polyelectrolyte [56].

Other sensitizers that have been attached to this polymer bead include eosin Y (18a), fluorescein (18b), chlorophyllin (18c) and hematoporphyrin (18d) [51]. The results from photooxidation of 2,3 diphenyl-1,4-dioxene with these photosensitizers are positive.

While considerably less efficient than the P–RB, the P-chlorophyllin and P-hematoporphyrin photosensitizers could be particularly valuable in investigations of biological oxidations [51]. All these immobilized photosensitizers can be used for photooxidations in water.

Another way leading to a water-soluble polymer-bound RB, is to functionalize the poly(ethyleneglycol) monomethyl ether with this dye. This polyglycol-RB (19) is soluble in solvents from toluene to water. Essentially it is a RB esterified with a low molecular weight polyglycol through the C-2 carboxylate function [57].

(19)

The quantum yield value Φ observed for $^1O_2$ formation from poly(ethyleneglycol) bound RB ($\Phi = 0.76$) indicates that in polar and protic solvents Φ is comparable with that obtained from RB in methanol. In nonpolar solvents Φ varies from 0.30 in toluene to 0.81 in CH_2Br_2 [57].

Another hydrophilic polymeric-bound RB HP–RB can be prepared by copolymerization of chloromethylstyrene or monomethacrylate ester of ethylene glycol with the bis-methacrylate ester of ethylene glycol as cross-linking agent 20 [58]. This heterogeneous sensitizer HP–RB is wetted by water and swells by 40% in this solvent [58]. The effectiveness of this HP–RB in sensitizing $^1O_2$ formation is illustrated by the successful photooxygenation of such substrates as tetramethylethylene known to react with $^1O_2$ [58].

$$\text{Ar}(CH=CH_2)(CH_2Cl) \quad CH_2=C(CH_3)COOCH_2CH_2OH \quad [CH_2=C(CH_3)COOCH_2]_2$$

(20)

As a hydrophilic specific photosensitizer for oxidation of histidine and aromatic amino acid residues in proteins, RB was also coupled to an aminohexyl-glycophase [59] and used to oxidize histidine. Sepharose shows advantages in porosity and hydrophilic nature: this involve a better distribution of the hydrophilic amino acid in the neighborhood of the immobilized RB. In addition, the high transparency of the sepharose suspension permits a good access of the available light to the chromophore.

2.8.2 Benzophenones

Ortho- and para(benzyloxy)benzaldehyde have been grafted on linear chloromethylated polystyrene [60]. These photosensitizers can be used to transfer energy to norbornadiene which isomerizes to quadricyclene [60].

Benzophenone can be linked to polystyrene without change of its quantum efficiency [61]. It was also bound to a polypeptide leading to poly-γ-p-benzoylbenzyl-L-glutamate (21). Two types were synthesized with an average molecular weight of 2600 and 11 000 [62].

$$\left(Ph-C(=O)-C_6H_4-CH_2-O-C(=O)-CH_2-CH_2-CH(COO-)(NH-) \right)_n \quad (21)$$

These polymers photosensitize the isomerization of *trans*-stilbene in CH_2Cl_2. The quantum yield (Φ) of the *trans*- to *cis*-photoisomerization is 0.55. When 4-methylbenzophenone is the sensitizer the Φ value is 0.52–0.56. Searle et al. using polyvinyl benzophenone as a sensitizer obtained the same quantum yield [61].

If the polymer supporting the photosensitizer is a polypeptide which has ketone chromophores in its side chain, one can wonder whether or not any side reaction such as H abstraction occurs between the excited carbonyl and other parts of the polypeptide. The results observed tend to prove that the chromophores are brought into the necessary arrangement in the side chains of the polypeptide. It makes it impossible for the carbonyl of the benzophenone to interact with the hydrogens of the peptide bonds, methylene groups and asymmetric α carbons [62].

2.8.3 Electron-transfer Photosensitizers

Anthracene or pyrene bonded to polyionene or to copolymers of 3-(1-pyrenyl) propyl methacrylate (22) behave as polymer-bound electron-transfer photosensitizers [63]. Their photoredox ability to reduce dimethyl viologen and crystal violet was shown spectroscopically.

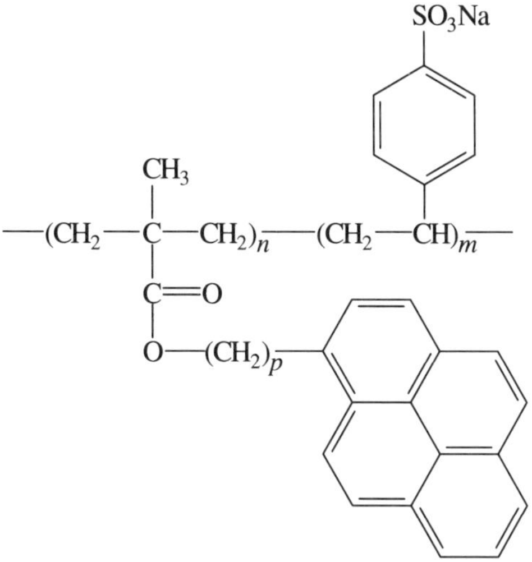

(22)

Here α, ω-dianthryl terminated poly(N-tertbutyliminoethylene) (23) p-TBIE has been synthesized by end capping of bifunctionally living p-(TBIE) using 9-amino(methyl)anthracene as a capping agent. It can be used as a polymeric photosensitizer for the photochemical conversion of tachysterol compounds into provitamin D2 and D3 [64].

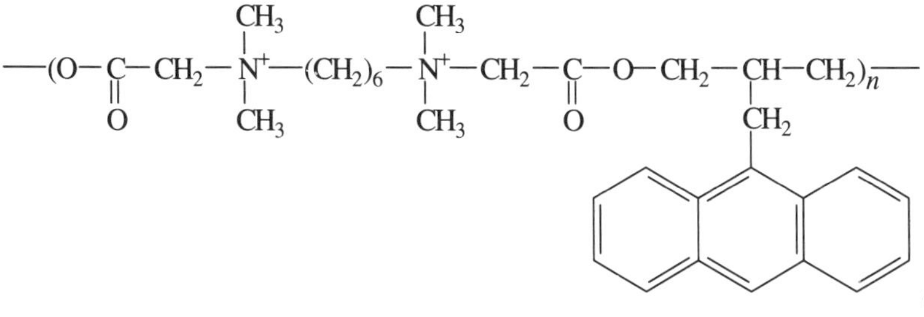

(23)

2.8.4 Porphyrins

The use of hematoporphyrin as photosensitizer is limited by the insolubility in neutral aqueous solution and in organic solvents. This can be overcome because hematoporphyrin having two carboxylic groups can be coupled with α-(3-aminopropyl-ω-methoxypoly(oxyethylene) through the acid-amide bond formed with carbodiimide (24) [65].

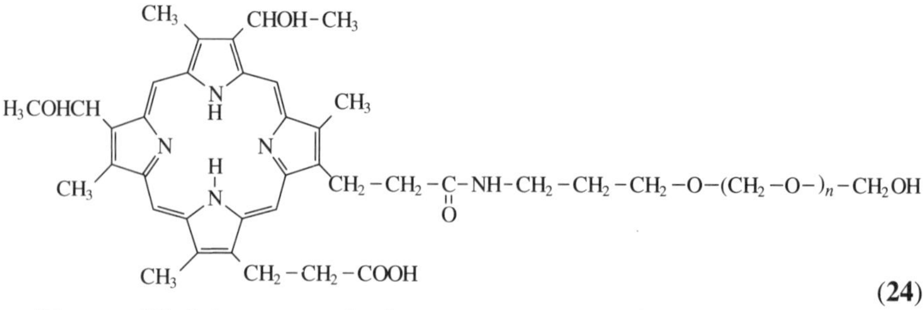

(24)

The modified hematoporphyrin acts as a true photooxidative sensitizer: imidazole and indole are oxidized in organic solvent. The same behavior is observed for uric acid in neutral aqueous solution [65].

Metalloporphyrins in enzymatic systems are embedded in a polymeric structure. Thus synthetic polymer-bound metalloporphyrins with a modified environment due to the polymer chain should be better model systems than the simple low molecular weight compounds.

Such are substituted derivatives of tetraphenylporphyrine bound to different polymers: poly(styrene-co-chloromethylstyrene), poly(vinylbenzyltriethyl ammonium chloride), poly(methacrylic acid and poly(N-vinylpyrrolidone-co-methacrylic acid) [66].

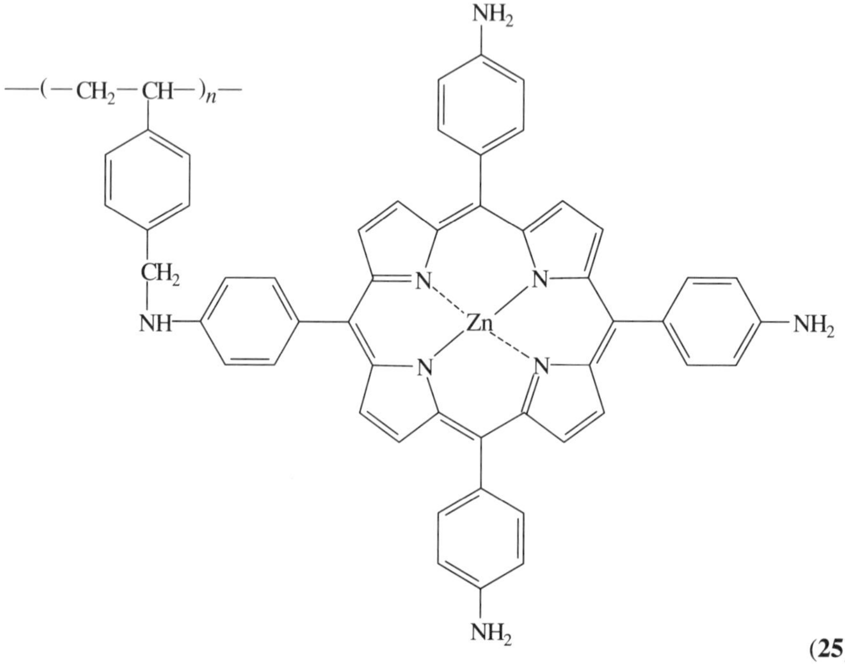

(25)

Similarly, some mesoporphyrin diesters have been grafted to insoluble chloromethylated polystyrene and α-(4-bromomethyl-3-nitrobenzamido)benzylpoly(styrene-co-divinylbenzene) [67]. Polystyrenes more heavily loaded in porphyrin behave as poorer sensitizers as already observed with soluble polystyrene bound RB systems of relatively high dye loading. Moreover the nitro group probably acts as a good electron acceptor deactivating the dye triplets through an electron-transfer process. For $^1O_2$ generation, an important decrease is observed on going from the soluble diesters to the insoluble polymer-bound species except when these are very lightly loaded in dye [67].

2.9 SILICA-BOUND PHOTOSENSITIZERS

2.9.1 Rose Bengal

Silica gel has attracted attention as a low-cost support for anchoring of metal catalysts. Accordingly, silica-bound RB P–Si–RB was developed [68]. It was prepared by the following reactions:

$$Cl_3Si(CH_2)_3Cl + EtOH \xrightarrow[C_6H_6]{Et_3N} (EtO)_3Si(CH_2)_3Cl$$

$$[P-Si-OH] + (EtO)_3Si(CH_2)_3Cl \xrightarrow[150\,°C]{xylene} [P-Si-O]_3-Si-(CH_2)_3Cl \quad (26)$$

$$[P-Si-O]_3-Si-(CH_2)_3Cl + RB \xrightarrow[135\,°C]{DMF} [P-Si-O]_3-Si-(CH_2)_3-RB$$

P–Si–RB was tested as a source of $^1O_2$ in some diagnostic reactions. In general, it was observed that it is as effective as RB supported on styrene–divinylbenzene copolymer beads. Moreover, P–Si–RB can function in aqueous solution as effectively as does RB itself. It must be underlined that P–Si–RB compared with RB in polar solvents is described as not being able to form superoxide ion. Another advantage is its versatility: it can be used in both nonpolar and polar solvents [68].

2.9.2 Electron-transfer Photosensitizers

There are two general ways to prepare silica with a chemically bonded compound other than RB.

1. Starting silica is previously modified with a sufficiently reactive group which is utilized in the next step of the synthesis. Amino group is often used for this purpose [69–75]. Aminopropyl silica is commercially available as a starting material.
2. In a second method, a silane is first prepared and then allowed to react with the silica surface yielding a chemically bonded functional group [76–82].

Aminopropyl silica as a starting material

The link between the sensitizer and the support is created by substitution of the amino group acting as a nucleophile on the carbon atom of the photosensitizer bearing a suitable nucleofugic group [83]. The photosensitizer is then bonded to the support by a short chain, namely $-(CH_2)_3NH-$.

$$\text{Si}-(CH_2)_3-NH_2 + \text{Photosens-X} \longrightarrow \text{Si}-(CH_2)_3-NH-\text{Photosens} \quad (27)$$

Some electron acceptors and electron donors photosensitizers have been prepared by this method [69,84].

TRNF-Si, **TTNF-Si**, **F5B-Si**, **CNFB-Si**, **CFB-Si**, **CNB-Si**, **FB-Si**, **TRNB-Si**, **TRNBS-Si**, **TRNBC-Si** (28)

IMMOBILIZED PHOTOSENSITIZERS AND PHOTOCATALYSIS 241

MABC-Si, **AOB-Si**, **AOBS-Si**, **AOC-Si**, **M3B-Si**, **M3B3-Si**, **MANS-Si**, **MA2B3-Si**

(**28** continued)

Bulky reagents are unable to react with the less accessible sites on the surface of the support [85], thus residual hydroxyl groups probably remain on the silica surface after the grafting reaction has been carried out. Reactions may also be performed starting from a sulfonyl chloride [86–89] (case of MANS–Si), sulfonic acids (case of TRNBS-Si, AOBS-Si) or aromatic acids (case of TRNBC-Si, AOC-Si, MABC-Si).

Silica as a starting material

Unmodified silica bears OH groups or silanols. Electron-acceptor or electron-donor molecules may be condensed on these OH only if they have previously been transformed into mono-, di- or trihalosilanes or mono-, di- or trialkoxysilanes. Monochlorosilanes usually give monomeric structures [90].

$$\text{Photosens}-CH_2-CH=CH_2 + H-Si(CH_3)_2-Cl \xrightarrow[\text{anti-Markovnikov}]{K_2PtCl_4} \text{Photosens}-(CH_2)_3-Si(CH_3)_2-Cl \quad (29)$$

$$\text{Photosens}-(CH_2)_3-Si(CH_3)_2-Cl + \underline{Si}-OH \xrightarrow{\text{silanization}} \text{Photosens}-(CH_2)_3-Si(CH_3)_2-O-\underline{Si} + HCl$$

The trinitrobenzene TRNB-Si and the 2,4,7-trinitrofluorenone TRNF-Si silica supported photosensitizers have been synthesized by this second method. The chlorosilane is itself obtained in three steps [69,84,91].

$$(29)$$

(30)

IMMOBILIZED PHOTOSENSITIZERS AND PHOTOCATALYSIS

[Scheme showing reaction of 2,4-dinitrophenyl ether silane with Si–OH under anhydrous p-xylene, 48 h reflux, yielding the silica-grafted dinitroaryl product + HCl] **(30** *continued*)

The first method involving aminopropylsilica is more convenient. It does not require special synthetic skills, the supported photosensitizer can be obtained in one step after 24 h. The second method is more time consuming and involves three or four steps.

Various applications of redox photosensitization have been reviewed [92–94]. To check the photosensitizing ability and efficiency of the silica grafted photosensitizers, the activation of methyl aromatics and benzyl bromides toward dioxygen was selected as model reactions [84].

The immobilization of some phthalocyanines Pc by covalent binding to silica is a prospective approach to achieve heterogeneous photocatalysis. Thus the Zn complexes of the tetracarboxylic acid chloride Pc and the tetraamino Pc have been covalently bound to aminopropyl or 3-chloropropyl silica (31) [95].

[Structure of silica-immobilized Zn tetracarboxy phthalocyanine] **(31)**

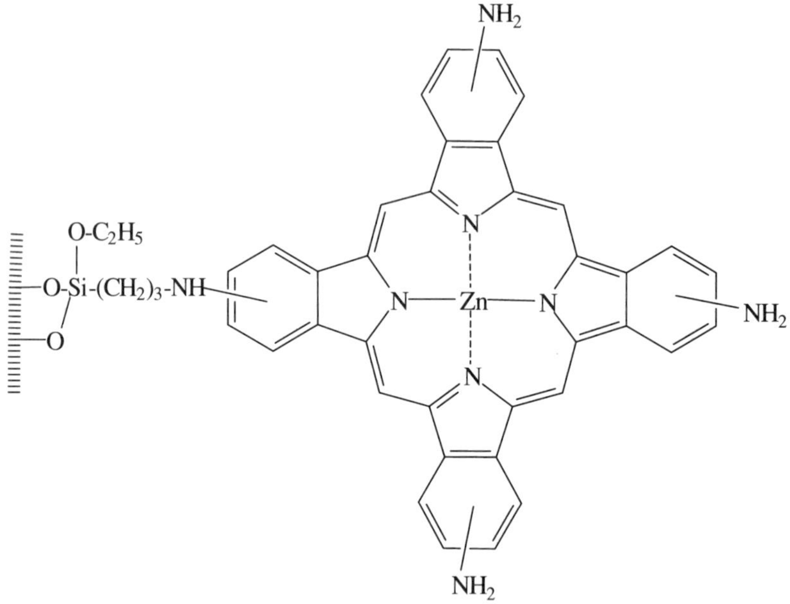

(**31** *continued*)

Finally, several features make 4-(*N*,*N*-dimethylamino)benzophenone an attractive photosensitizer for immobilization. Its high extinction coefficient permits relatively low loading. Several synthetic techniques allowed to graft it onto silica (32) [96].

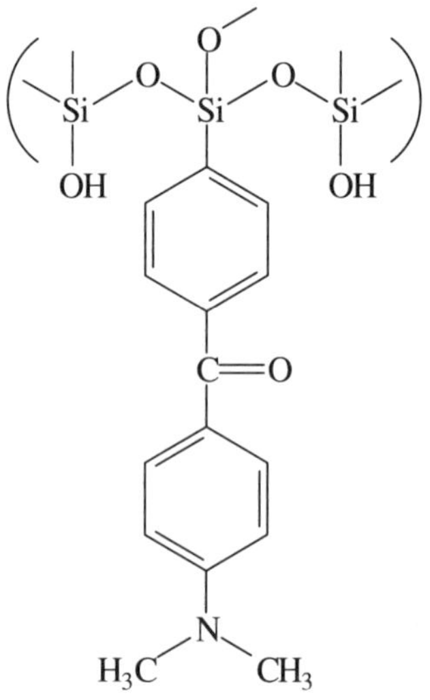

(**32**)

3 ADVANTAGES AND DRAWBACKS OF IMMOBILIZED PHOTOSENSITIZERS

3.1 SILICA-ADSORBED PHOTOSENSITIZERS

Among the benefits which can be realized from adsorption of a photosensitizer onto an insoluble silica gel matrix one example may be selected.

Ir(bipy)$_3$OH^{2+} adsorption onto the surface of silica gel affords a highly efficient heterogeneous photosensitizer for the endergonic valence isomerization of norbornadiene NBD to quadricyclene Q [9].

The absorption and the luminescence spectra show that the adsorption of the photosensitizer has little effect on the energies of its low-lying electronic states. The irradiation of an unstirred slurry of immobilized photosensitizer at 366 nm in pure NBD results in the production of Q. The quantum yield Φ is 0.7–0.8, close to the 0.72 value observed with the homogeneous photosensitizer in CH$_3$CN solution. This correspondence implies that essentially all of the adsorbed iridium photosensitizer is accessible to the NBD solution phase. Moreover the iridium complex suffers no detrimental effects from being in close proximity to the highly polar silica surface in contrast to the behavior of an organic photosensitizer such as 4-(N,N-dimethylamino)benzophenone [96].

A second kind of benefit may be obtained with photosensitizers adsorbed on silica gel particles attached on a plane surface, the photosensitizer side down, above a solution containing the substrate to be photooxidized. For example in the case of RB, ^{1}O$_2$ generated by illumination of the surface can still diffuse through a short distance (1–2 mm) in air and react with a substrate in the solution. Since the photosensitizer and the substrate are physically separated, all electron and H transfer reactions are suppressed, making it possible to perform the ^{1}O$_2$ reaction alone [97]. This method could be useful in photobiology for studying the killing of cells by ^{1}O$_2$ as the sole damaging species.

3.2 ION-EXCHANGE SUPPORTED PHOTOSENSITIZERS

3.2.1 The Two Phases Model

As ^{1}O$_2$ generator, Dowex-exchanged Ru(bpy)$_3$$^{2+}$ was comparable to homogeneous RB, and the optical properties of Ru(bpy)$_3$$^{2+}$ were not greatly affected by binding to the Dowex resin [17]. Interestingly Dowex-complexed RB is also able to efficiently generate ^{1}O$_2$ in protic solvents, including water. The resin rapidly scavenges water, but 32% weight is the highest water concentration that yields clear solutions.

A two-solvent phases model may explain the behavior of this photosensitizing resin when the ratio of water increases [17]. Dowex 50W-X1 is a porous, sulfonic acid cation exchanger and Ru (bpy)$_3$$^{2+}$ is closely associated with the acid groups, yielding a dispersed sensitizer in a permeable matrix. In pure CH$_3$OH, the porous resin solvates with solvent which allows O$_2$ and substrate to diffuse throughout it. The distance before encountering is the same as in homogeneous medium.

When water is introduced, the strongly hydrophilic resin hydrates. There is a hydrated hydrophilic region around the photosensitizer and an essentially pure CH_3OH-substrate bulk solvent region well removed from the polar regions of the resin.

In this model, the water layer over the photosensitizer thickens with increased H_2O loading; $^1O_2$ formed in the water layer must diffuse across this layer to be trapped by the substrate. Thus the quantum yield depends on water layer thickness and the $^1O_2$ lifetime in this layer. It is assumed that $^1O_2$ is formed uniformly on a surface and to diffuse across a uniform-thickness barrier while undergoing a simultaneous first-order decay. The mean diffusion path of $^1O_2$ was estimated as 100–300 nm. These large values reveal a shortcoming in the model. It seems highly unlikely that a 100–300 nm water layer over the surface of the resin could be maintained. More likely than a clear H_2O-CH_3OH boundary, a smooth transition from pure H_2O to pure CH_3OH may be responsible for the 15% quantitative difference observed when comparing calculated and experimental quantum yield values [17].

3.2.2 Stereoselective System

One of the difficulties in developing a photochemical system displaying stereoselectivity often lies in the fact that the steric control by a single excited molecule cannot be large enough to direct a reaction in a stereoselective way [98–99]. Nevertheless, optically active sulfoxide can be produced by photosensitizing a sulfide on the surface of a clay optically active chelate as $Ru(bpy)_3^{2+}$ [18], whereas stereoselectivity is not induced when the reaction is carried out with homogeneous $Ru(bpy)_3^{2+}$. The observed enantiomeric excess provides the possibility that stereoselectivity in a photochemical process can be enhanced when a reactant or a catalyst is adsorbed on a solid surface. One of the reasons for such an enhancement is that the reaction at the surface may proceed under the cooperative interactions of two or three chelates. Adsorption of the pure enantiomer of a tris-chelated complex with bulky ligands such as $Ru(phen)_3^{2+}$ or $Ru(bpy)_3^{2+}$ leaves half of the surface sites unoccupied so that a sulfide molecule may be adsorbed in the empty space, photoactivated, and then attacked by an oxygen molecule [18].

3.2.3 Light Absorption in a Heterogeneous System

Even though many advantages have been pointed out in a heterogeneous system, only a few kinetic analyses which treat quantitatively the photosensitizer activity and the photosensitizer concentration have been reported. One of the reasons may be the difficulty in evaluating accurately the light absorption rate in a heterogeneous reaction system. This was done in the case of photooxygenation of 2,3-dimethyl-2-butene with RB and eosin supported on Amberlite beads [100]. The light absorption rate was assessed by a Monte Carlo simulation. The reaction rate was found proportional to the nth order of light absorption rate where n is a little less than unity and to a 0th order of butene concentration in a feed range of $5 \times 10^1 - 2 \times 10^2$ mol/m^3. An higher level dye loading acts to lower the photooxygenation rate owing to a probable dye–dye interaction [100].

3.3 GLASS BEADS SUPPORTED PHOTOSENSITIZERS

Dye-sensitized photooxidations are used to modify selectively amino acid residues in polypeptides, leading to a clarification of the roles played by those amino acids in protein function. One application of this method is the photooxidation of methionine using free methylene blue as the sensitizer [101]. In this case, the extent of oxidation of methionine is complicated by the presence of the free dye in solution. This problem is overcome by immobilizing methylene blue on glass beads. The selective photooxidation of the methionyl residues of lysozyme which has been shown to be highly sensitive to the pH and to the solvent used is successfully carried out using immobilized methylene blue without adsorption of the enzyme onto the beads. Moreover, glass beads are stable in variety of solvents, in acidic or basic media and they may be stored and reused [19].

Nevertheless a common problem in using glass beads to immobilize reactive materials is the nonspecific adsorption of compounds onto the bead. A suitable diazotization of the beads prevents nonspecific adsorption while achieving stable coupling between the beads and a photosensitizer such as methylene blue.

(33)

3.4 POLYMERIC PHOTOSENSITIZERS

3.4.1 Energy Migration Along the Chain

One of the most interesting results obtained with polymeric photosensitizers is that intramolecular energy migration along the polymer chain seems to be the rule rather than the exception in polymers containing chromophores. This migration may increase the half-lifetime of the triplet excited state of the chromophore bound to the polymer. This is interpreted by the change of the nature of the triplet electronic state in the polymer from the $n - \pi^*$ of the benzoyl group to the $\pi - \pi^*$ state of the naphthoyl one due to energy transfer. For example, phenyl vinyl ketone-2-vinylnaphthalene copolymers show high sensitizing efficiency for the photoisomerization reaction of stilbene [23]. A polymeric photosensitizer in which the excitation energy can be transferred intramolecularly would be different from a monomeric sensitizing of the same chromophore in two aspects [30]:

1. First, the rate of energy transfer to an acceptor might be greater. If intramolecular energy migration is sufficiently facile, the migration distance may begin to approximate the polymer chain length. In this case the entire macromolecule can be considered as the reactive species. The net result is an increase in the effective collisional radius for reactions with small-molecule substrates. In principle, this could lead to increased efficiencies in photosensitized reactions when using polymeric rather than small-molecule photosensitizers [102].
2. Second, the polymer might interact sterically with an acceptor and thus influence the mode of decay of the acceptor. The latter interaction could be pictured as a sort of cage effect in which the cage is the polymeric sensitizer molecule instead of several solvent molecules [30].

Attempts to demonstrate the feasibility of this approach have centered about the photosensitized isomerization of *cis*- and *trans*-stilbene, but the result is not clear. Poly(phenyl vinyl ketone) is no more efficient as a sensitizer than are small-molecule triplet energy donors such as acetophenone [22,102–104]. A similar conclusion arises from a study carried out on poly-(*p*-vinylbenzophenone) [105]. Copolymers of phenyl vinyl ketone with a small amount of 2-vinyl naphthalene are, however, better photosensitizers for stilbene isomerization than ketone itself [23].

To conclude, it seems difficult to quantify the occurrence of intramolecular energy migration along a polymer chain. Good evidence for this process is the observation in dilute solutions of a P-type delayed fluorescence resulting from triplet–triplet annihilation of poly(vinyl-naphthalene) [106].

3.4.2 Photosensitizing Efficiency

Some differences in the photosensitizing efficiency may result from the polymeric medium. This situation is sometimes advantageous when there is a competition to form normal products and unwanted ones resulting from side reactions of the excited photosensitizer or the products. For example, in the case of the photoisomerization of norbornadiene to quadricyclene, oxetane formation is much more prominent with free benzophenone than with polymeric benzophenone [107].

If the polymer is not swelled significantly in the solvent, the energy transfer process between the polymer-bound photosensitizer and the reactants is likely to be less efficient. Conversely, the steric hindrance of secondary reactions results, for some cases, in an increased efficiency of the polymer-sensitized reaction for the photooxidation of secondary alcohols [27].

Nevertheless, this polymer effect is not obvious when the sensitization step proceeds at a rate which is considerably below the diffusion-controlled limit [102]. Thus, for the ketyl radical formation from cumene via hydrogen abstraction by excited benzophenone or poly(vinylbenzophenone), the rate constants are 1.1×10^6 and 0.81×10^6 $M^{-1}s^{-1}$ respectively. Conversely in donor-quencher interactions that begin to approach the diffusion-controlled limit, the intramolecular energy migration leads to an enhancement of photosensitization efficiencies. The rate constants for energy transfer to 1-methylnaphthalene are 0.94×10^9 $M^{-1}s^{-1}$ for benzophenone and 1.85×10^9 $M^{-1}s^{-1}$ for poly(vinylbenzophenone) [102].

Aiming to examine the positive effect of molecular size on the ability of polymeric photosensitizers to transfer their energy to acceptor molecules, several systems were examined but the response is not clear.

When different samples of polyvinylbenzophenone were involved to sensitize the photoisomerization of cis- and trans-stilbene, no difference with the monomer was observed in sensitizing effect, optical behavior and quantum efficiency [39]. For the same reaction, polystyrenes naphthoylated with α- or β-naphthoyl chloride were found to have the same quantum efficiencies as their corresponding model compounds except for the isomerization of trans-stilbene by the α-naphthoylated polymer which showed a better yield [22].

The photosensitizing ability of poly(4-vinylbenzophenone) was also checked: the rate constant for hydrogen abstraction from tetrahydrofuran was roughly twice as large as for benzophenone itself [29]. The photosensitized dimerization of cyclohexene and the oxidation of secondary alcohols lead to photoadducts with a yield less than threequarters of that obtained when using benzophenone [27]. Finally in the case of the photoisomerization of (E)-1,3 pentadiene, the photostationary state depends on the degree of functionalization of the polymers via steric hindrance to the encounter between the donor and the quencher [108].

3.4.3 Electron-transfer Photosensitizers

The radical ion separation is favoured when using a polymeric rather than a low molecular weight photosensitizer [36]. This is a point of interest because competition between separation and back electron transfer is crucial in determining the overall efficiency of photoinduced electron transfer reaction. But this advantage can be cancelled by secondary reactions of the chromophore with its support. For example, poly(ethyleneterephthalate) plays the role of an effective electron transfer photosensitizer, causing almost the same amount of dimerization of phenyl vinyl ether as does the same weight of dissolved dimethyl terephthalate [36]. It can be reused, but its effectiveness decreases strongly after each run. This is attributed to irreversible reaction of the terephthalic chromophore at the surface of the polymer [36]. Nevertheless, monomeric esters are also consumed through side reactions when used as electron-transfer photosensitizers.

3.4.4 Multifunctional Polymeric Photosensitizers

In this field, the efficiency of the photosensitizing system is strongly affected by the content of the photosensitizing group in the bead because large amounts either reduce the degree of introduction of the substrate attracting group in the polymer bead or decrease the energy transfer by self-quenching between the photosensitizing groups.

The photosensitizing efficiency is also strongly affected by the bulkiness or the hydrophilicity of the substrate-attracting group on the bead. Basically, a multifunctional polymeric photosensitizer bead having tiny or hydrophilic quaternary salt as a substrate-attracting group should have higher efficiency than the polymer bead having a bulky or hydrophobic quaternary salt. This was shown for the photochemical *cis-trans*-isomerization of potassium sorbate or potassium cinnamate [38,40].

The rates of the valence isomerization of 2,5-norbornadiene-2-carboxylate photosensitized by multifunctional polymeric photosensitizers containing nitroaryl oxygroups as photosensitizing moiety, and quaternary ammonium or phosphonium salts as substrate-attracting groups, are significantly higher than those with low molecular weight photosensitizers. As expected, the efficiency depends on the content of the pendant photosensitizing group whose the best was 11.8 mol% in the polymer chain. The benzyltributylphosphonium was found to be the better attracting group [41]. Other multifunctional photosensitizers containing pendant benzophenone moiety and pendant quaternary phosphonium or ammonium salts have been evaluated for their photosensitizing efficiency in the *trans-cis*-isomerization of potassium sorbate [60]. The same study was performed on potassium cinnamate with pendant 4-nitrophenoxybutyrate and pendant benzyltributyphosphonium chloride [40].

3.4.5 Degradation of Polymeric Photosensitizers

In some cases, if the normal products of the reaction are observed, the recovered photosensitizing polymer cannot be reused. For example when polymeric benzophenone has photosensitized the dimerization of cyclohexene or the oxidation of secondary alcohols, it no longer shows the original ketone carbonyl band in its infrared spectrum after reaction [27].

In the case of benzoylated polystyrene-divinylbenzene (P-benzoyl), the photodecarboxylation of the styryl benzoate ester gives benzophenone as an observed product, whereas hydrogen abstraction from the polystyrene by the triplet aromatic carbonyl group leads to direct cross-linking of the polymer backbone to the carbonyl group. These degradation reactions significantly decrease the energy-transfer efficiency [26,107].

Poly(p,p'-divinylbenzophenone) which has a more rigid structure, also degrades by hydrogen atom abstraction from the polymer backbone, but more slowly [107]. The degradation can be prevented by hindering abstraction of atoms from the polymer backbone by replacing the hydrogens with fluorines [107]. Thus poly(p-trifluorovinyl)benzophenone TEF–BP and poly(p-trifluorovinyl) acetophenone TEF–AP are among the most photostable aryl ketone photosensitizers. They can be used for several triplet-sensitized reactions such as the dimerization of indene, the (2 + 2) cycloaddition of benzo-b-thiophene to dichloroethylene or acetylenedicarboxylate and the isomerization of norbornadiene [107]. The relative quantum yields of these polymer-sensitized reactions, compared with benzophenone itself under similar conditions, are usually between 0.5 and 0.8 probably due to their bimolecular nature. The effective encounter of the reactants with a sterically crowded chromophore is limited by the restricted accessibility.

3.4.6 Photophysics of Polymeric Photosensitizers

In the field of photophysical studies, the preparation and handling of polymeric samples is critical in order to obtain reproducible results. The quenching constants obtained may be of the same order of magnitude, but spectrophotometric determination must be considered cautiously because of the possible degradation undergone by the sample during the measurements [109].

Triplet–triplet absorption studies undertaken to determine the actual spectroscopic triplet lifetime of TEF–BP give a value of 10^{-5} s in benzene [107], an almost identical value with that of benzophenone. The triplet lifetime of TEF–AP is, like that of acetophenone, too short to be measured by triplet–triplet absorption technique in fluid solution.

In the same field, poly(phenyl vinyl ketone) PPVK has been the subject of several studies. Variations of lifetime of the triplet state were attributed to the incorporation into the fragmented macromolecules of unsaturated end-groups which can act as internal quenchers [110]. The mechanism of energy migration along the chain allows the energy of the excited triplet to arrive intramolecularly at the incorporated quencher. Therefore samples of differently photodegraded PPVK give rise to triplets with different lifetimes [22,109]. As a result, the energy-transfer efficiency of these carbonyl chromophores is higher when they are in free state that when they are polymer bound [22].

Finally, as already mentioned, efficient intramolecular self-quenching processes may occur in the functionalized polymers because of the large local concentration of chromophore [111]. An increase of the chromophore content in the polymer leads to lowering its sensitizing efficiency. That is the case for polymeric photosensitizers with various amounts of 4-nitro-1-naphthyl carbamoyl groups used as triplet energy donors against *trans*-stilbene [111].

3.5 PHOTOSENSITIZERS INCORPORATED INTO POLYMERIC FILMS

As a rule, the polymer film retards the transport of solvent, dissolved oxygen, and/or substrate to the sites occupied by the photosensitizer.

The four photosensitizers immobilized into polymer film: RB/cellulose acetate (RB/CA), RB/PVC, Ru(bpy)$_3^{2+}$/PVC and Safranine/PVC photosensitize the oxidation of di-n-butyl sulfide into di-n-butyl sulfoxide. The Bu$_2$S photooxidation rate is higher for RB/CA films, but RB/CA films, are far more susceptible to photobleaching than are RB/PVC films [43]. In fact they display different susceptibilities to photodegradation: RB/CA, RB/PVC, Safranine/PVC and Ru(bpy)$_3^{2+}$/PVC are 50% photodegraded after irradiation for 0.15, 5, 7 and 320 h respectively. This provides a rationale for the high stability of Ru(bpy)$_3^{2+}$/PVC to photodegradation. In fact its photoinduced electron-donor ability leads to Ru(bpy)$_3^{3+}$ a strong reducing agent which may be expected to return to the 2+ oxidation state by reaction with a reductant rather than undergoing subsequent reactions [43].

In the context of a system designed to mediate the photooxidation of organic substrates, the incorporation of photosensitizers into solution-cast polymer films provides an alternative to covalently bound photosensitizers. Using a cast film as a support material permits the use of photosensitizers that could not, otherwise, be supported owing to preparative difficulties. An example of this is the efficient photosensitizer Ru(bpy)$_3^{2+}$ whose stability with respect to photodegradation is good. Cast films containing Ru(bpy)$_3^{2+}$ can be made on a large scale, they are more amendable to investigations in which determination of quantum efficiencies is desired. However, cast sensitizer films that are highly permeable to solvents are also prone to leaching of the sensitizer into solution. A compromise between quantum efficiency and resistance to sensitizer leaching is therefore necessary in their use [43].

3.6 PHOTOSENSITIZERS INCORPORATED INTO MEMBRANES

The widely used RB photosensitizer, when immobilized in a poly(methylmethacrylate) film, is not a particularly efficient photosensitizer [112].

Zinc tetraphenylporphyrin incorporated into cellulose acetate membrane exhibits fluorescence with a short lifetime < 10 ns and a spectrum consistent with that of ZnTPP [44]. Usually, the ultimate goal of the ZnTPP modified fibers is the accomplishment of photosensitized redox reactions in solution. Electron donors such as EDTA, triethanolamine or tertiary amines, and electron acceptors such as viologens were tested. The irradiations with microheterogeneous ZnTPP indicate a reduced efficiency caused by a reduced access of the solution-phase donors or acceptors to the sensitizer or by a reduction of the excited state lifetime of the photosensitizer [44].

Conversely, methylene blue incorporated into a Nafion membrane exhibits a higher fluorescence yield, longer fluorescence lifetime and a blue shift of the absorption spectrum compared to an aqueous solution. The photoxidation of anthracene is also enhanced with this polymer as sensitizer [113].

3.7 PHOTOSENSITIZERS INCORPORATED INTO ZEOLITES

In homogeneous medium, a photoinduced electron transfer is often followed by an efficient back transfer. Zeolites possess a stable and rigid framework which provides the opportunity to assemble redox systems and prevent back reaction. This is observed with the system Ru(bpy)$_3^{2+}$-dimethylviologen-zeolite Y [47].

IMMOBILIZED PHOTOSENSITIZERS AND PHOTOCATALYSIS 253

Moreover, the spectroscopic properties of the intrazeolitic complex Ru(bpy)$_3^{2+}$ are similar to those exhibited by the complex in aqueous solution [46].

Some protective effect may occur. Thus 2,4,6-triphenylpyrylium cation encapsulated in zeolite Y is able to promote the *cis-trans*-photoisomerization of stilbene. The reaction is not affected by the presence of oxygen as shown by the absence of byproducts arising from oxidative cleavage. This contrasts with the extensive photooxygenation produced by the parent homogeneous photosensitizer [48].

3.8 POLYMER-BOUND PHOTOSENSITIZERS
3.8.1 Rose Bengal

Although polymer-bound RB has been demonstrated to be useful in most organic solvents, it is a poor photosensitizer in aqueous systems. The reason for this limited effectiveness is related to the observations that the hydrophobic polymer is difficult to suspend in aqueous media, is not wetted and does not swell in water. The hydrophilic poly(ethylene glycol) bound RB HP—RB [58] provides an alternative for photooxygenation in aqueous media. It is stable under photooxidative conditions and can be reused without apparent loss of efficiency. It was compared in a variety of solvents to polystyrene-divinylbenzene-bound RB marketed as Sensitox. It behaves very similarly in polar solvents such as acetone, CH_2Cl_2 or CH_3OH [58]. The quantum yield for $^1O_2$ formation is 0.48 for HP—RB and 0.43 for Sensitox. In contrast to Sensitox, HP—RB is not useful in nonpolar media such as dioxane, toluene or octane in accord with its hydrophilic nature. In water, the sodium salt of 9,10-anthracene dipropionic acid is photoxidized with 63% yield in 1 min, whereas only 4% reaction occurs with Sensitox. Compared with P—RB based on poly(styrene-co-divinylbenzene) beads [57], in P—RB the nonpolar polymer chain carries a polar dye into a nonpolar solvent. In this case, hundreds of RB molecules are attached to one polymer chain, whereas in the case of HP—RB each RB possesses its own small polymer residue. The first one is a heterogeneous photosensitizer and is truly an example of heterogeneous photochemical catalysis. The second is an example of a microheterogeneous system where the dye is essentially dispersed, preventing aggregation which enhances self-quenching [57].

A third way is to bind RB to copolymers soluble in water where these copolymers adopt a compact conformation resulting in the formation of microdomains which are capable of solubilizing organic compounds sparingly soluble in water. The antenna chromophores absorb light and the excitation energy is transferred to substrate molecules solubilized within the polymer coil [56]. These "photozymes" behave as efficient generators of $^1O_2$.

3.8.2 Photosensitizing Efficiency

The quantum yield Φ for the production of $^1O_2$ with the polymer-bound RB is 0.43, whereas for RB itself it is reported that $\Phi^1O_2 = 0.76$ in methanol [114].

When coupled to aminohexyl-glycophase, RB is two to three times less efficient as a photodynamic agent than in solution [59]. This could be because the activation energy for the photooxidation process is 15.1 kJ/mol for RB, whereas a value of 30.6 kJ/mol is calculated for the immobilized sensitizer. The difference might be associated with chemical modification of the carboxylic group of the dye upon immobilization [59].

These values suggest also that when a small amount of RB is attached to a polymeric backbone the lowering of the actual quantum yield of $^1O_2$ formation seems neither sensitizer nor solvent controlled, but due to a decrease in the mobility of the molecules [5].

When thionine is covalently bound to linear copoly(styrene-p-vinylbenzyl chloride) or to linear copoly(acrylic acid-2-ethylhexyl acrylate) the resulting immobilized photosensitizers present quantum yields of $^1O_2$ generation lower than their corresponding low molecular weight photosensitizers thionine hydrochloride or acetylthionine. High chromophore concentrations within the volume encompassed by each macromolecule in the solution can explain this fall in efficiency [115].

Finally, as expected, the photosensitizing efficiency is lower when the emission is higher: *ortho* and *para*(benzyloxy)benzaldehyde grafted onto chloromethylated polystyrene are less efficient energy donors toward norbornadiene than monomeric model compounds but, on the contrary, polymer binding enhances the phosphorescence of both substituted chromophores [60].

3.8.3 Electron-transfer Polymer-bound Photosensitizers

The best example of photosensitized electron transfer inside a polymeric system is found in photosynthesis (Chapter 10) where the photons absorbed are driven to the reaction center by means of energy transfer along the array of antenna pigments. Quenching is made possible by electron transfer across thylacoid membrane. A reasonable approach to this problem is to study the behavior of electron-transfer photosensitizers bonded to polymers, to determine the factors that control the heterogeneous electron transfer at the polymer–solution interface. Extensive studies on the excited state behavior of polymer-bound chromophores [116,117] indicate that the mobility of the chromophores is reduced and that the excimer/exciplex formation is enhanced [116,117].

There are, in fact, limited examples of small and positive polymer effects in the field of photoinduced electron transfers. The greatest merit of the polymer-bound photosensitizers is the interfacial electron transfer and the photoredox coupling across the polymer membrane leading to facile product separation. If the interfacial propertics are well analyzed and the solution–solid interface is properly designed, the efficiency of the polymer-bound photosensitizers could approach the value of small molecular models in homogeneous systems [63].

As a rule, the photoinduced electron-transfer efficiency is reduced if compared to homogeneous photosensitizer. This is caused by:

IMMOBILIZED PHOTOSENSITIZERS AND PHOTOCATALYSIS

1. A loss of segment mobility;
2. Enhanced excimer formation (energy trap);
3. Enhanced side reactions;
4. The reduction of the solvent polarity in the microenvironment of the polymer chain.

Slow solvation and stabilization of ion radicals in the polymer environment and, consequently, a larger fraction of ion-radical dissipation from the nonrelaxed state in polymeric systems may also be responsible for the lower yields of ion formation [63].

Conversely, porphyrins are known to work as photosensitizers in the presence of donors and acceptors for the conversion of visible light into chemical energy. In order to convert more light into chemical energy, the simultaneous use of various photosensitizers absorbing at different wavelengths in the visible region of light would be important, but a mixture of monomeric porphyrins leads to a decrease in photocatalytic activity. The employment of different porphyrins covalently bound to a polymer prevents this intermolecular interaction. The overall electron-transfer efficiency with electron donors or electron acceptors is higher with polymers. This increase is in the following order: monomeric porphyrins < porphyrins on uncharged polymers < porphyrins on negatively charged polymers < porphyrins on positively charged polymers [66]. Moreover, in general, a covalently bound porphyrin exhibits an enhanced triplet lifetime in comparison with the monomeric photosensitizer [66].

3.8.4 Bleaching of Polymer-bound Photosensitizers

As a rule, polymer-bound photosensitizers are significantly more stable toward bleaching than are the free sensitizers [50,118].

Evaluation of insoluble photosensitizing materials for further technological development must include tests of their energy-transfer efficiency and also of their light stability. A method of quantifying the light stability of a photoactive surface is reported [119]. It uses the same industrial testing procedures as those applied on dyestuffs, pigments and polymers and consists in the preparation of planar surfaces of convenient size, irradiated in a standard xenotest equipment and subsequently analyzed using a reflection spectrophotometer. Four polymer-based photosensitizers were tested:

1. Sensitox I or RB esterified to chloromethylated beads [50–51];
2. Sensitox II or RB bound to a copolymer of chloromethylstyrene and ethylene glycol monomethacrylate [58];
3. Rose Bengal bound to a polyepoxy resin of Araldite type;
4. Hematoporphyrin bound to the same Araldite support.

There is practically no bleaching of pure Sensitox pellets. The colour changes are found only when the photosensitizer is mixed with spacers such as KBr.

But there is neither a chromophore nor a support specific stabilizing effect of KBr [119].

3.8.5 Degradation of Polymer-bound Photosensitizers

Commercial polymer-bound RB Sensitox can be used for six to eight successive photooxygenations of 2–4 h duration, with a typical visible radiation source and extensive stirring, before the beads become so mechanically distorted and damaged RB that they can no longer be separated by filtration [25].

When and toluidine blue bound to trihydroxymethylisocyanuric acid are used as photosensitizing agents to oxidize α-pinene, a 10% loss of photosensitizer is observed after 3 h irradiation [120].

3.8.6 Photophysics of Polymer-bound Photosensitizers

In some cases, one can predict the absorption properties of an heterogeneous photosensitizer from the absorption spectrum of its soluble equivalent. Thus, the diffuse reflectance spectrum of the polymer-based RB shows that the chromophore is not affected by its binding [50]. Polymeric $Ru(bpy)_3^{2+}$, in which one of the rings is covalently bound to insoluble cross-linked polystyrene [121], behaves spectroscopically, similarly as in a homogeneous solution: emission from the MLCT complex is observed. Nevertheless, experiments using methyl viologen show that luminescence quenching is about 40 times slower than in homogeneous aqueous solutions because the hydrophobic nature of the polymer impedes the approach of the quencher to the ruthenium complex [122].

Other examples exist where, at low concentrations of dye on the polymer chain, the absorption spectrum resembles that of the dilute solution. But when the distance between the chromophores on the polymeric backbone decreases, there is then the possibility of aggregation of the dye molecules along the polymer chain allowing one dye molecule to influence another. In the higher concentration cases, both the form of the absorption curve and the molar absorptivity no longer appear as that of the simple photosensitizer.

Water-soluble phthalocyanines and porphyrins also aggregate strongly in aqueous solutions. This can be seen by a shift of the Q band to shorter wavelengths. Interestingly, the polymer backbone may inhibit strong intermolecular dye–dye interactions [66]. Then the cationically charged polymers show no aggregation. The other polymers slightly aggregate in water.

If the direct measurement of the absorption spectrum of a heterogeneous photosensitizer is impossible because of light dispersion from the polymer surface, one can measure the emission excitation spectrum as exemplified by P–RB in a frozen nonpolar solvent [5].

3.9 SILICA-BOUND PHOTOSENSITIZERS

Rose Bengal immobilized on silica gel has the virtue of being solvent versatile, but is less photostable than the original polymer supported reagent [25];

4-(N,N-dimethylamino)benzophenone grafted onto silica photosensitizes the isomerization norbornadiene-quadricyclene used for solar energy storage. The homogeneous system is efficient (100%), whereas the silica functionalized photosensitizer is significantly less effective (24%) [96]. This lower efficiency is attributed to a decrease in triplet yield and to a lowering of the triplet energy from 66 kcal/mole to 57 kcal/mole caused by the polar silica surface. The phosphorescence spectrum shows substantial loss of structure characteristic of a more polar environment.

Conversely, photooxidative activation of aromatics towards dioxygen is optimized by bonding the photosensitizer to SiO_2. The use of supported electron-acceptor photosensitizers improves the yields and the selectivities in aldehyde formation, whereas with homogeneous photosensitizers the reaction leads also to the corresponding acid. One explanation may be the internal filter effect of the photosensitizer which absorbs light as far as the visible region [69,91].

In the same way, zinc phthalocyanines bound to modified silica behave as a homogeneous dispersion of the Pc on the surface of silica. No evidence for aggregation is seen in the visible reflectance spectra [95]. Thus, when used in photocatalytic reactions such as oxidation of thiols to disulfides they exhibit better activities and O_2 consumption than the monomeric photosensitizers.

4 CONCLUDING REMARKS

Immobilized photosensitizers have several properties which differ from classical soluble photosensitizers.

In principle the scope of this approach is unlimited, but the use of insoluble photosensitizers has some disadvantages as follows:

1. Difficulty of achieving quantitative measurements, particularly those concerning light absorption.
2. Difficulty of determining the concentration of the species involved, especially excited species.
3. Possible restricted accessibility of the chromophore, resulting in a less effective encounter with the acceptor and thus limited photosensitizing efficiency.
4. When incorporated, usually the polymeric film retards transport of solvent and dissolved reactants to the sites occupied by the photosensitizing chromophore.
5. Some support-chromophore couples undergo secondary cross-linking reactions and lead to degradation. The photosensitizing efficiency then decreases with time.
6. As a rule, a large amount of the photosensitizing group in the immobilizing system must be avoided in order to prevent self-quenching of the photosensitizer.

7. Depending on the immobilization method, the sensitizer can be desorbed from the support by polar solvents.

The primary advantage of this approach is that the sensitizer can also be highly resistant to separation from the support material, thus permitting use of the photosensitizer in a wide variety of solvent systems.

1. For carrying out preparative photosensitized reactions, the sensitizer can be easily eliminated at the end of the reaction by filtration.
2. As a rule, immobilized photosensitizers are more stable toward bleaching than are free molecules.
3. If the ratio of chromophore is not too high, support acts like a microheterogeneous system preventing aggregation: self-quenching is not observed, the absorption spectrum resembles that of the photosensitizing monomer solution and triplet lifetime may even be enhanced, leading to greater efficiency.
4. For photosensitized electron-transfer reactions, the rigid framework of the support seems to provide opportunity to reduce or prevent back electron transfer, the radical ions separation may also be favoured.
5. Encapsulation may hinder secondary oxidative cleavage reactions.

Finally, the support is not just a vehicle for the immobilized photosensitizer but it is an important part of the reagent. It is responsible for the stability of the photosensitizing chromophore, the accessibility of the acceptor and the solvent compatibility of the system. The surface of the support or the membrane properties are more influential in determining the photosensitizer behavior than are the properties of the surrounding fluid medium. Nevertheless, greater efficiency demands that the quencher fluid medium be readily accessible to the chromophore.

REFERENCES

[1] O. Valdesaguilera and D. C. Neckers, Acc. Chem. Res., **22**, 171 (1989).
[2] D. Xu and D. C. Neckers, J. Photochem. Photobiol. A: Chem., **40**, 361 (1987).
[3] H. Kautsky, Trans. Faraday Soc., **35**, 216 (1939).
[4] R. B. Merrifield, J. Am. Chem. Soc., **85**, 2149 (1963).
[5] D. C. Neckers, React. Polym. Ion Exch. Sorbents, **3**, 277 (1985).
[6] B. Paczkowska, J. Paczkowski and D. C. Neckers, Macromolecules, **19**, 863 (1986).
[7] R. Nilson and D. R. Kearns, Photochem. Photobiol., **19**, 181 (1974).
[8] H. Kautsky, Trans. Faraday Soc., **35**, 3795 (1939).
[9] P. A. Grutsch and C. Kutal, J. Chem. Soc. Chem. Commun., 893 (1982).
[10] A. Isbell and D. T. Sawyer, Anal. Chem., **41**, 1381 (1969).
[11] W. R. Midden and S. Y. Wang, J. Am. Chem. Soc., **105**, 4129 (1983).
[12] J. R. Williams, G. Orton and L. R. Unger, Tetrahedron Lett., **46**, 4603 (1973).
[13] J. R. Williams, G. Orton and L. R. Unger, Tetrahedron Lett., **14**, 4603 (1973).
[14] L. Rixin, Z. QinQin, C. Guozhu and W. Shikang, Kexue Tong Bao, **28**, 1197 (1983).

[15] T. Yokota, Y. Takahata, Y. Hosoya, K. Suzuki and K. Takahashi, J. Chem. Eng. Japan, **22**, 543 (1989).
[16] A. P. Schaap, A. L. Thayer and D. C. Neckers, J. Am. Chem. Soc., **97**, 3741 (1975).
[17] S. L. Buell and J. N. Demas, J. Phys. Chem., **87**, 4675 (1983).
[18] T. Hikita, K. Tamaru, A. Yamagishi and T. Iwamoto, Inorg. Chem., **28**, 2221 (1989).
[19] C. Lewis and W. H. Scouten, Biochim. Biophys. Acta, **444**, 326 (1976).
[20] R. R. Hautala and J. L. Little, Adv. Chem. Ser., 184 (1980).
[21] D. C. Neckers, Chem. Tech., **8**, 108 (1978).
[22] F. Catalina, R. Martinez-Utrilla and R. Sastre, Polymer Bulletin, **8**, 369 (1982).
[23] S. Irie, M. Irie, Y. Yamamoto and K. Hayashi, Macromolecules, **8**, 424 (1975).
[24] S. N. Gupta, L. Thijs and D. C. Neckers, Macromolecules, **13**, 1037 (1980).
[25] D. C. Neckers, Nouv. Journal de Chimie, **6**, 645 (1982).
[26] E. C. Blossey and D. C. Neckers, Tetrahedron Lett., 323 (1974).
[27] J. L. Bourdelande, J. Font and F. Sanchez-Ferrando, Can. J. Chem., **61**, 1007 (1983).
[28] J. L. Bourdelande, J. Font and F. Sanchez-Ferrando, Tetrahedron Lett., 21, 3805 (1980).
[29] M. Kamachi, Y. Kikuta and S. Nozakura, Polym. J., **11**, 273 (1979).
[30] H. A. Hammond, J. C. Doty, T. M. Laakso and J. L. R. Williams, Macromolecules, **3**, 711 (1970).
[31] K. Fukunaga, K. Mitarai and H. Mitoguchi, Nippon Kagaku Kaishi, 1064 (1992).
[32] F. R. Diaz, L. H. Tagle, F. Garcia, R. Sastre, M. Conde, J. L. Mateo and F. Catalina, J. Polymer Sci. Part A: Chemistry, **28**, 3499 (1990).
[33] M. Nowakowska, E. Sustar and J. E. Guillet, J. Photochem. Photobiol. A: Chem., **80**, 369 (1994).
[34] F. Almasy, J. Chim. Phys., **30**, 528 (1933).
[35] G. R. De Maré and J. R. Fox, Eur. Polym. J., **17**, 315 (1981).
[36] A. Albani and S. Spreti, J. Chem. Soc. Chem. Commun., **18**, 1426 (1986).
[37] J. Mattay, M. Vondenhof and G. Trampe, EPA Newsl., **32**, 23 (1988).
[38] T. Nishikubo, T. Kondo and K. Inomata, Macromolecules, **22**, 3827 (1989).
[39] K. Inomata, T. Kawashima, Y. Suzuki and T. Nishikubo, Nippon Kagaku Kaishi, 223 (1991).
[40] T. Nishikubo, J. Uchida, K. Matsui and T. Iizawa, Macromolecules, **21**, 1583 (1988).
[41] T. Nishikubo, T. Kawashima, K. Inomata and A. Kameyama, Macromolecules, **25**, 2312 (1992).
[42] A. Zweig and W. A. Henderson J. Polym. Sci. Polym. Chem. Ed., **13**, 717 (1975).
[43] R. A. Kenley, N. A. Kirschen and T. Mill Macromolecules, **13**, 808 (1980).
[44] C. C. Wamser, M. Calvin and G. A. Graf, J. Membr. Sci., **28**, 31 (1986).
[45] L. R. Faulkner, S. L. Suib, C. L. Renschler, J. M. Green and P. R. Bross, ACS Symp. Ser., **207** 99 (1982).
[46] T. L. Pettit and M. A. Fox, J. Phys. Chem., **90**, 1353 (1986).
[47] P. K. Dutta and J. A. Incaco, J. Phys. Chem., **91**, 4443 (1987).
[48] A. Corma, V. Fornes, H. Garcia, M. A. Miranda, J. Primo and M. J. Sabater, J. Am. Chem. Soc., **116**, 2276 (1994).
[49] R. B. Merrifield, Science, **150**, 178 (1965).
[50] A. P. Schaap, A. L. Thayer, E. C. Blossey and D. C. Neckers, J. Am. Chem. Soc., **97**, 3741 (1975).
[51] E. C. Blossey, D. C. Neckers, A. L. Thayer and A. P. Schaap, J. Am. Chem. Soc., **95**, 5820 (1973).

[52] J. J. M. Lamberts, D. R. Schumacher and D. C. Neckers, J. Am. Chem. Soc., **106**, 5879 (1984).
[53] J. J. M. Lamberts and D. C. Neckers, Z. Naturf., **39b**, 474 (1984).
[54] E. C. Blossey, D. C. Neckers, A. L. Thayer and A. P. Schaap, J. Am. Chem. Soc., **95**, 5820 (1973).
[55] D. C. Neckers, Chemtech., Feb. 108 (1978).
[56] M. Nowakowska, E. Sustar and J. E. Guillet, J. Photochem. Photobiol. A: Chem., **80**, 369 (1994).
[57] J. Paczkowski and D. C. Neckers, J. Photochem., **35**, 283 (1986).
[58] A. P. Schaap, A. L. Thayer, K. A. Zablika and P. C. Valenti, J. Am. Chem. Soc., **101**, 4016 (1979).
[59] F. I. Llorca, J. L. Iborra and J. A. Lozano, Photobiochem. Photobiophys., **5**, 105 (1983).
[60] R. Martinez-Utrilla, F. Catalina and R. Sastre, J. Photochem. Photobiol. A Chem., **44**, 187 (1988).
[61] R. Searle, J. L. R. Williams, J. C. Doty, D. E. De Meyer, S. H. Merrill and T. M. Laakso, Makromol. Chem., **3**, 711 (1967).
[62] A. Ueno, F. Toda and Y. Iwakura, Polymer Letters, **12**, 287 (1974).
[63] S. Tazuke, R. Takasaki, Y.Iwaya and Y.Suzuki, J. Am. Chem. Soc., **8**, 187 (1984).
[64] K. Kowalski, E. J. Goethals and R. B. Koolstra J. Macromol. Sci. Chem., **A28**, 249 (1991).
[65] Y. Kodera, A. Ajima, K. Takahashi, A. Matsushima, Y. Saito and Y. Inada, Photochem. Photobiol., **47**, 22 (1988).
[66] D. Wohrle, J. Gitzel, G. Krawczyk, E. Tsuchida, H. Ohno, I. Okura and T. Nishisaka, J. Macromol. Sci. Chem., **A25**, 1227 (1988).
[67] J. M. Ribo, M. L. Sese and F. R. Trull, Reactive Polymers, **10**, 239 (1989).
[68] S. Tamagaki, E. Liesner and D. C. Neckers, J. Org. Chem., **45**, 1573 (1980).
[69] M. Julliard, C. Legris and M. Chanon, J. Photochem. Photobiol. A: Chem., **61**, 137 (1991).
[70] C. H. Lochmüller, R. R. Ryall and C. W. Amoss, J. Chromatogr., **178**, 298 (1979).
[71] W. E. Hammers, A. G. M. Theenwes, W. K. Brederode and C. L. J. de Ligny, J. Chromatogr., **243**, 321 (1982).
[72] L. Nondek and R. Ponec, J. Chromatogr., **294**, 2434 (1984).
[73] M. Verzele and N. Van de Velde, Chromatographia, **20**, 239 (1985).
[74] C. H. Lochmüller and C. W. Amoss, J. Chromatogr., **108**, 85 (1975).
[75] L. Nondek and J. Malek, J. Chromatogr., **155**, 187 (1978).
[76] H. Hemetsberger, H. Klaar and H. Ricken, Chromatographia, **13**, 277 (1980).
[77] T. H. Mourey and S. Siggia, Anal. Chem., **51**, 763 (1979).
[78] C. H. Lochmüller, A. S. Colborn, L. M. Hunnicut and J. M. Harris, Anal. Chem., **55**, 1344 (1983).
[79] S. A. Matlin, W. L. Lough and D. G. Bryan, J. High. Resolut. Chromatogr. Chromatogr. Commun., **3**, 33 (1980).
[80] G. Felix, C. Bertrand and F. Van Gastel, Chromatographia, **20**, 155 (1985).
[81] G. Felix and C. Bertrand, J. High. Resolut. Chromatogr. Chromatogr. Commun., **7**, 714 (1984).
[82] G. Felix and C. Bertrand, J. High. Resolut. Chromatogr. Chromatogr. Commun., **8**, 362 (1985).
[83] G. Eppert and I. Schinke, J. Chromatogr., **260**, 305 (1983).
[84] M. Julliard, New J. Chem., **18**, 243 (1994).
[85] A. Pryde, J. Chromatogr. Science, **12**, 486 (1974).
[86] L. Nondek and J. Malek, J. Chromatogr., **155**, 187 (1978).
[87] L. Nondek and R. Ponec, J. Chromatogr., **294**, 175 (1984).

[88] L. Nondek, M. Minarik and J. Malek, J. Chromatogr., **178**, 427 (1979).
[89] L. Nondek and J. Malek, Czech. Pat., 194 081 (1982).
[90] H. Engelhardt and P. Orth, J. Liq. Chromatogr., **10**, 1999 (1989).
[91] M. Julliard and M. Chanon, Bull. Soc. Chim. France, **129**, 242 (1992).
[92] M. Julliard and M. Chanon, Chem. Rev., **83**, 461 (1983).
[93] G. J. Kavarnos and N. J. Turro, Chem. Rev., **86**, 401 (1986).
[94] M. Chanon and L. Eberson, in *Photoinduced Electron Transfer*; edited by M. A. Fox and M. Chanon; (Elsevier, New York 1988), pp 409–581.
[95] G. Schneider, D. Wöhrle, W. Spiller, J. Stark and G. Schulz-Ekloff, Photochem. Photobiol., **60**, 333 (1994).
[96] R. R. Hautala and J. L. Little, Adv. Chem. Ser., **184**, 1 (1980).
[97] C. Murali Krishna, Y. Lion and P. Riesz, Photochem. Photobiol., **45**, 1 (1987).
[98] J. P. Pete, Asymmetric photoreactions of conjugated enones and esters in *Adv. in Photochem.* (Wiley, New York, 1995) **21**, 135 (1995).
[99] Y. Inoue, Chem. Rev., **92**, 741 (1992).
[100] T. Yokota, Y. Takahata, J. Hosoya, K. Suzuki and K. Takahashi, J. Chem. Eng. Japan, **22**, 543 (1989).
[101] G. Jori, G. Galiazzo and E. Scoffone, Biochemistry, **8**, 2868 (1969).
[102] E. H. Urruti and T. Kilp, Macromolecules, **17**, 50 (1984).
[103] R. E. Moser and H. G. Cassidy, J. Polym. Sci. Polymer Lett., **2**, 545 (1964).
[104] S. Irie, M. Irie, Y. Yamamoto, K. Hayashi, Macromolecules, **8**, 424 (1975).
[105] R. Searle, J. L. R. Williams, J. C. Doty, D. E. Demeyer, S. H. Merrill and T. M. Laakso, Makromol. Chem., **107**, 246 (1967).
[106] R. F. Cozzens and R. B. Fox, J. Chem. Phys., **50**, 1532 (1969).
[107] N. Asai and D. C. Neckers, J. Org. Chem., **45**, 2903 (1980).
[108] J. L. Bourdelande, J. Font and F. Sanchez-Ferrando, Polym. Photochem., **2**, 383 (1982).
[109] R. Martinez-Utrilla, F. Catalina and R. Sastre, Polym. Photochem., **4**, 361 (1984).
[110] M. V. Encinas, K. Funabashi and J. C. Scaiano, Macromolecules, **12**, 1167 (1979).
[111] R. Sastre, J. L. Mateo, P. Bosch, F. Catalina, F. Amatguerri and F. R. Diaz, Eur. Polymer J., **29**, 539 (1993).
[112] R. A. Kenley, N. A. Kirshen and T. Mill, Macromolecules, **13**, 808 (1980).
[113] E. Niu, K. P. Ghiggino, A. W. H. Mau and W. H. F. Sasse, J. Luminescence, **40, 41**, 563 (1988).
[114] K. K. Gollnick and G. O. Schenck, in *1-4-cycloaddition Reactions;* J.Hamer, Ed.; (Academic Press, New York, NY 1967); p 255.
[115] F. Amatguerri, J. M. Botija and R. Sastre, J. Polym. Sci. A Polymer Chem., **31**, 2609 (1993).
[116] S. Tazuke, J. Syn. Org. Chem. Jpn., **40**, 806 (1982).
[117] H. L. Yuan and S. Tazuke, Polym. J., **15**, 125 (1983).
[118] A. P. Schaap, A. L. Thayer, K. A. Zaklika and P. C. Valenti, J. Am. Chem. Soc., **101**, 4016 (1979).
[119] E. Oliveros, M. T. Maurette, E. Gassmann and A. M. Braun, Dyes and pigments, **5**, 457 (1984).
[120] K. Dorgham, B. Richard, M. Richard and M. Lenzi, J. Chim. Phys., **85**, 579 (1988).
[121] P. Bosch, C. Campa, J. Camps, J. Font, P. de March and A. Virgili, An. Quim. scr., **C 81**, 162 (1985).
[122] J. L. Bourdelande, C. Campa, J. Camps, J. Font, P. de March, F. Wilkinson and C. J. Willsher, J. Photochem. Photobiol. A: Chem., **44**, 51 (1988).

8 Water Splitting: from Molecular to Supramolecular Photochemical Systems

EDMOND AMOUYAL
Laboratoire de Physico-Chimie des Rayonnements, Université Paris-Sud, Orsay France

1	Introduction	264
2	Photolysis of water into hydrogen and oxygen: the fascinating challenge	265
3	Required general functions	266
4	Homogeneous and quasi-homogeneous molecular systems	267
	4.1 General schemes	267
	4.2 First model systems for H_2 production	271
	4.2.1 First model systems	271
	4.2.2 Photosensitizers	273
	4.2.3 Electron relays	273
	4.2.4 Electron donors	274
	4.2.5 Microheterogenous catalysts	274
	4.2.6 Homogeneous catalysts and homogeneous systems for H_2 production	277
	4.3 The classical model system	279
	4.3.1 Introduction	279
	4.3.2 Mechanism	281
	4.3.3 Optimization	282
	4.4 Model systems for O_2 production	284
	4.4.1 First model systems	284
	4.4.2 Other microheterogeneous molecular systems	285
	4.4.3 Homogeneous systems	286
	4.4.4 Confined systems	287
5	Supramolecular systems	288
	5.1 Supramolecular systems for charge separation	288
	5.2 Hydrogen-generating supramolecular systems	294
6	Constrained systems	295

Homogeneous Photocatalysis. Edited by M. Chanon
© 1997 John Wiley & Sons Ltd

7 Complete water splitting into hydrogen and oxygen 297
8 Conclusion 300
References 301

1 INTRODUCTION

According to J. M. Lehn, supramolecular chemistry can be defined as "chemistry beyond the molecule" [1]. In this relatively young area, the term "supermolecule" was originally used to describe an assembly of two or more individual molecules, called receptors and substrates, linked by weak intermolecular (non-covalent) interactions such as van der Waals' forces, electrostatic interactions and hydrogen bonding (Figure 8.1a). This association, leading to a defined supramolecular structure, gives rise to a specific function, characteristic of the supermolecule or to new and more elaborate physical or chemical properties compared to the individual molecular subunits. This definition has now been extended, in particular in photochemistry [2] to systems involving covalent links between the components provided that the chemical identity of each molecular subunit is preserved in the supramolecular structure (Figure 8.1b). In other words, the intrinsic properties of each component which can be easily identified are only weakly perturbed by the other components. This means that the intramolecular interactions between

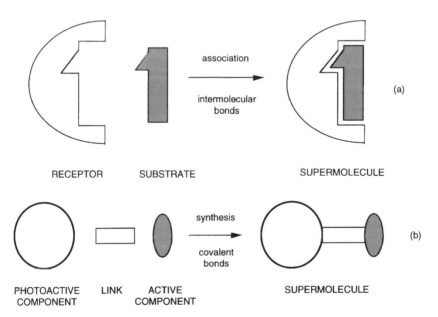

Figure 8.1. (a) Classical definition of supermolecule; (b) extended definition of supermolecule (polyad)

individual subunits are weak in the supramolecular system in its ground state. Systems involving several components covalently linked to one another are called dyad, triad, tetrad, pentad, ... according to whether the number of components is two, three, four, five, ... respectively. I shall use the generic term of polyad for these multicomponent molecular systems (Figure 8.1b).

One of the most interesting aspects of the chemistry of supramolecular systems is their interaction with light. Indeed, supramolecular photochemistry offers a great variety of processes: photoinduced electron and proton transfer, charge separation, energy migration, etc. which can be modulated by the arrangement of the linked subunits. From a practical point of view, it would be very important to use this possibility of modulation for homogeneous photocatalysis, in particular in the design of model systems for photochemical conversion and storage of solar energy. It should be noted that this storage is achieved in green plant photosynthesis and bacterial photosynthesis. This is the reason why these systems are also called artificial photosynthetic systems. This chapter describes the evolution of homogeneous artificial systems from the molecular to the supramolecular approach. In fact, homogeneous, quasi-homogeneous and microheterogeneous model systems in which the illumination of the photoactive component gives rise to photoinduced redox processes are presented. Microheterogeneous systems involving the formation of electron–hole pairs through direct excitation or photosensitization of semiconductor particles are not considered. Several reviews on these semiconductor-based systems are available [3–5].

2 PHOTOLYSIS OF WATER INTO HYDROGEN AND OXYGEN: THE FASCINATING CHALLENGE

The direct conversion of solar energy into chemical energy leading to the production of renewable and non-polluting fuels remains a great and fascinating challenge for the end of this century [6]. Among various interesting reactions (water-splitting, carbon dioxide reduction, nitrogen reduction), the splitting of water (reaction 1) into molecular hydrogen and molecular oxygen by visible light (sunlight) is potentially one of the most promising ways for the photochemical conversion and storage of solar energy [7–9].

$$H_2O + \text{sunlight} \longrightarrow H_2 + \tfrac{1}{2}O_2 \qquad (1)$$

Indeed hydrogen is a valuable fuel: the free enthalpy needed in reaction 1 to produce one mole of H_2, i.e. the energy stored per mole, is $\Delta G^0_{298} = 237.2 \text{ kJ mol}^{-1}$. Due to its small weight, the energy storage capacity of H_2 per gram, $119\,000 \text{ J g}^{-1}$, is very high (Table 8.1). It is, for example three times higher than the storage capacity of oil ($40\,000 \text{ J g}^{-1}$, Table 8.1). Moreover, the water-splitting process has two other advantages. First, the raw material, i.e. water, is abundant and cheap. Second, the combustion of H_2 in air (reverse of reaction 1) again gives water. In other words, the overall process is cyclic and non-polluting.

Table 8.1. Energy storage capacity of several fuels

Fuel	Energy (J g^{-1})
Hydrogen	119 000
Methane	49 000
Oil	43 000
Coal	~ 28 000
Ethanol	28 000
Methanol	21 000
Dry wood	~ 17 000
Municipal waste	~ 7 000

The main disadvantage is that H_2 can react explosively with O_2. The crash of the hydrogen-driven zeppelin *Hindenburg* in 1937 remains a terrifying memory for mankind. However, the explosive limit of H_2 in air is 4.00%, less than that of butane, 1.86%. Hence the utilization of hydrogen is no more dangerous than that of natural gas, and like natural gas hydrogen can be quite easily stored and transported [10]. Besides, probably because hydrogen technology is now considered as safe enough, car makers (Daimler-Benz, Mazda) launch into the development of hydrogen-powered cars. Other reasons are that the reserves of fossil fuels, especially crude oil, should decrease in the first decades of the next century, and that local air pollution, in particular by automobiles, becomes more and more a serious matter for concern.

3 REQUIRED GENERAL FUNCTIONS

Since water does not absorb visible light, it is easily understood that, as in natural photosynthesis, one or several intermediates are needed to achieve the water photocleavage via a cyclic pathway (reaction 1). These intermediates have ideally the following functions (Figure 8.2):

1. Visible light absorption;
2. Conversion of the excitation energy to redox energy (charges);
3. Concerted transfer of several electrons to water, leading to the formation of H_2 as energy-storage compound and/or to the formation of O_2.

Indeed, one of the main difficulties in achieving the splitting of water by means of light-induced redox processes is that hydrogen requires two electrons (reaction 2) while oxygen requires four electrons (reaction 3).

$$2H_2O + 2e^- \longrightarrow H_2 + 2OH^- \quad E°(pH\ 7) = -0.41V \text{ vs NHE} \quad (2)$$

$$2H_2O \longrightarrow O_2 + 4H^+ + 4e^- \quad E°(pH\ 7) = +0.82V \text{ vs NHE} \quad (3)$$

WATER SPLITTING 267

This number of charges corresponds to the most favorable thermodynamic conditions for reaction (1). In other words, this reaction is a multielectron transfer process which requires 1.23 eV per electron transferred. Hence, photons with $\lambda < 1008$ nm corresponding to a minimum energy of 1.23 eV can induce the cleavage of water.

4 HOMOGENEOUS AND QUASI-HOMOGENEOUS MOLECULAR SYSTEMS

4.1 GENERAL SCHEMES

Photochemical systems involving several compounds were proposed in a first approach. In these multimolecular systems, each function is fulfilled by one molecule (Figure 8.2) namely:

1. A photosensitizer PS able to absorb visible light to generate excited species PS* with useful redox properties (reaction 4);

$$PS \xrightarrow{h\nu} PS^* \qquad (4)$$

2. A second compound R which can be reduced or oxidized by quenching of the excited species PS* in electron-transfer reactions leading to the formation

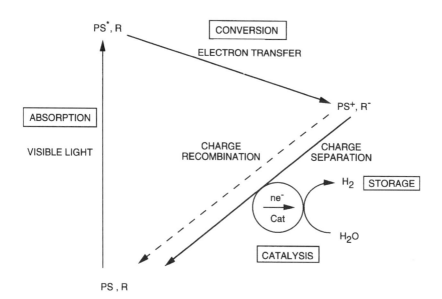

Figure 8.2. Schematic representation of required general functions for photochemical conversion and storage of solar energy via photoredox processes (PS = photosensitizer, R = electron relay, Cat = catalyst)

of charge pairs, such as (PS^+, R^-) in the case of the oxidative quenching of PS (reaction 5);

$$PS^* + R \longrightarrow PS^+ + R^- \tag{5}$$

3. A third compound able to collect several electrons to facilitate the exchange of two (reaction 6) or four electrons with water. This multielectron collection and transfer can be realized by a specific redox catalyst, Cat:

$$2R^- + 2H^+ \xrightarrow{Cat} 2R + H_2 \tag{6}$$

In such a system, the second compound R acts as an electron relay between the photosensitizer PS and the catalyst Cat mediating the electron collection. The redox potential of its reduced species R^- must be less than -0.41 V (vs NHE, pH 7) to take part in reaction (2).

In practice, difficulties arise from a fast recombination of charge pairs (reaction 7).

$$PS^+ + R^- \longrightarrow PS + R \tag{7}$$

The main problem, for these multimolecular systems and more generally for photochemical systems, is how to retard this back electron transfer reaction in order to obtain a charge separation with a long lifetime.

In the case of multimolecular systems, the back reaction should be prevented by using a fourth compound, an electron donor D, which scavenges the oxidized photosensitizer PS^+ in a competitive electron transfer reaction to give the initial PS and a donor oxidation product D^+ (reaction 8):

$$PS^+ + D \longrightarrow PS + D^+ \tag{8}$$

The latter rapidly decomposes irreversibly (reaction 9), and such systems have been qualified as "sacrificial"; D is the only compound, apart obviously from water (H^+), which is consumed. The other compounds PS, R and Cat follow catalytic cycles.

$$D^+ \longrightarrow products \tag{9}$$

Two schemes for cyclic production of hydrogen from water can be envisaged [11]. The first is called the "oxidative quenching mechanism" because it involves oxidation of the excited photosensitizer PS^* to PS^+ by the electron relay R (Figure 8.3). It corresponds to reactions (4)-(9). The second scheme involving reduction of the excited state photosensitizer PS^* by D is called the "reductive quenching mechanism" (Figure 8.4).

$$PS^* + D \longrightarrow PS^- + D^+ \tag{10}$$

This primary reaction (reaction 10) yields the reduced photosensitizer PS^- and the oxidized donor D^+ which decomposes irreversibly (reaction 9). In this way,

WATER SPLITTING

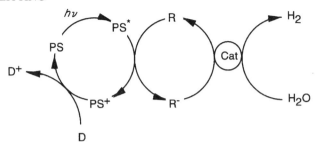

Figure 8.3. Schematic representation of the redox catalytic cycles in the photoreduction of water to hydrogen by visible-light irradiation of a four-component model system PS/R/D/Cat: oxidative quenching mechanism

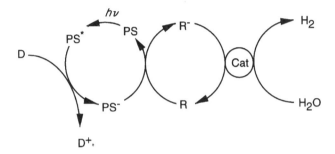

Figure 8.4. Schematic representation of the redox catalytic cycles in the photoreduction of water to hydrogen by visible-light irradiation of a four-component model system PS/R/D/Cat: reductive quenching mechanism

PS^- can accumulate and react with an electron relay R to regenerate PS and to yield R^- (reaction 11):

$$PS^- + R \longrightarrow PS + R^- \qquad (11)$$

In the presence of a suitable catalyst, R^- can lead to the formation of hydrogen as in the first scheme (reaction 6).

It should be remarked that PS^- is a more powerful reducing species than R^-. Hence, the reduction of water to H_2 can be achieved directly by PS^- itself in the presence of a suitable catalyst. As a consequence, this scheme involves only three components (PS, D, Cat) and the mechanism become simplified (Figure 8.5).

Similar three-component systems for O_2 production from water have been proposed (Figure 8.6). These systems require the formation, following visible-light excitation of the photosensitizer PS, of a strong oxidizing species PS^+, having a redox potential $E^\circ(PS^+/PS)$ greater than 0.82 V (vs NHE, pH 7). This can be achieved by using an electron acceptor A as quencher which, once reduced to A^- (reaction 12), decomposes irreversibly (reaction 13).

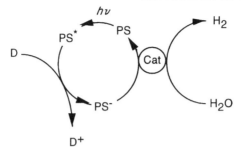

Figure 8.5. Schematic representation of the redox catalytic cycles in the photoreduction of water to H_2, via reductive quenching mechanism, for a three-component model system PS/D/Cat

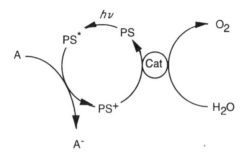

Figure 8.6. Schematic representation of the redox catalytic cycles in the photooxidation of water to oxygen by visible-light irradiation of a three-component model system PS/A/Cat

$$PS^+ + A \longrightarrow PS^+ + A^- \qquad (12)$$

$$A^- \longrightarrow \text{decomposition products} \qquad (13)$$

The oxidized PS^+ can thus accumulate and lead to oxygen evolution in the presence of a suitable catalyst capable of facilitating the exchange of four electrons with water (reaction 14):

$$4PS^+ + 2H_2O \xrightarrow{\text{Cat}} 4PS + 4H^+ + O_2 \qquad (14)$$

Another approach (Figure 8.7) consists in using the photosensitizer PS as an antenna and transferring the excitation energy to a receptor molecule R_{en} (reaction 15):

$$PS^* + R_{en} \longrightarrow PS + R_{en}^* \qquad (15)$$

The receptor can subsequently react with the electron relay R via electron transfer (reaction 16) to give a charge pair (R_{en}^+, R^- for example).

$$R_{en}^* + R \longrightarrow R_{en}^+ + R^- \qquad (16)$$

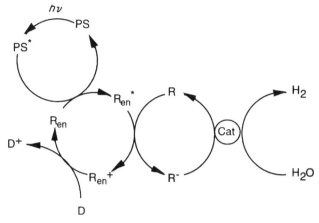

Figure 8.7. Schematic representation of the redox catalytic cycles in the photoreduction of water to H_2, via energy transfer (antenna effect), by visible-light irradiation of a five-component model system PS/R_{en}/R/D/Cat

The reduction of water to H_2 can be achieved in the presence of a sacrificial electron donor D and a suitable catalyst (Figure 8.7) as in the first scheme (Figure 8.3). In this five-component system PS/R_{en}/R/D/Cat, the energy-transfer photosensitizer PS is not involved in any redox processes as are the antenna molecules in natural photosynthesis, and the receptor R_{en} acts as an energy-electron relay.

It should be noted that, although the multimolecular approach is the simplest way to achieve cyclic photochemical water cleavage, its accomplishment necessitates overcoming several difficulties. Indeed, the different components of such systems must fulfil spectral, photophysical, thermodynamic and kinetic conditions. Some of them have been mentioned here. The other requirements can be found in earlier comprehensive reviews [4,5,8,9,12-15].

4.2 FIRST MODEL SYSTEMS FOR H_2 PRODUCTION

4.2.1 First Model Systems

Several sacrificial model systems of H_2 production from water have been proposed since 1977. The first ones are listed in Table 8.2. They used acridine dyes such as acridine yellow [16] as PS. But transition metal complexes, in particular $[Ru(bpy)_3]^{2+}$ [11,17], then appeared to be remarkable photosensitizers with respect to visible light absorption, excited state properties, redox potentials and kinetic requirements [14]. The electron relay species first investigated include Eu^{3+} and V^{3+} salts [16], a transition metal complex $[Rh(bpy)_3]^{3+}$ [17] which can transfer two electrons, and methyl viologen MV^{2+} the most commonly used electron relay [11,16]. Cysteine [16] and especially tertiary amines such as EDTA

Table 8.2. First hydrogen-generating model systems

System	Photosensitizer (PS)	Electron relay (R)	Electron donor (D)	Catalyst (Cat)
Shilov group [16]	Acridine yellow	Eu^{3+} or V^{3+}	Cysteine	PtO_2
Lehn group [17]	$[Ru(bpy)_3]^{2+}$	$[Rh(bpy)_3]^{3+}$	TEOA	K_2PtCl_6
Orsay groups [11]	$[Ru(bpy)_3]^{2+}$	MV^{2+}	EDTA	Colloidal Pt (PVA)

(ethylenediamine tetra-acetic acid) [11,16] and triethanolamine TEOA [16,17], which are rapidly decomposed when oxidized, were used as sacrificial electron donors. Platinum compounds [11,16,17] turned out to be suitable catalysts.

The first system (Table 8.2) has been described by Shilov and coworkers [16]. It consists of acridine yellow AY as PS, cysteine as sacrificial electron donor, salicylate complexes of Eu^{3+} and V^{3+} as R and Adams' catalyst (PtO_2) as Cat. They also used EDTA, TEOA or H_2S as D, MV^{2+} as R and K_2PtCl_6 as catalyst. The mechanism assumed was of the "reductive" type (Figure 8.4), and the quantum yield of H_2 production in the case of AY/Eu^{3+}/cysteine/PtO_2 model system was of the order of 1%. This quantum efficiency was too low so that the validity of the system was questioned at that time. Indeed, in the early 1970s, binuclear metal complexes were considered as more promising candidates for PS than organic compounds [18]. The two following systems, that of Lehn and Sauvage [17,19] (Table 8.2) and our own [11,20] (Table 8.2) used a metal complex, $[Ru(bpy)_3]^{2+}$, as PS. The hydrogen quantum yields $\oslash$ $\left(\frac{1}{2} H_2\right)$ were much higher (>10%) than that of Shilov's system [16]. Consequently, it was easier to produce and characterize H_2, and even to detect the formation of H_2 bubbles with the naked eye. In addition, in the case of the Orsay system (Table 8.2), the oxidative quenching mechanism for H_2 production (Figure 8.3) by laser flash spectroscopy was clearly established [11,21]. So it is not surprising that these two systems [11,17] drew attention to the interest of the multimolecular approach and contributed largely to the popularity of the domain.

It should be noted that in the Shilov system [16] and in the Lehn system [17], it was assumed that Pt particles are formed *in situ* through the photosensitized reduction of K_2PtCl_6, while in Orsay system [11] it was demonstrated for the first time that colloidal metals (Pt, Au), chemically prepared and stabilized by polymers (polyvinyl alcohol PVA), can be used successfully as catalysts in a photochemical model system.

Grätzel and coworkers [22] described at the same time a system, $[Ru(bpy)_3]^{2+}$/MV^{2+}/TEOA/PtO_2, similar to the Orsay system (Table 8.2) [11,21], but using platinum oxide PtO_2 (Adams' catalyst) instead of colloidal platinum, and triethanolamine TEOA or cysteine instead of EDTA. With these components (PtO_2, TEOA) [22], the H_2 yields are very low [21,23].

It is hardly possible to give an exhaustive list of all the multimolecular systems of H_2 production from water described in the literature and based on the general schemes (Figures 8.3–8.5, 8.7). However, I have categorized the different constituents used in these systems as PS, R, D and Cat, and I shall describe only a few systems, in particular the Orsay system, which is also called the classical model system.

4.2.2 Photosensitizers

The principal classes of photosensitizers PS are: transition metal complexes of Ru, Cr, Os... [24,25], metalloporphyrins and metallophthalocyanines [25,26] and acridine dyes [16,27–29], $[Ru(bpy)_3]^{2+}$ being the most investigated (Table 8.3). This complex is essentially involved in the oxidative quenching mechanism. For the reductive mechanism, one of the best candidates as PS is $[Ru(bpz)_3]^{2+}$ (bpz = 2, 2'-bipyrazine) [30,31]. Hydrogen is produced with good yields when it is used in a photochemical system with TEOA as D, MV^{2+} as R and a platinum compound as catalyst [32,33]. Another attractive complex is $[Ru(bpy)_2dppz]^{2+}$ (dppz = dipyrido [3,2-a: 2',3'-c] phenazine) which can be involved as PS either in an oxidative mechanism, with EDTA as D, or in a reductive mechanism, with TEOA as quencher [34]. Interestingly, a rigid copper(I) complex $[Cu(dpp)_2]^+$ (dpp = 2, 9-diphenyl-1,10-phenanthroline) has been used as energy-transfer photosensitizer (Figure 8.7) in a five-component system [35]. It should be noted that high quantum yields for H_2 formation with an optimum $\Phi\left(\frac{1}{2}H_2\right) = 0.6$ [36] have been found when aqueous solutions of a water-soluble zinc porphyrin, Zn(II) tetrakis (N-methyl-4-pyridyl) porphyrin $ZnTMPyP^{4+}$ [26,36–38], are irradiated with 550 nm light in the presence of MV^{2+}, EDTA and colloidal Pt. However, these yields decrease dramatically within an irradiation time of 4 hrs [26,36]. More recently, cage complexes [39,40], cyanine dyes [41] and organic compounds absorbing visible light such as poly(pyridine-2,5-diyl) [42] have been tested as PS.

4.2.3 Electron Relays

Bipyridinium ions, also called viologens, are the main compounds used as electron relays R (Table 8.4), the most popular being methyl viologen MV^{2+}. They also provide an extended range of redox potentials [43] (Table 8.5).

Table 8.3. Principal classes of photosensitizers

Photosensitizer (PS)
Metal complexes of Ru, Cr, Os, Ir, Pt ...: $[Ru(bpy)_3]^{2+}$
Metal porphyrins of Zn, Mg, Ru ...: $ZnTMPyP^{4+}$
Metal phtalocyanines of Zn, Co, Mg ...
Acridine dyes: acridine yellow, proflavine.
Xanthene dyes: fluorescein, eosin Y
Cyanine dyes
Organic compounds: poly(pyridine-2,5-diyl)

Table 8.4. Principal classes of electron relays

Electron relay (R)
Bipyridinium ions: MV^{2+}
Phenanthrolinium ions
Metal ions: Eu^{3+}, V^{3+}, Cr^{3+}
Metal complexes of Rh, Co ...: $[Rh(bpy)_3]^{3+}$, $[Co(Sep)]^{3+}$
Proteins: cytochrome c_3

Several homogeneous series of 4,4'-bipyridinium (or paraquat), 2,2'-bipyridinium (or diquat), and 1,10-phenanthrolinium ions as mediators for water photoreduction were investigated [43–46] the most efficient electron relays being MV^{2+} and 1,1'-dimethylene-4,4'-dimethyl-2,2'-bipyridinium ion [43–45]. Transition metal complexes are also interesting mediators (Table 8.4), in particular $[Rh(bpy)_3]^{3+}$ which can transfer two electrons [17,47,48] and cage complexes such as $[Co(sep)]^{3+}$ (sep = sepulchrate) [49–52]. $[Co(sep)]^{3+}$ is, contrary to viologens, insensitive to hydrogenation, an undesired side reaction which can occur at the catalyst surface. A natural electron mediator, cytochrome c_3, which unlike MV^{2+} is not at all toxic, has been tested in model systems in association with hydrogenase as catalyst [29,53]. It is of interest to note that the only compound used as energy-electron relay R_{en} in a five-component system (Figure 8.7) is 9-carboxylate anthracene anion [35,54].

4.2.4 Electron Donors

Krasna [27] thoroughly tested several classes of organic compounds as electron donors D (Table 8.6), with proflavine as PS, MV^{2+} as R and the enzyme hydrogenase or Pt asbestos as Cat. With this model system, the most effective donors were EDTA and 1,2-diaminocyclohexane tetraacetic acid. Whitten and coworkers [55] found that triethylamine TEA which is not at all effective in Krasna's systems, leads to high H_2 yields (0.53) in a three-component system $[Ru(bpy)_3]^{2+}/TEA/PtO_2$, i.e. in the absence of MV^{2+}, but in 25% water–acetonitrile mixtures. Coenzymes such as NADH and NADPH have been tested as sacrificial electron donors [53,56]. However, the photoinduced regeneration of these natural reductants can also be achieved [56,57]. In photochemical systems involving a sacrificial electron donor, it is interesting, from a practical point of view, to find and use electron donors such as H_2S [56] which are easily available as waste industrial products.

4.2.5 Microheterogeneous Catalysts

As regards the catalyst (Table 8.7), and since the first report on the catalytic activity of colloidal Pt in a photochemical system of H_2 production [11], colloidal metals (of groups VIII and IB essentially), metals deposited on solid supports

Table 8.5. Molecular structure and reduction potentials for viologen compounds

Molecular structure						Usual notation	Reduction potential (V vs NHE)
R_1	R_2	R_3	R_5	R_6			
CH_3	H	H	H	H		MV^{2+}, PQ^{2+}	−0.44
CH_3	CH_3	H	H	H		TMV^{2+}	−0.51
CH_3	CH_3	H	H	CH_3		HMV^{2+}	−0.64
CH_3	H	CH_3	H	H			−0.83
$CH-(CH_3)_2$	H	H	H	H			−0.45
CH_2-CH_2OH	H	H	H	H			−0.40
CH_2-Ph	H	H	H	H		BV^{2+}	−0.33
CH_2-CN	H	H	H	H			−0.15
$(CH_2)_6-CH_3$	H	H	H	H			−0.41
$(CH_2)_3-SO_3^-$	H	H	H	H		SPV, PVS	−0.41
$(CH_2)_3-SO_3^-$	H	CH_3	H	H		MPVS	−0.79
							−0.31

(continued overleaf)

Table 8.5. continued

Molecular structure	n	R_3	R_4	R_5	R_6	Usual notation	Reduction potential (V vs NHE)
						DQ^{2+}	−0.72
	2	H	H	H	H	DQ_2^{2+}	−0.37
	3	H	H	H	H	DQ_3^{2+}	−0.55
	4	H	H	H	H	DQ_4^{2+}	−0.65
	2	H	CH_3	H	H		−0.49
	3	H	CH_3	H	H		−0.70
	4	H	CH_3	H	H		−0.78
	2	CH_3	H	H	H		−0.59
	2	H	H	CH_3	H		−0.48
	2	H	H	H	CH_3		−0.48

Table 8.6. Principal classes of electron donors

Electron donor (D)
EDTA and glycine derivatives (NPG)
Amines: TEOA, TEA.
Sulphur compounds: cysteine, thiols (mercaptoethanol), H_2S
Urea derivatives; allylthiourea
Amino acids
Carbon compounds: ascorbate, ethanol
Metal ions: Eu^{2+}
Coenzymes: NADH, NADPH

Table 8.7. Principal classes of catalysts

Catalyst (Cat)
Colloidal metals: group VIII metals (Ir, Pt, Ni...), Au, Ag
Pt salts: K_2PtCl_6, K_2PtCl_4
Metal powders: Pt, Ru, Ni.
Supported metals: Pt–TiO_2, Rh–$SrTiO_3$, Ni–TiO_2
Metal oxides: RuO_2, PtO_2, IrO_2, PdO_2, TiO_2, Fe_2O_3.
Supported metal oxides: RuO_2 + IrO_2/zeolite
Colloidal metal systems: Ni–Pd
Enzymes: hydrogenase, nitrogenase.

(semiconductor, zeolite), metal and metal oxide powders have been systematically investigated [58,59].

Iridium and platinum hydrosols are extremely efficient [58,59] (Table 8.8). Platinum supported on TiO_2 [33,58–60] leads to similar high H_2 yields. Ruthenium oxides, known to be good catalysts for O_2 generation from water (Section 4.4), efficiently mediate the photoreduction of water to H_2 without catalyzing the undesired hydrogenation of the electron relay [61–63]. RuO_2 and IrO_2 codeposited on zeolite give the highest H_2 yields [58]. More recently, we have observed an improvement of catalytic activity by an alloying effect in the case of bimetallic sols of Ni-Pd [64]. It is of interest to note that the enzyme hydrogenase which contains at least Fe_4S_4-type clusters has been used as a natural catalyst [27,29,63], but it is unstable and the H_2 yields are lower than those obtained with Pt compounds [27].

4.2.6 Homogeneous Catalysts and Homogeneous Systems for H_2 Production

The great majority of catalysts described in the literature are heterogeneous. Very few homogeneous catalysts and hence homogeneous systems for H_2 production (Table 8.9) have been reported (see Section 4.4 for homogeneous catalysts for O_2 production). In these systems, following visible-light excitation of the photosensitizer PS, one of the components is transformed into an unstable intermediate, for

Table 8.8. Efficiency of different classes of catalysts. Quantum yields for hydrogen formation $\Phi\left(\frac{1}{2} H_2\right)$ [58,84] from the irradiation ($\lambda_{exc} = 453$ nm) of pH 5 aqueous solutions of $[Ru(bpy)_3]^{2+}$ 5.65×10^{-5}M, MV^{2+} 3×10^{-3}M, EDTA 0.1 M and various catalysts

Catalyst	Particle diameter (Å)	Metal concentration[a] ($\times 10^5$ M)	Amount of catalyst added[b] (mg)	$\Phi\left(\frac{1}{2} H_2\right)$[c]
Colloidal metals				
Ir	40	2		0.173
Ir	12	2		0.173
Ir	<8	2		0
Pt	16–1000	1.92		0.171
Os	<50	5		0.160
Ru	<50	6		0.139
Rh	<50	4		0.080
Co	<50	2		0.066
Ni	<50	4		0.060
Pd	<50	50		0.056
Ag	<50	2		0.050
Au	<50	10		0.042
Cu, Cd, Pb	<50	1-100		0
Metal-supported powders				
0.5 wt % Pt-TiO$_2$	20	6.15	12	0.160
1.4 wt % Pt-SiO$_2$		17.2	12	0.154
0.6 wt % Pt-Al$_2$O$_3$	30–35	7.38	12	0.120
4.83 wt % Ni-TiO$_2$	135	132	8	0.108
Metal oxide powders				
(RuO$_2$ + IrO$_2$) zeolite			12	0.142
RuO$_2$-TiO$_2$			5	0.121
RuO$_2$, xH$_2$O			3	0.117
PtO$_2$, xH$_2$O			22	0.072
TiO$_2$ P25			12	0.068
Fe$_2$O$_3$			4	0.056
MnO$_2$			3	0.042
WO$_3$			6	0.039

[a] Optimum concentration for colloidal metals. Metal concentration in at-g l^{-1} for metal-supported powders.
[b] Amount added in 5 ml samples.
[c] Corrected for light-scattering effects by the colloidal metal particles, but uncorrected for catalyst powders. In the latter case, it corresponds to a lower limit.

instance a metal hydride, which in turn decomposes to yield H$_2$ and the starting component. Therefore, this component acts as homogeneous catalyst. Good candidates are inorganic compounds, such as homogeneous hydrogenation catalysts, which have a metallic site able to present different oxidation states during the catalytic cycle, and which can form an intermediate hydride, unstable in solution, to provide a pathway for H$_2$ release. Homogeneous catalysts (Table 8.9) which were first reported in 1979 are a macrocyclic cobalt(II) complex, [Co(Me$_6$[14]diene N$_4$)(H$_2$O)$_2$]$^{2+}$ [65], and [Rh(bpy)$_3$]$^{3+}$ [47,48]. Other Co(II) complexes have

WATER SPLITTING

Table 8.9. Components, quantum yields and mechanism for H_2-generating homogeneous systems

Photosensitizer	Electron donor	Homogeneous catalyst	pH	$\Phi(H_2)$	Mechanism[a]	Ref.
$[Ru(bpy)_3]^{2+}$	Eu^{2+}	$[Co(Me_6[14]diene N_4)(H_2O)_2]^{2+}$	5	0.05	Red	[65]
$[Ru(bpy)_3]^{2+}$	Ascorbate	$[Co(Me_6[14]diene N_4)(H_2O)_2]^{2+}$	5	0.0005	Red	[65]
$[Ru(bpy)_3]^{2+}$	EDTA	$[Rh(bpy)_3]^{3+}$	5.2	0.04 [41]	Ox	[47]
$[Ru(bpy)_3]^{2+}$	TEOA	$[Rh(bpy)_3]^{3+}$	5	0.02	Ox	[48]
$[Ru(bpy)_3]^{2+}$	Ascorbate	$[Co(bpy)_n]^{2+}$	5	0.03	Red	[66]
$[Ru(bpy)_3]^{2+}$	TEOA	$[Co(dimethyl glyoxime)_2]$	8.7[b]		Red	[69]
$[Ru(4,7-(CH_3)_2phen)_3]^{2+}$	TEOA	$[Co(bpy)_3]^{2+}$	8	0.29[c]	Ox	[68]

[a]Reductive (red) or oxidative (ox) quenching of PS*.
[b]In DMF/H_2O or in neat organic media (DMF, acetone, acetonitrile ...).
[c]In 50% aqueous acetonitrile.

been proposed, namely $[Co(bpy)_3]^{2+}$ [66-68], cobaloxime and other macrocyclic complexes [69]. These homogeneous systems (Table 8.9) consist of three components, the photosensitizers PS being essentially $[Ru(bpy)_3]^{2+}$ [47,48,65-67,69] or $[Ru(4,7-(CH_3)_2 phen)_3]^{2+}$ (phen = 1, 10-phenanthroline) [68], and the electron donors being EDTA [47,48], Eu(II) [65], ascorbate [65-67] or tertiary amines like TEOA [48,68,69]. As in the case of microheterogeneous systems, the mechanism of H_2 production involves either a reductive [65-67,69] (reaction 10) or an oxidative [47,48,68] (reaction 5) quenching of PS*. These systems are efficient, especially in organic media [68,69]. In the case of the $[Ru(4,7-(CH_3)_2phen)_3]^{2+}$ /TEOA/$[Co(bpy)_3]^{2+}$ system proposed by Sutin and coworkers [68], the H_2 yields increase from about 0.02 in H_2O to 0.29 in 50% CH_3CN-H_2O (Table 8.9).

4.3 THE CLASSICAL MODEL SYSTEM

4.3.1 Introduction

The classical system proposed in 1978 for the first time by the Orsay groups [11,21] (Table 8.2) comprises $[Ru(bpy)_3]^{2+}$ as PS, MV^{2+} as R, EDTA as D, and colloidal Pt as catalyst (Figure 8.8). It has been known since 1934 [70] that electron exchange between MV^{2+} and H_2 is catalyzed by Pt. But for the first time, it has been proved that catalyst hydrosols [71] used in a model photosystem [11] mediate visible-light H_2 generation from water. In previous studies [16,17] the formation *in situ* of such colloids had been assumed through the reduction of Pt salts. This system produces H_2 very efficiently, leads to reproducible results, and is well characterized as regards H_2 formation quantum yields [58,72] and detailed mechanism [21,43]. These are the reasons why this system has been

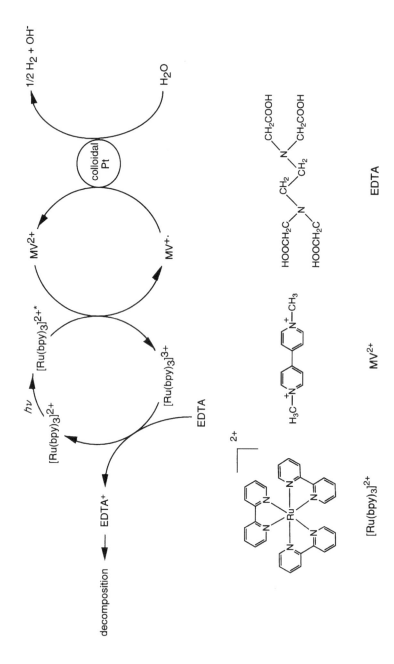

Figure 8.8. (Upper) Schematic representation of the redox catalytic cycles in the photoreduction of water to H_2 by visible-light irradiation of the Orsay model system $[Ru(bpy)_3]^{2+}/MV^{2+}/EDTA/$colloidal Pt proposed by Moradpour, Amouyal et al. [11]. (Lower) Molecular structure of the components

WATER SPLITTING

thoroughly studied by several groups [23,32,60,73–80] and is still considered as a reference for testing new PS, R, D and catalysts (Table 8.3–8.8), and for evaluating solar photochemical reactors on a pilot level [81]. The system has been described in many reviews and books. I now wish to recall some of its important features.

4.3.2 Mechanism

When an aqueous solution containing $[Ru(bpy)_3]^{2+}$, MV^{2+}, EDTA, and colloidal Pt is irradiated with visible light (400 nm $< \lambda <$ 600 nm), an important H_2 evolution is observed (Figure 8.8) according to the following mechanism established from laser flash spectroscopy experiments [11,21]. In deaerated solutions, the main reactions are as follows:

$$[Ru(bpy)_3]^{2+} \xrightarrow{h\nu} [Ru(bpy)_3]^{2+*} \qquad (17)$$

$$[Ru(bpy)_3]^{2+*} \xrightarrow{k_0} [Ru(bpy)_3]^{2+} \qquad (18)$$

$$[Ru(bpy)_3]^{2+*} + MV^{2+} \xrightarrow{k_q} [Ru(bpy)_3]^{3+} + MV^{+\cdot} \qquad (19)$$

$$[Ru(bpy)_3]^{3+} + MV^{+\cdot} \xrightarrow{k_b} [Ru(bpy)_3]^{2+} + MV^{2+} \qquad (20)$$

$$[Ru(bpy)_3]^{3+} + EDTA \xrightarrow{k_{ox}} [Ru(bpy)_3]^{2+} + EDTA^{+} \qquad (21)$$

$$MV^{+\cdot} + H^+(H_2O) \xrightleftharpoons{Pt} MV^{2+} + \tfrac{1}{2}H_2 \qquad (22)$$

$$\text{Net: EDTA} + H^+ \xrightarrow{h\nu, Pt} EDTA^+ + \tfrac{1}{2}H_2 \qquad (23)$$

The proposed mechanism for EDTA degradation (reaction 24) and for the catalytic process on metallic particles (reaction 22) have been described in detail in previous reports [21,43,59,82].

$$EDTA^+ \longrightarrow \text{products (mainly glyoxylic acid)} \qquad (24)$$

Besides the EDTA consumption (reaction 24), difficulties arise from undesirable reactions such as $MV^{+\cdot}$ dimerization [82] (reaction 25)

$$2MV^{+\cdot} \rightleftharpoons (MV^{+\bullet})_2 \qquad (25)$$

and the irreversible hydrogenation of methyl viologen [21,43,82] (reaction 26, (Figure 8.9)):

$$MV^{+\cdot}(\text{or } MV^{2+}) \xrightarrow{H_2, Pt} MV^+H \xrightarrow{H_2, Pt} \text{products} \qquad (26)$$

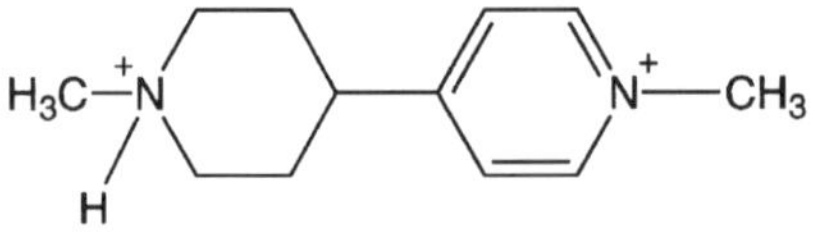

Figure 8.9. MV^{2+} hydrogenation products: 1-methyl-4-[4'-(1'-methylpiperidyl)] pyridinium (a) and 1,1'-dimethyl-4,4'-bipiperidinium (b)

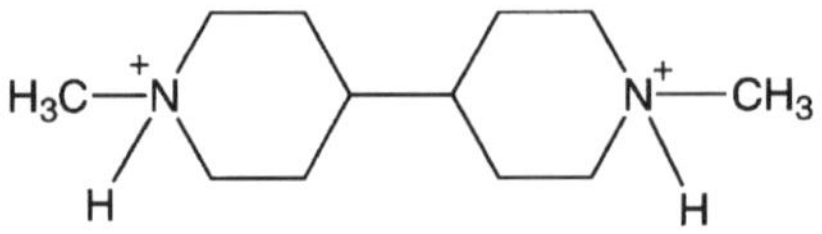

The rate constant values, as determined by flash photolysis [21,65] are the following:

$$k_0 = 1.45 \times 10^6 \text{ s}^{-1}$$
$$k_q = 1.03 \times 10^9 \text{ M}^{-1}\text{s}^{-1} \text{ (for } \mu = 0.018 \text{ M and dried } MV^{2+})$$
$$k_b = 2.8 \times 10^9 \text{ M}^{-1}\text{s}^{-1}$$
$$k_{ox} = 1.1 \times 10^8 \text{ M}^{-1}\text{s}^{-1}$$

k_q increases with the ionic strength μ of the solution [83] with $k_q^o = 2 \times 10^8$ M^{-1}s^{-1} extrapolated for zero μ [43].

In non-deaerated solutions, i.e. in the presence of molecular oxygen, H_2 formation rates and yields decrease due mainly to reaction (27) with $k = 4.6 \times 10^9$ M^{-1}s^{-1} [84] and to reaction (28) with $k = 8 \times 10^8$ M^{-1}s^{-1} [85].

$$[Ru(bpy)_3]^{2+*} + O_2 \longrightarrow [Ru(bpy)_3]^{2+} + {}^1O_2 \qquad (27)$$

$$MV^{+\cdot} + O_2 \longrightarrow MV^{2+} + O_2^- \qquad (28)$$

This reaction leads to H_2O_2 as a stable intermediate and to MV^{2+} degradation products [86,87].

$$2O_2^- + 2H^+ \longrightarrow 2HO_2^{\cdot} \longrightarrow O_2 + H_2O_2 \qquad (29)$$

$$MV^{+\cdot} + O_2^- (HO_2^{\cdot}) \longrightarrow MVO_2 \longrightarrow \text{products} \qquad (30)$$

4.3.3 Optimization

The mechanism shows that H_2 production depends on light intensity, pH and concentration of the four constituents of the system. An optimal quantum yield:

$$\Phi \left(\tfrac{1}{2}H_2\right) = 0.171$$

WATER SPLITTING

was found for the following optimized concentrations [58]:

$$pH = 5,$$
$$[Ru(bpy)_3]^{2+} = 5.65 \times 10^{-5} \text{ M},$$
$$MV^{2+} = 3 \times 10^{-3} \text{ M},$$
$$EDTA = 0.1 \text{ M},$$
$$\text{Colloidal Pt} = 1.92 \times 10^{-5} \text{ M}.$$

For the same optimized conditions but in the absence of Pt, the $MV^{+\cdot}$ quantum yield was

$$\Phi(MV^{+\cdot}) = 0.181$$

in good agreement with the value calculated from the relationship (31):

$$\Phi(MV^{+\cdot}) = \phi_T \phi_q \phi_{ce} \tag{31}$$

provided that the cage-escape efficiency $\phi_{ce} = 0.30$ [58] (ϕ_T is the quantum yield for intersystem crossing, ϕ_q is the efficiency of the quenching reaction (19), ϕ_{ce} is the efficiency of net formation of $[Ru(bpy)_3]^{3+}$ and $MV^{+\cdot}$ from the solvent cage before the ions undergo back electron transfer [88]). The irreversible hydrogenation of $MV^{+\cdot}$ (or MV^{2+}) catalyzed by colloidal Pt (reaction 26) is a main limiting factor to the longevity of the system [21,89,90].

The catalytic turnover numbers TN for H_2 production, defined as the ratio of the total H_2 amount to the initial amount of the considered component, reflect the stability of each constituent of the system, and particularly MV^{2+}. They are for example [21]:

$$TN([Ru(bpy)_3]^{2+}) > 290$$
$$TN(MV^{2+}) = 115$$
$$TN(\text{colloidal Pt}) > 2900$$

Effectively, H_2 generation stops when the viologen is completely destroyed via hydrogenation. Meanwhile $[Ru(bpy)_3]^{2+}$ is slightly decomposed and colloidal Pt remains apparently intact. Thus these calculated TN values are lower limits of the true TN for $[Ru(bpy)_3]^{2+}$ and Pt. Indeed, an extrapolated TN [47] of 6000 has been found [74a] for $[Ru(bpy)_3]^{2+}$. It should also be stressed that these TN were not determined under optimal experimental conditions for H_2 production [89]. The inhibition of the catalytic hydrogenation is thus of major importance. Such an undesired reaction may be prevented:

1. By finding electron relays whose structure would be less sensitive to hydrogenation such as HMV^{2+} (1,1',2,2',6,6'-hexamethyl-4,4'-bipyridinium ion) [91,92] and $[Co(sep)]^{3+}$ [49–52];

2. By using more specific catalysts such as RuO_2 [61,62]; and/or
3. By adding to the solution hydrogenation poisons such as sulphur compounds, glutathione GSH in particular [58,59,74].

Indeed, in the presence of GSH, colloidal Pt is operating with 100% efficiency. In particular, for $[Pt] > 1.92 \times 10^{-5}$ M, we found that:

$$\Phi \left(\tfrac{1}{2} H_2 \right) = \Phi(MV^{+\cdot}) = 0.18$$

Not only H_2 formation rates but also H_2 amounts are considerably improved in the presence of a hydrogenation poison. With GSH, the MV^{2+} turnover reaches a value $TN(MV^{2+}) = 750$ [92].

In a five-component modified version (Figure 8.7) $[Ru(bpy)_3]^{2+}/AC^-/MV^{2+}/$EDTA/colloidal Pt using 9-carboxylate anthracene anion AC^- as energy-electron relay R_{en}, Sasse and coworkers [54b] have achieved the highest quantum yield for H_2 production [54b,82]:

$$\Phi \left(\tfrac{1}{2} H_2 \right) = 0.85$$

This result reflects a cage escape efficiency ϕ_{ce} (eq. 31) of about 100% for the charge pair $(AC\cdot, MV^{+\cdot})$ in reaction (16). Recently, a high turnover number TN (MSPV) = 358 has been calculated for a zwitterionic viologen MSPV (3,3'-dimethyl-1,1'-bis(3 sulfonato-propyl)-4,4'-bipyridinium) in a modified system $[Ru(bpy)_3]^{2+}/MSPV/TEOA/TiO_2$-Pt at pH 7.6 [33]. This result reinforces the interest of using microheterogeneous catalysts such as TiO_2–Pt which was found to be as efficient as colloidal Pt for H_2 generation from water [58] (Table 8.8).

4.4 MODEL SYSTEMS FOR O_2 PRODUCTION

4.4.1 First Model Systems

Contrary to half photosystems of H_2 production, the number of model systems of water photooxidation is much more limited. The first model systems for O_2 production from water by visible-light radiation have been proposed in 1979 by Lehn et al. [93] and a little later by Kalyanasundaram and Grätzel [94]. Also based on the multimolecular approach, they consist of three components including $[Ru(bpy)_3]^{2+}$ as PS, $[Co(NH_3)_5Cl]^{2+}$ as sacrificial electron acceptor, and RuO_2 as catalyst (Figure 8.10).

The main reactions (32–35) leading to O_2 evolution are the following:

$$[Ru(bpy)_3]^{2+} \xrightarrow{h\nu} [Ru(bpy)_3]^{2+*} \tag{32}$$

$$[Ru(bpy)_3]^{2+*} + [Co(NH_3)_5Cl]^{2+} \longrightarrow [Ru(bpy)_3]^{3+} + [Co(NH_3)_5Cl]^{+} \tag{33}$$

$$[Co(NH_3)_5Cl]^{+} + 5H^{+} \longrightarrow Co_{aq}^{2+} + 5NH_4^{+} + Cl^{-} \tag{34}$$

$$4[Ru(bpy)_3]^{3+} + 2H_2O \xrightarrow{RuO_2} 4[Ru(bpy)_3]^{2+} + 4H^{+} + O_2 \tag{35}$$

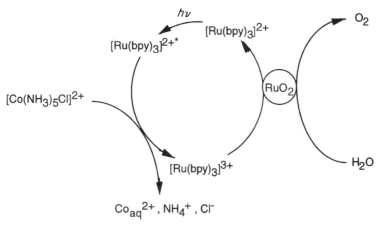

Figure 8.10. Schematic representation of the redox catalytic cycles in the photooxidation of water to oxygen by visible-light irradiation of the $[Ru(bpy)_3]^{2+}/[Co(NH_3)_5 Cl]^{2+}/RuO_2$ model system [93,94]

The mechanism involves an oxidative quenching of the excited state of $[Ru(bpy)_3]^{2+}$ by a Co(III) complex, $[Co(NH_3)_5Cl]^{2+}$ (reaction 33). The rapid and irreversible aquation of $[Co(NH_3)_5Cl]^{+}$ (reaction 34) allows the accumulation of $[Ru(bpy)_3]^{3+}$ which is an oxidant strong enough to oxidize water to O_2 in the presence of a suitable redox catalyst like RuO_2 (reaction 35).

4.4.2 Other Microheterogeneous Molecular Systems

Modified versions of this system have been proposed with practically only $[Ru(bpy)_3]^{2+}$ as PS because it is the most efficient, although $[Ru(bpz)_3]^{2+}$ [95] and metalloporphyrins such as $ZnTMPyP^{4+}$ [96] also appear suitable for O_2 generation. However, several sacrificial electron acceptors have been examined including $[Co(NH_3)_5Cl]^{2+}$ [93-95,97-99], $[Co(NH_3)_5Br]^{2+}$ [94], $[Co(C_2O_4)_3]^{3-}$ [94], Tl^{3+} [94,98], $S_2O_8^{2-}$ [95,98-100], Ag^+ [96,101], Fe^{3+} [96,102] and a Mn(IV) pyrophosphate complex [103]. Because redox catalysts represent an important key to the water cleavage problem, they were investigated even more widely. Thus catalysts which are active in photochemical systems include RuO_2 powder [93,94], colloidal RuO_2 [94], RuO_2 supported on alumina [93] or TiO_2 [96], $CoSO_4$ [99] and Prussian blue [104-106]. Quantum yields for O_2 generation from the irradiation of aqueous solutions of $[Ru(bpy)_3]^{2+}/[Co(NH_3)_5Cl]^{2+}$/catalyst systems at pH 5 were found to be 0.03, 0.02 and 0.003 with colloidal RuO_2, $CoSO_4$ and RuO_2 powder respectively [99]. Lehn et al. [107] and Harriman et al. [108] tested several metal oxide catalysts (RuO_2, IrO_2, PtO_2 MnO_2, Mn_2O_3, Rh_2O_3, Co_3O_4, $NiCo_2O_4$). Iridium and ruthenium oxides supported on Y zeolites are much more efficient than RuO_2 powder or RuO_2 supported on γ-alumina and silica, the most efficient being a mixture of IrO_2 and RuO_2 deposited on Y zeolite [107]. Pure RuO_2 (stoichiometric) and PtO_2 are inactive for O_2 generation [107]. It is of interest to note, because of the involvement of Mn ions in the oxygen-evolving center in green-plant photosystem II [109–111], that MnO_2 [103,107], Mn_2O_3 [108] and a Mn(IV) cluster bound to a lipid vesicle [112] are also efficient catalysts. Shafirovich et al. [97] have claimed that visible-light irradiation of aqueous solutions of a two-component homogeneous system $[Ru(bpy)_3]^{2+}/[Co(NH_3)_5Cl]^{2+}$ resulted in continuous O_2 generation at pH 7. They assumed that $[Co(H_2O)_6]^{2+}$, the hydrolysis product of $[Co(NH_3)_5Cl]^+$, acts as a redox catalyst. However, other laboratories [99,113] did not observe any photochemical evolution of oxygen, although it has been reported that hydroxo complexes of Co(II), Fe(II), Fe(III), Ni(II) and Cu(II) are active catalysts for non-photochemical generation of O_2 from water in alkaline solutions using $[M(bpy)_3]^{3+}$ (M = Fe, Ru, Os) or $[IrCl_6]^{2-}$ as oxidants [97,114].

4.4.3 Homogeneous System

In a different approach, μ-oxo dinuclear complexes of transition metals have been examined not only for their possible activity as homogeneous catalysts but also because they may provide models for the oxygen-evolving site of photosystem II in natural photosynthesis [109–111]. For this purpose, the di-μ-oxotetrakis (2,2'-bipyridyl) dimanganese(III,IV) perchlorate (Figure 8.11) has been proposed as model system by Calvin [115]. But, contrary to this first report, Cooper and Calvin [116] did not succeed in producing O_2 photochemically. However, it was shown that dinuclear complexes of Mn as well as μ-oxo diruthenium complexes (Figure 8.12) as first examined by Meyer and coworkers

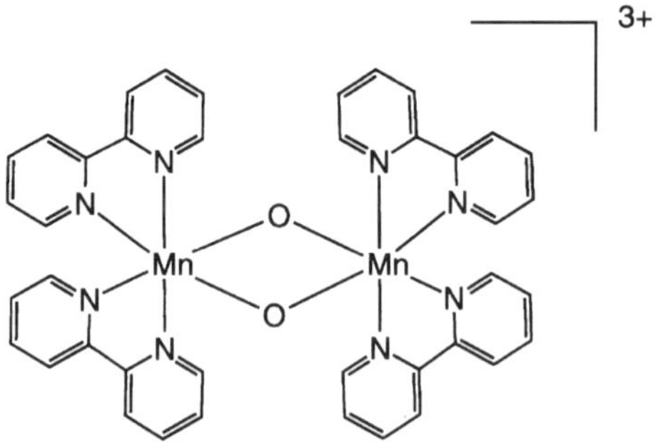

Figure 8.11. Di-μ-oxotetrakis (2,2'-bipyridyl) dimanganese(III,IV) complex: $[(bpy)_2 Mn^{III} (\mu$-O$)_2 Mn^{IV} (bpy)_2]^{3+}$, $3ClO_4^-$

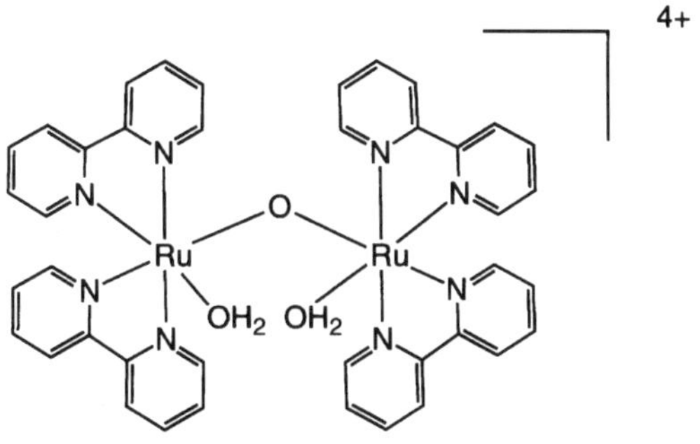

Figure 8.12. μ-oxo dinuclear complex of ruthenium (III): $[(bpy)_2(H_2O)Ru^{III}ORu^{III}(H_2O)(bpy)_2]^{4+}$, $4ClO_4^-$

[117], are good catalysts for O_2 generation from water in a non-photochemical process [117–119]. It was postulated that O_2 formation involves higher oxidation states of ruthenium, probably a Ru(V), Ru(V) dinuclear complex [117,118]. More recently, similar water-soluble μ-oxo dinuclear complexes were proposed as homogeneous catalysts for thermal water oxidation [120,121]. Moreover, photoinduced generation of oxygen from water was also observed, according to the general scheme (Figure 8.6), by using these complexes as catalysts, Ru(II) tris(4,4'-dicarbethoxy-2,2'-bipyridine) as PS and $S_2O_8^{2-}$ as a sacrificial electron acceptor [120,121].

4.4.4 Confined Systems

Another interesting approach involves a restricted geometry such as zeolite cages [122–126]. Calzaferri and coworkers [123–126] have shown that irradiation of silver zeolites A dispersed in aqueous solution leads to the reduction of silver cations Ag^+ and to the production of oxygen with high yields. The process may be schematically represented by reaction (36):

$$(Ag^+, H_2O)_{zeolite} \xrightarrow{h\nu} (Ag^0, H^+)_{zeolite} + O_2 \qquad (36)$$

The system which is initially insensitive to visible light needs to be illuminated with near-UV light (about 370 nm) before becoming active in the visible up to about 600 nm. Clearly, O_2 is produced by self-sensitization. This phenomenon is interpreted as a quantum size effect (p. 190). It can be explained by the photoinduced formation of partially reduced silver clusters (probably Ag_6^{m+}, with $m < 6$, in zeolite A) which absorb at longer wavelength and act as new chromophores.

Similar results have been obtained with monograin layers of silver zeolite A [125,126a,b]. More recently, Calzaferri and coworkers have found that oxygen evolves from thin AgCl layers on SnO_2-coated glass plates, in the presence of a small excess of Ag^+ ions in the aqueous solutions, with a maximum evolution rate at pH 4-5. As for silver zeolites, oxygen is produced by self-sensitization with high quantum yields of 0.8 upon irradiation with UV or blue light (420-480 nm) and 0.5 in the 500-540 nm excitation range.

5 SUPRAMOLECULAR SYSTEMS

5.1 SUPRAMOLECULAR SYSTEMS FOR CHARGE SEPARATION

The lifetime of the charge pair produced initially (reaction 5 p. 268), following light excitation, is very short because of the occurence of a back electron-transfer reaction (reaction 7). As a result, the charge recombination in the multimolecular approach, is too rapid to let the ions react with water at the catalyst surface. In all the homogeneous and microheterogeneous systems with three, four or five components described before, efficient charge separation is achieved by means of an electron donor which is "sacrificed". To overcome this problem, another approach has been explored which consists in the utilization of supramolecular structures. Excellent reviews describing photoactive supramolecular systems for visible-light induced charge separation appeared recently [1,2,127-136]. In this section, I shall consider only a few supramolecular systems, in particular among those involving several components covalently linked to one another, in other words polyads (Section 1).

In a first approach, the simplest polyads, of PS-A type, involve porphyrins as photosensitizers (and electron donors) and quinones as electron acceptors. They were synthesized in order to mimic the primary electron-transfer events leading to charge separation in natural photosynthesis. The first porphyrin-quinone dyads were proposed in 1978 by Kong et al. [137] (Figure 8.13) and in 1979 by Tabushi et al. [138]. Afterwards, numerous porphyrin-quinone dyads were investigated to study the dependence of electron-transfer processes on donor-acceptor distance, orientation and energetics, on solvent and on the nature of the link [2,127,129-132]. Transition metal complexes of Ru and Os [2,128,132-135,139], and viologens [2,128,134,135,139,140] have also been extensively explored as photosensitizers and as electron acceptors respectively. In all these dyads, photoinduced charge separation and charge recombination occur on a picosecond time-scale, which is too fast for practical applications to solar energy conversion and storage. Consequently, more sophisticated supramolecular systems have been developed to achieve long-lived charge separation. As will be seen later, the simple incorporation of an electron donor or another electron acceptor to form a triad increases the lifetime of the charge-separated state.

The first triads were proposed in 1983 by Gust, Moore and coworkers [141] and by Nishitani, Mataga and coworkers [142].

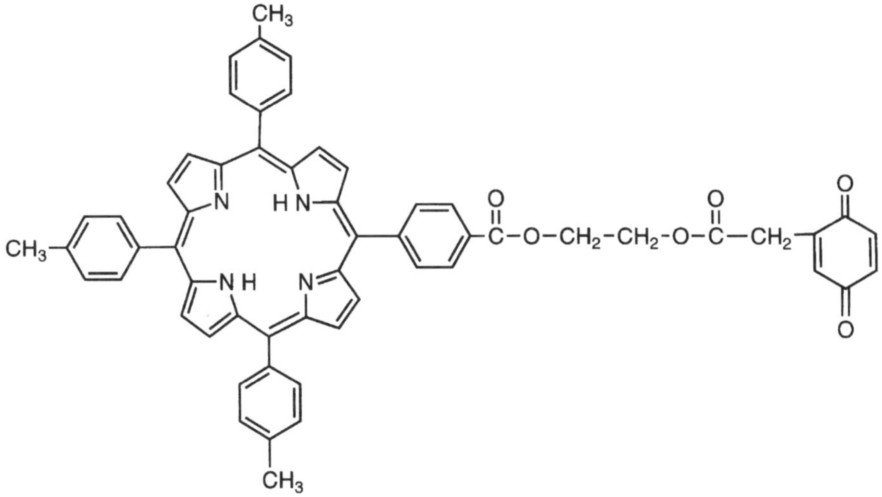

Figure 8.13. Porphyrin–quinone dyad of Kong and Loach [137]

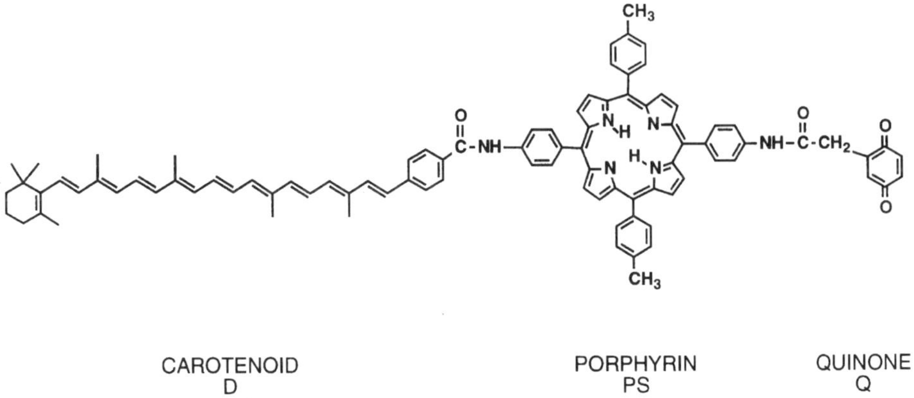

Figure 8.14. Carotenoid–porphyrin–quinone triad of Gust, Moore et al. [141]

In the triad of Gust and Moore (Figure 8.14), of D–PS–A type, a donor D (carotenoid Car) and an electron acceptor A (benzoquinone BQ) are covalently linked to a photosensitizer PS (porphyrin P), whereas in the triad of Nishitani and Mataga (Figure 8.15), of PS–A_1–A_2 type, the photosensitizer PS (porphyrin P) is covalently linked to two successive acceptors A_1 and A_2, with A_2 (trichlorobenzoquinone ClBQ) being a stronger electron acceptor than A_1 (benzoquinone BQ). Light excitation of PS leads to the formation of charge-separated states

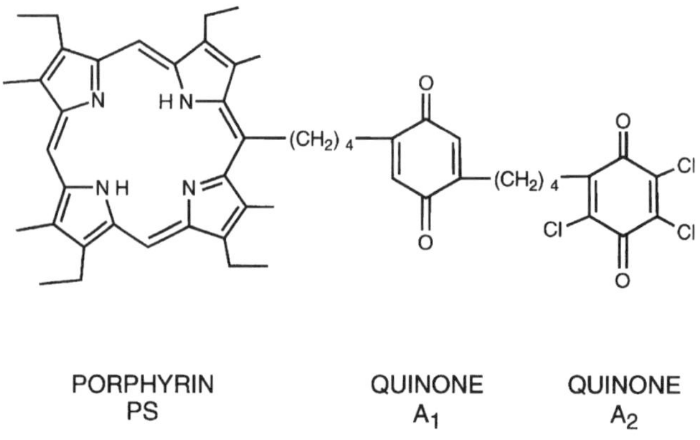

Figure 8.15. Porphyrin–quinone–chloroquinone triad proposed by Nishitani, Mataga and coworkers [142]

D^+-PS-A^- (reactions 37) and PS^+-A_1^--A_2^- (reactions 38) respectively, via two consecutive intramolecular electron transfer steps:

$$D\text{-}PS\text{-}A \xrightarrow{h\nu} D\text{-}PS^*\text{-}A \xrightarrow{e^-} D\text{-}PS^+\text{-}A^- \xrightarrow{e^-} D^+\text{-}PS\text{-}A^- \quad (37)$$

$$PS\text{-}A_1\text{-}A_2 \xrightarrow{h\nu} PS^*\text{-}A_1\text{-}A_2 \xrightarrow{e^-} PS^+\text{-}A_1^-\text{-}A_2 \xrightarrow{e^-} PS^+\text{-}A_1\text{-}A_2^- \quad (38)$$

Figure 8.16 gives an illustration of such an electron transfer cascade in the case of a D–PS–A triad. In Table 8.10, different polyads are listed with the corresponding lifetimes τ_{cs} and quantum yields of formation Φ_{cs} of charge-separated states, as well as the energy stored E_{st} (Figure 8.16). When the components are linked by flexible chains, as in the case of the triad of Nishitani and Mataga [142], the lifetime τ_{cs} is very short (100 ps). With more rigid spacers, i.e. when there is a smaller chance for the oxidized donor and the reduced acceptor to be

Table 8.10. Lifetimes τ_{cs} and quantum yields of formation of charge-separated states Φ_{cs} and energy stored E_{st} for several polyads

Polyad	Solvent	τ_{cs}	Φ_{cs}	E_{st} (eV)	Ref.
P ~ BQ		100 ps			[143]
P ~ BQ ~ ClBQ	Dioxane	300 ps			[142]
	THF	400 ps			
Car–P–BQ	CH_2Cl_2	300 ns	0.04	1.1	[141]
	butyronitrile	2 µs			
DMA–P–NQ	Butyronitrile	2.45 µs	0.71	1.39	[144]
P ~ $(MV^{2+})_2$	DMSO	150 ns			[145]
P ~ $(MV^{2+})_4$	DMSO	6.4 µs	0.33		[146]
$(PTZ)_2$ ~ $[Ru(bpy)_3]^{2+}$ ~ DQ^{2+}	CH_2Cl_2	160 ns	0.26	1.29	[147]
lysine ~ $[Ru(bpy)_3]^{2+}$, PTZ, AQ	CH_3CN	174 ns	0.26	1.54	[148]

WATER SPLITTING

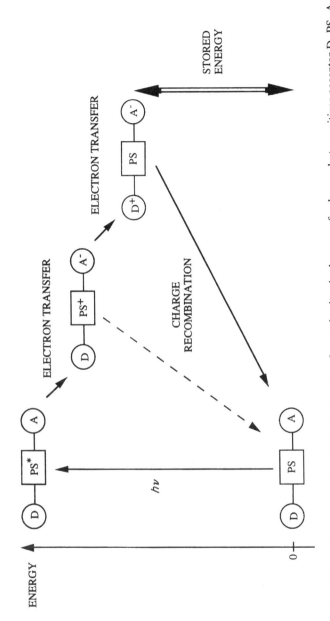

Figure 8.16. Schematic representation of the electron transfer mechanism in the case of a donor-photosensitizer-acceptor D-PS-A triad

in contact, lifetimes in the nanosecond and microsecond time-scales have been determined (Table 8.10). The triad of Gust, Moore and coworkers [141], which presents a long lifetime $\tau_{cs} = 2$ μs in butyronitrile, played a great part in the development of the supramolecular approach for solar energy conversion and for the design of molecular systems that mimic the functions of natural photosynthesis. Wasielewski et al. [144] obtained the best charge separation data with a rigid triad dimethylaniline–porphyrin–naphthoquinone (DMA–P–NQ) using triptycene as rigid spacers. In butyronitrile, the charge-separated state is formed in two successive steps of 9 ps and 71 ps (reactions 39):

$$\text{DMA-P-NQ} \xrightarrow{h\nu} \text{DMA-P*-NQ} \xrightarrow{e^-} \text{DMA-P}^+\text{-NQ}^-$$
$$9 \text{ ps} \qquad\qquad 71 \text{ ps}$$
$$\downarrow e^- \qquad (39)$$
$$\text{DMA}^+\text{-P-NQ}^-$$
$$2.45 \text{ μs}$$

with a quantum efficiency of 71% and it lasts 2.45 μs, the energy stored being estimated at 1.4 eV (Table 8.10). Table 8.11 collects some of the elegant polyads prepared by Gust, Moore and coworkers [130,141,149–153]. This table shows that τ_{cs} can be increased by increasing the number of components of the polyad. Thus a lifetime of 340 μs has been measured in CH_2Cl_2 in the case of a pentad carotenoid–porphyrin–porphyrin–naphthoquinone–benzoquinone (Car–P_1–P_2–NQ–BQ) [130,153].

Table 8.11 also shows that charge separation data can be optimized by a substituent effect (F instead of CH_3 on the porphyrin). In particular, the quantum yield of formation of the final charge-separated state changes from 4% for Car–P–BQ to 30% for the fluorinated Car–P(F)–BQ triad (Table 8.11). It is interesting to note that in polyads of P–$(MV^{2+})_n$ type (Table 8.10) consisting

Table 8.11. Lifetimes τ_{cs} and quantum yields of formation of charge-separated states Φ_{cs} for polyads prepared by Gust, Moore and coworkers

Polyad	Solvent	τ_{cs}	Φ_{cs}	Ref.
Car–P–BQ	CH_2Cl_2	300 ns	0.04	[141]
	Butyronitrile	2 μs		
Car–P–BQ	Benzonitrile	370 ns	0.13	[149]
Car–P(F)–BQ	CH_2Cl_2	455 ns	0.30	[150]
Car–P–NQ	CH_2Cl_2	70 ns	0.04	[151]
Car–P–NQ–BQ	CH_2Cl_2	460 ns	0.23	[151]
	CH_3CN	4 μs		
Car–P_1–P_2–NQ	Anisole	2.9 μs	0.25	[152]
Car–ZnP_1–P_2–NQ–BQ	$CHCl_3$	55 μs	0.83	[153]
	CH_2Cl_2	200 μs	0.60	
Car–P_1–P_2–NQ–BQ	CH_2Cl_2	340 μs	0.15	[153]

WATER SPLITTING

of a porphyrin covalently linked to two [145] or four [146] viologens via a flexible chain, the efficiency of charge separation increases with the number n of viologen molecules. The lifetimes of charge-separated states are 160 ns and 6.4 μs for $n = 2$ and $n = 4$ respectively (Table 8.10).

Artificial supramolecular systems which can be defined as systems in which the photosensitizer is not a porphyrin, have also been described. The photosensitizer can be either a coordination complex of transition metals [2,128,132,134] or a molecule completely organic [2,129,134,136,154–156]. Figure 8.17. gives an example of a flexible polyad [147] $(PTZ)_2 \sim [Ru(bpy)_3]^{2+} \sim DQ^{2+}$ consisting of a diquat DQ^{2+} as electron acceptor and two phenothiazine molecules PTZ as electron donors covalently linked via polymethylene chains to the photoactive site $[Ru(bpy)_3]^{2+}$. For this polyad, the charge-separated state lasts 160 ns and the energy stored is about 1.29 eV (Table 8.10).

More recently, Meyer and coworkers [148,157] proposed a new interesting type of polyad in which a photosensitizer $[Ru(bpy)_3]^{2+}$, an electron donor (phenothiazine PTZ) and an electron acceptor (anthraquinone AQ [148] (Figure 8.18) or MV^{2+} [157]) are attached to an amino acid (L-lysine). Although these polyads are flexible, lifetimes of the order of a 100 ns (Table 8.10) have been obtained for the charge-separated states.

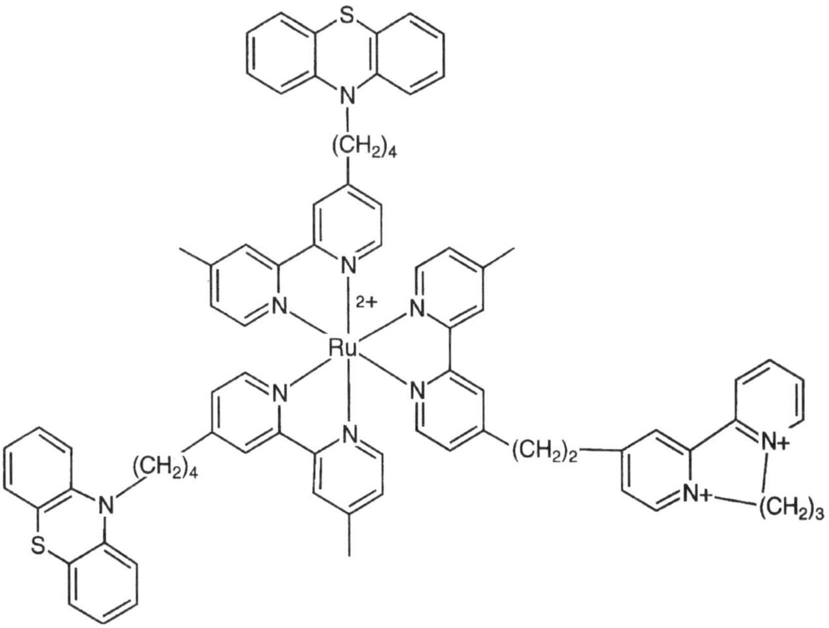

Figure 8.17. Polyad consisting of a ruthenium complex as photosensitizer, a diquat as electron acceptor and two phenothiazine molecules as electron donors [147]

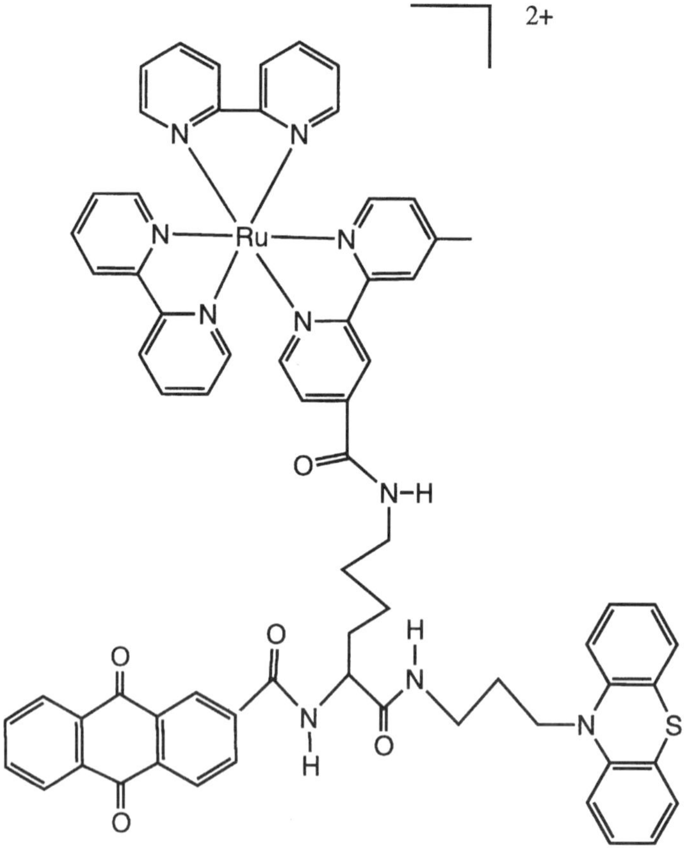

Figure 8.18. Polyad consisting of a ruthenium complex, phenothiazine and anthraquinone attached to L-lysine [148]

5.2 HYDROGEN-GENERATING SUPRAMOLECULAR SYSTEMS

Several supramolecular systems exhibiting efficient charge separation have been proposed, but few of them have been tested for photochemical H_2 or O_2 production from water [41,135,158–160]. For instance, Okura and coworkers [158] synthesized a series of dyads involving a porphyrin and MV^{2+} covalently linked by flexible chains of different length $((CH_2)_n, n = 2\text{-}6)$. Visible-light irradiation of these dyads in the presence of NADPH as electron donor and colloidal Pt or hydrogenase as catalyst gives rise to H_2 evolution only in the case of the dyad with the shortest chain ($n = 2$). It is of interest to note that, although the system only works with UV light ($\lambda < 400$ nm), a dyad proposed by Mau et al. [159,160] and consisting of anthracene as PS and a Co(III) cage complex as electron relay, leads to H_2 generation in the presence of EDTA as D and colloidal Pt

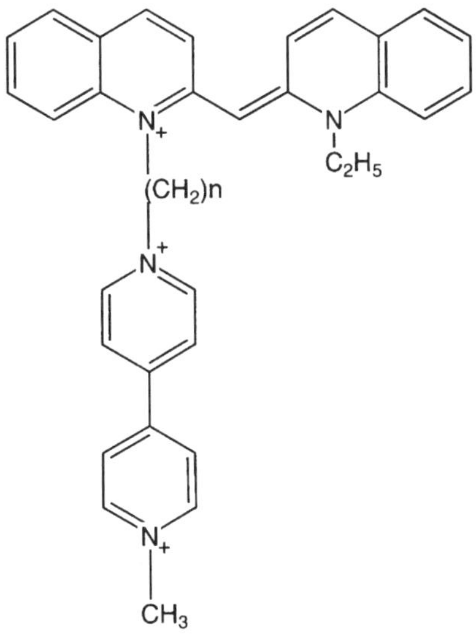

Figure 8.19. Cyanine-viologene dyad [41]

as catalyst. High H_2 quantum yields $\Phi\left(\frac{1}{2} H_2\right) = 0.14$ were determined at pH 6.5 [160]. More recently, Königstein and Bauer [41] proposed a system involving a cyanine-MV^{2+} dyad (Figure 8.19) which can form J-aggregates.

When this dyad is illuminated by visible light ($\lambda > 400$ nm) in the presence of EDTA and a platinum catalyst, H_2 is generated. H_2 yields are low (<1%) and optimum for a J-aggregate containing two cyanine-dye molecules per one cyanine-MV^{2+} dyad [161]. In this system, J-aggregates act as an antenna allowing light harvesting and migration until the excitation energy reaches the dyad. This induces a primary electron transfer reaction which leads to an efficient intramolecular charge separation. According to Königstein and Bauer [41], charge migration and charge distribution in the J-aggregate are probably responsible for the retardation of the back reaction.

6 CONSTRAINED SYSTEMS

The utilization of organized assemblies, constrained and confined media or the combination of these media with, for instance, supramolecular systems, constitutes another approach which has attracted much interest with a view to retarding the back electron-transfer reaction. An example of this is the system of Königstein and Bauer [41] based on the property of some cyanine dyes spontaneously to form J-aggregates, in other words organized molecular systems. I shall limit this overview to a few other examples of model systems in constrained media such as sol-gel silica glasses and zeolites. Concerning photochemical conversion in organized media (micelles, vesicles, monolayers, bilayers, Langmuir-Blodgett films, polymers, polyelectrolytes...), the reader is referred to papers and review articles that have appeared recently [4,128,162–171].

Slama-Schwok et al. [172] described a two-component donor–acceptor system consisting of an Ir(III) complex, [Ir(bpy)$_2$(C^3,N′)bpy]$^{3+}$, and 1,4-dimethoxybenzene (DMB). The excited state of Ir(III) trapped in a transparent, inert porous SiO$_2$ glass by the "sol-gel" process [173], can be reductively quenched by DMB dissolved in the water-filled pores of the glass. The back reaction of the photoinduced charge-separated pair (Ir(II), DMB$^+$) is retarded by four orders of magnitude with respect to homogeneous solutions. This strong retardation is attributed to the combined effects of trapping Ir(II) and adsorption of DMB$^+$, i.e. to the restriction of the diffusional processes of the charge pair, leading to long-lived charge separation at spatially separated sites. The back reaction is sufficiently inhibited so that under acidic pH (1.3), Ir(II) is able to react with water on the glass surface to produce H$_2$ with a quantum yield of about 1%. The turnover number for Ir(III), which acts both as a photosensitizer and a catalyst, is estimated to be higher than 100 [172].

The same group proposed recently [174] a three-component system of PS/A$_1$/A$_2$ type involving pyrene (Py) as PS, N,N'-tetramethylene-2,2′-bipyridinium ion (DQ$_4^{2+}$) as primary electron acceptor A$_1$, and MV^{2+} as secondary electron acceptor A$_2$. Pyrene and MV^{2+} are both immobilized in a porous sol-gel silica glass and the redox reaction is carried out by the mediation of DQ$_4^{2+}$ acting as a mobile charge carrier in the intrapore space (reactions 40–42):

$$(Py)_{trapped} \xrightarrow{h\nu} (Py^*)_{trapped} \tag{40}$$

$$(Py^*)_{trapped} + (DQ_4^{2+})_{mobile} \longrightarrow (Py^+)_{trapped} + (DQ_4^{+\bullet})_{mobile} \tag{41}$$

$$(DQ_4^{+\bullet})_{mobile} + (MV^{2+})_{trapped} \longrightarrow (DQ_4^{2+})_{mobile} + (MV^{+\bullet})_{trapped} \tag{42}$$

Under these conditions, the unidirectionality of the shuttling reaction is due to the difference of redox potentials of DQ$_4^{2+}$ and MV^{2+}, the latter being a better electron acceptor than the former (Table 8.5 p. 275). The spatial separation of Py$^+$ and MV$^{+\bullet}$ strongly inhibits the back electron-transfer reaction between these two species. Low yields of about 5% but very long lifetimes τ_{cs} of at least 4 h were found for the photoinduced charge-separated pair (Py$^+$, MV$^{+\bullet}$) [174].

In a similar way, but with zeolites as constraining media, Mallouk and coworkers [175] proposed an interesting three-component system, of PS–A$_1$...A$_2$ type, which consists of a [Ru(R-bpy)$_3$]$^{2+}$–DQ$_n^{2+}$ flexible dyad (R = H or CH$_3$ and DQ$_n^{2+}$ = N,N'-dialkyl-2,2′-bipyridinium ion with $n = 2$ or 3) and benzylviologen BV^{2+} as a secondary electron acceptor (A$_2$), BV^{2+} being contained within the zeolite framework. The [Ru(R-bpy)$_3$]$^{2+}$ moiety of the dyad is too large to enter through the 7.1 Å window of zeolite L. Thus only the DQ$_n^{2+}$ moiety can enter the zeolite anionic structure. This spatial arrangement (Figure 8.20) restricts the motion of the flexible ethylene chain and allows the contact of DQ$_n^{2+}$ with BV^{2+}. When the system is irradiated by visible light, an

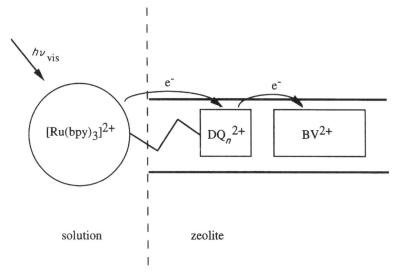

Figure 8.20. Schematic representation of unidirectional electron transfer cascade in a zeolite-based triad proposed by Mallouk and coworkers [175]. (Reproduced from ref. 176; Copyright 1995 Elsevier)

electron cascade occurs from the excited state of $[Ru(R\text{-}bpy)_3]^{2+}$ to DQ_n^{2+} and then to the mobile BV^{2+} according to a mechanism similar to reactions (38). A charge-separated state having a long lifetime $\tau_{cs} = 37$ µs (see Tables 8.10 p. 290 and 8.11 for comparison with supramolecular systems in non-constrained media) is formed with a quantum yield of 17% [175].

In the trimolecular system, of $PS/A_1/A_2$ type, developed recently by Borja and Dutta [177] and based on the same approach as that of Ottolenghi and coworkers [174], PS and A_1 are encapsulated in zeolite Y, whereas A_2 is in the surrounding solution. Here PS is $[Ru(bpy)_3]^{2+}$, A_1 is DQ_4^{2+} and A_2 is the zwitterionic sulfonatopropyl viologen SPV (Table 8.5 p. 275). Since SPV is neutral, it does not replace DQ_4^{2+} in the zeolite. A unidirectional electron-transfer cascade takes place from the excited state of $[Ru(bpy)_3]^{2+}$ to DQ_n^{2+} in zeolite, and then across the zeolite–solution interface from the reduced $DQ_4^{+\bullet}$ to SPV, according to a mechanism similar to reactions (38) p. 290. However, the long-lived $SPV^{-\bullet}$ is formed in very low yields (5×10^{-4}) [177]. New developments concerning this system appeared recently [178].

7 COMPLETE WATER SPLITTING INTO HYDROGEN AND OXYGEN

The incorporation of at least one component of a photochemical system in organized assemblies and the utilization of constrained/confined media in order to inhibit the back reaction is not really a new approach. As early as 1976, Whitten and coworkers [179] reported on the complete cleavage of water into H_2 and O_2 by visible light, using monolayer assemblies of surfactant derivatives of $[Ru(bpy)_3]^{2+}$. Although this observation could not be reproduced by the same group [180] and by many other laboratories [181–185], it has drawn attention towards organized media (monolayers, micelles, vesicles, microemulsions...) and towards the idea that microheterogeneous environments can strongly modify photochemical behaviour and reactivity of compounds incorporated in these media.

Another system was developed by Van Damme, Fripiat and coworkers [186] in 1983. These authors used as support solid-state materials such as clays and related minerals. Photoinduced water splitting was achieved by compartmentalizing the system into two subsystems [186,187] (Figure 8.21). In the first of these, $[Ru(bpy)_3]^{2+}$ as PS and RuO_2 as catalyst for O_2 evolution, were deposited on negatively charged colloidal particles of a fibrous clay (sepiolite), whereas in

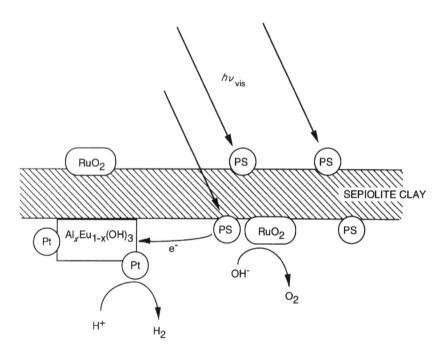

Figure 8.21. Schematic representation of the system $[Ru(bpy)_3]^{2+}/Eu^{3+}/Pt/RuO_2$ proposed by Van Damme, Fripiat and coworkers [186] for the simultaneous photogeneration of H_2 and O_2 from water by visible-light irradiation of this system incorporated in colloidal suspensions of a fribrous clay (sepiolite). PS is $[Ru(bpy)_3]^{2+}$ and the electron relay is Eu^{3+} which is embedded into Al hydroxide colloids. (Reproduced from ref. 176; Copyright 1995 Elsevier.)

the second colloid, which is positively charged, Pt particles, as catalyst for H_2 evolution, were deposited on amorphous aluminium–europium mixed hydroxide $Al_xEu_{1-x}(OH)_3$, Eu^{3+} acting as electron relay (Figure 8.21). The two colloidal subsystems were prepared separately and were then mixed. Their coupling leading to a supramolecular edifice (Figure 8.21) occurs spontaneously through interparticular association by electrostatic interactions. Visible-light irradiation of aqueous suspensions of the system leads to the oscillatory generation of H_2 and O_2 via oxidative quenching of $[Ru(bpy)_3]^{2+*}$ by Eu^{3+}, but with a turnover number for the Eu^{3+} relay of only 5. Although the catalytic nature of the process is not clear, this confined-system approach deserves more consideration.

A system similar to that of Van Damme and Fripiat, but without any organized solid support, was proposed before by Kalyanasundaram and Grätzel in 1979 [94]. This four-component system, based on the multimolecular approach (Section 4), consists of $[Ru(bpy)_3]^{2+}$ as PS, MV^{2+} instead of Eu^{3+} as electron relay, colloidal Pt and RuO_2 as catalysts for H_2 and O_2 generation respectively. The idea was to couple the two half photosystems (reactions 6 and 14) in order to avoid the use of any sacrificial electron donor or acceptor. It was claimed that visible-light irradiation of this system at pH 4.7 leads to simultaneous generation of H_2 and O_2. But these observations were not reproduced by several laboratories [21,81,102,104,188]. Nevertheless, this system is described even in recent books and review articles without specifying that it has given rise to much controversy. In fact, this system does not work for at least one reason, viz. the obligation to use very specific catalysts in order to avoid short-circuit processes, principally the back reaction (reaction 20). In the case of RuO_2 which was considered at that time as a highly specific catalyst for O_2 evolution, it was proved, for the first time that RuO_2 was also a good catalyst for H_2 evolution in a four-component sacrificial system $[Ru(bpy)_3]^{2+}/MV^{2+}/EDTA/RuO_2$ [61–63].

Several other questionable attempts to generate H_2 and O_2 simultaneously have been reported. For instance, in 1982, Kaneko and coworkers [104,105] described a microheterogeneous system involving $[Ru(bpy)_3]^{2+}$ as PS, Prussian blue colloids acting both as electron mediator and catalyst, and K^+ or Rb^+ cations. Prussian blue (PB), which is a mixed-valence iron complex $\{Fe_4^{III}[Fe^{II}(CN)_6]_3\}$, is known to form a three-dimensional polymeric structure of zeolitic nature. Visible-light irradiation of aqueous solutions of this system results in the formation of H_2 and O_2 with low yields at the optimum pH 2 and with a turnover number for $[Ru(bpy)_3]^{2+}$ of 11 after an irradiation time of 90 h. But under the same conditions, one can calculate a turnover number for PB catalyst much lower than 1. In addition, Harriman and coworkers [106] could not confirm the simultaneous production of H_2 and O_2, but they did show that PB can catalyze O_2 generation at pH 7 with a quantum efficiency of 2% in a three-component system $[Ru(bpy)_3]^{2+}/Na_2S_2O_8/PB$ using sodium persulfate as sacrificial electron acceptor. More recently, in 1992, Katakis et al. [189] described a two-component homogeneous system for water splitting based

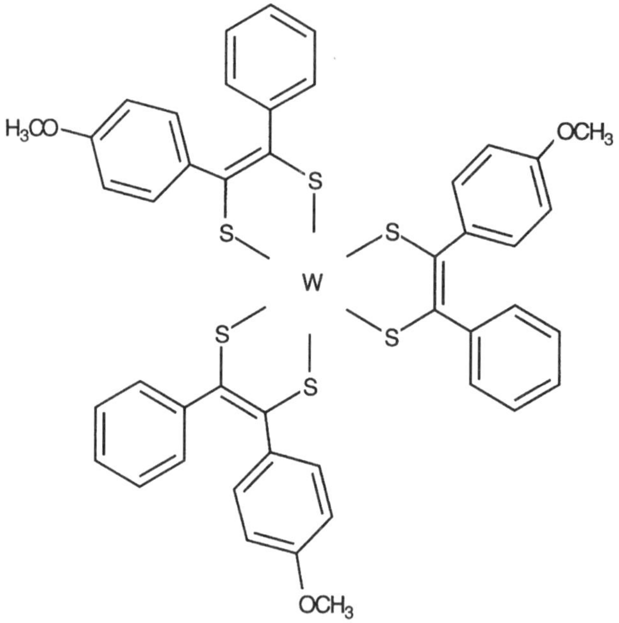

Figure 8.22. Tris-[-1-(4-methoxyphenyl)-2-phenyl-1,2-ethylenodithiolenic-S,S'] tungsten complex. [189]

on a dithiolene complex of tungsten, tris-[1-(4-methoxyphenyl)-2-phenyl-1,2-ethylenodithiolenic-S,S'] tungsten (Figure 8.22), acting as both photosensitizer and catalyst. In the presence of MV^{2+}, the authors claimed that the visible-light irradiation in the 350–500 nm range of acetone-water (70:30 v/v) solutions of the tungsten dithiolene results in H_2 and O_2 evolution in stoichiometric proportions with an average quantum yield for O_2 production (in equivalents) of 4%, and turnover numbers for the tungsten dithiolene greater than 1000. We know from experience that the greatest care must be taken with such water-splitting results. Therefore, before concluding as to the catalytic nature of this system, the observations of Katakis et al. [189–190] should be repeated by other groups.

8 CONCLUSION

In this chapter, we have tried to illustrate different approaches for visible-light-induced water splitting into H_2 and/or O_2, and their evolution from multimolecular to supramolecular systems, and from homogeneous or quasi-homogeneous to organized, constrained or confined media. These approaches give rise to numerous sacrificial model systems capable of producing H_2 and O_2 separately from water. It should be emphasized that, excepting the homogeneous system of Katakis et al. [189,190] which await confirmation, no reproducible, complete system generating H_2 and O_2 simultaneously is available up to now. Among the half photosystems, the four-component $[Ru(bpy)_3]^{2+}/MV^{2+}/EDTA/$colloidal Pt classical model system remains a reference for testing new photosensitizers, electron relays, electron donors and catalysts. A large selection of such components is now available which opens the possibility of designing *à la carte* photochemical systems, in particular supramolecular systems.

Thanks to these polyads, remarkable progress has been made in the effective retardation of the back electron-transfer reaction, and hence in the achievement of charge separation with a long lifetime. This was also accomplished by incorporating at least one component in organized molecular assemblies or in constrained/confined media, and by a combination of such multiphase media with supramolecular polyads. Moreover, such media introduce organized microenvironments which it would be judicious to use in order to compartmentalize the reducing and oxidizing processes. Indeed, it is not desirable, on the microscopic level, to produce H_2 and O_2 at the same catalytic site. Although enormous progress, essentially in fundamental research, has been made in less than 20 years, the development of practical devices for water splitting still requires facing several problems. For example:

1. The elaboration of new photosensitizers and/or the association of an ensemble of photosensitizers are needed in order to cover the whole solar spectrum:
2. The efficiency and selectivity of catalyst for O_2 generation must be highly improved:
3. The separation of H_2 and O_2 in the early stages of the reaction is required in order to avoid the catalytic regeneration of water.

Finally, the development of supramolecular systems coupled with constrained media has opened the door to the elaboration of efficient photochemical devices for solar energy storage. However, this needs further creative efforts and imagination. Altogether, at the present time, visible light water splitting into hydrogen and oxygen remains one of the most stimulating challenges for scientists, all the more so as economic and environmental considerations will become more and more pressing.

REFERENCES

[1] J. M. Lehn, Angew. Chem. Int. Ed. Engl. **27**, 89 (1988).
[2] V. Balzani and F. Scandola, *Supramolecular Photochemistry* (Ellis Horwood, Chichester, 1991).
[3] Z. W. Tian and Y. Cao, (editors) *Photochemical and Photoelectrochemical Conversion and Storage of Solar Energy* (International Academic Publishers, Beijing, 1993).

[4] M. Grätzel, *Heterogeneous Photochemical Electron Transfer* (CRC Press, Boca Raton, Florida, 1989).
[5] J. M. Lehn and J. S. Connolly, (editors). *Photochemical Conversion and Storage of Solar Energy* (Academic Press, New York, 1981) p 161.
[6] A. J. Bard and M. A. Fox, Acc. Chem. Res. **28**, 141 (1995).
[7] E. Pelizzetti and N. Serpone, (editors). *Homogeneous and Heterogeneous Photocatalysis* (D. Reidel, Dordrecht, 1986) C174.
[8] A. Harriman and M. A. West, (editors). *Photogeneration of Hydrogen* (Academic Press, London, 1982).
[9] S. Claesson and L. Engström, (editors). *Solar Energy: Photochemical Conversion and Storage* (National Swedish Board for Energy Source Development, Stockholm, 1977).
[10] P. Dantzer and H. Wipf, (editors). *Hydrogen in Metals III; Topics Appl. Phys.* (Springer-Verlag, Berlin, 1997) *in press*.
[11] A. Moradpour, E. Amouyal, P. Keller and H. Kagan, Nouv. J. Chim. **2**, 547. (1978).
[12] J. R. Bolton, Science **202**, 705. (1978).
[13] M. D. Archer and J. R. Bolton, J. Phys. Chem. **94**, 8028 (1990); ibid. **95**, 4172 (1991), and references therein.
[14] V. Balzani, A. Juris and F. Scandola, in reference [7], p. 1.
[15] J. Sykora and J. Sima, *Photochemistry of Coordination Compounds* (Elsevier, Amsterdam, 1990); Coord. Chem. Rev. **107**, 1 (1990).
[16] B. V. Koriakin, T. S. Dzhabiev and A. E. Shilov, Dokl. Akad. Nauk. SSSR., **233**, 620 (1977).
[17] J. M. Lehn and J. P. Sauvage, Nouv. J. Chim., **1**, 449 (1977).
[18] V. Balzani, L. Moggi, M. F. Manfrin, F. Bolletta and M. Gleria, Science **189**, 852 (1975).
[19] J. M. Lehn, J. P. Sauvage and M. Kirch, *Second International Conference on the Photochemical Conversion and Storage of Solar Energy* (Cambridge, Great Britain, 1978); late abstract.
[20] A. Moradpour, E. Amouyal, P. Keller and H. Kagan, *Second International Conference on the Photochemical Conversion and Storage of Solar Energy* (Cambridge, Great Britain, 1978); Book of Abstracts, p. 31.
[21] P. Keller, A. Moradpour, E. Amouyal and H. B. Kagan, Nouv. J. Chim. **4**, 377 (1980).
[22] K. Kalyanasundaram, J. Kiwi and M. Grätzel, Helv. Chim. Acta, **61**, 2720 (1978).
[23] M. Gohn and N. Getoff, Z. Naturforsch., **34a**, 1135 (1979).
[24] A. Juris, F. Barigelletti, S. Campagna, V. Balzani, P. Belser and A. von Zelewsky, Coord. Chem. Rev., **84**, 85 (1988).
[25] K. Kalyanasundaram, *Photochemistry of Polypyridine and Porphyrin Complexes* (Academic Press, London, 1992).
[26] J. Darwent, P. Douglas, A. Harriman, G. Porter and M. C. Richoux, Coord. Chem. Rev., **44**, 83 (1982).
[27] A. I. Krasna, Photochem. Photobiol. **29**, 267 (1979).
[28] O. I. Micic and M. T. Nenadovic, J. Chem. Soc. Faraday Trans. 1, **77**, 919 (1981).
[29] L. H. Eng, M. B. M. Lewin and H. Y. Neujahr, Photochem. Photobiol., **58**, 594 (1993).
[30] R. J. Crutchley and A. B. P. Lever, J. Am. Chem. Soc., **102**, 7129 (1980).
[31] G. Neshvad and M. Z. Hoffman, J. Phys. Chem. **93**, 2445 (1989). and references therein.
[32] H. Dürr, G. Dörr, K. Zengerle, E. Mayer, J. M. Curchod and A. Braun, Nouv. J. Chim., **9**, 717 (1985).

[33] H. Dürr, S. Bossmann and A. Beuerlein, J. Photochem. Photobiol. A: Chem., **73**, 233 (1993).
[34] E. Amouyal, A. Homsi, J. C. Chambron and J. P. Sauvage, J. Chem. Soc. Dalton Trans., 1841 (1990).
[35] A. Edel, P. A. Marnot and J. P. Sauvage, Nouv. J. Chim. **8**, 495 (1984).
[36] A. Harriman, G. Porter and M. C. Richoux, J. Chem. Soc. Faraday Trans. 2, **77**, 833 (1981).
[37] I. Okura and N. Kim Thuan, J. Mol. Catal. **5**, 311 (1979); ibid. **6**, 227 (1979).
[38] K. Kalyanasundaram and M. Grätzel, Helv. Chim. Acta. **63**, 478 (1980).
[39] F. Barigelletti, L. De Cola, V. Balzani, P. Belser, A. von Zelewsky, F. Vögtle, F. Ebmeyer and S. Grammenudi, J. Am. Chem. Soc. **111**, 4662 (1989).
[40] S. Bossmann and H. Dürr. New. J. Chem., **16**, 769 (1992).
[41] C. Königstein and R. Bauer, Sol. Energy Mater. Sol. Cells **31**, 535 (1994).
[42] S. Matsuoka, T. Kohzuki, Y. Kuwana, A. Nakamura and S. Yanagida, J. Chem. Soc. Perkin Trans. 2, 679 (1992)
[43] E. Amouyal and B. Zidler, Isr. J. Chem. **22**, 117 (1982).
[44] E. Amouyal, B. Zidler, P. Keller and A. Moradpour, Chem. Phys. Lett. **74**, 314 (1980).
[45] P. Keller, A. Moradpour, E. Amouyal and B. Zidler, J. Mol. Catal., **12**, 261 (1981).
[46] E. Amouyal, B. Zidler and P. Keller, Nouv. J. Chim., **7**, 725 (1983).
[47] M. Kirch, J. M. Lehn and J. P. Sauvage, Helv. Chim. Acta **62**, 1345 (1979).
[48] S. F. Chan, M. Chou, C. Creutz, T. Matsubara and N. Sutin, J. Am. Chem. Soc., **103**, 369 (1981).
[49] V. Houlding, T. Geiger, U. Kölle and M. Grätzel J. Chem. Soc. Chem. Commun., 681 (1982).
[50] M. A. Rampi Scandola, F. Scandola A. Indelli and V. Balzani, Inorg. Chim. Acta **76**, L67 (1983).
[51] P. A. Lay, A. W. H. Mau, W. H. F. Sasse, I. I. Creaser, L. R. Gahan and A. M. Sargeson, Inorg. Chem. **22**, 2347 (1983).
[52] A. Launikonis, P. A. Lay, A. W. H. Mau, A. M. Sargeson and W. H. F. Sasse, Sci. Pap. Inst. Phys. Chem. Res. (Jpn) **78**, 198 (1984).
[53] I. Okura, Coord. Chem. Rev. **68**, 53 (1985).
[54] (a) O. Johansen, A. W. H. Mau and W. H. F. Sasse, Chem. Phys. Lett. **94**, 107 (1983); (b) ibid., **94**, 113 (1983).
[55] P. J. Delaive, B. P. Sullivan, T. J. Meyer and D. G. Whitten, J. Am. Chem. Soc., **101**, 4007 (1979).
[56] A. Harriman, J. Photochem., **29**, 139 (1985).
[57] I. Willner, D. Mandler and R. Maidan Nouv. J. Chim., **11**, 109 (1987).
[58] E. Amouyal and P. Koffi, J. Photochem., **29**, 227 (1985); and references therein.
[59] E. Amouyal, in reference [7] p. 253.
[60] J. M. Lehn, J. P. Sauvage and R. Ziessel, Nouv. J. Chim., **5**, 291 (1981).
[61] E. Amouyal, P. Keller and A. Moradpour, J. Chem. Soc. Chem. Commun. 1019 (1980).
[62] P. Keller, A. Moradpour and E. Amouyal, J. Chem. Soc. Faraday Trans. 1, **78**, 3331 (1982).
[63] J. M. Kleijn, E. Rouwendal, H. P. van Leeuwen and J. Lyklema, J. Photochem. Photobiol. A: Chem., **44**, 29 (1988)
[64] E. Amouyal, M. Georgopoulos and M. O. Delcourt New. J. Chem., **13**, 501 (1989).
[65] G. M. Brown, B. S. Brunschwig, C. Creutz, J. F. Endicott and N. Sutin, J. Am. Chem. Soc., **101**, 1298 (1979).

[66] C. V. Krishnan and N. Sutin, J. Am. Chem. Soc., **103**, 2141 (1981).
[67] C. V. Krishnan, C. Creutz, D. Mahajan, H. A. Schwarz and N. Sutin, Isr. J. Chem., **22**, 98 (1982).
[68] C. V. Krishnan, B. S. Brunschwig, C. Creutz and N. Sutin, J. Am. Chem. Soc., **107**, 2005 (1985).
[69] J. Hawecker, J. M. Lehn and R. Ziessel, Nouv. J. Chim., **7**, 271 (1983).
[70] D. E. Green and L. M. Stickland, Biochem. J., **28**, 898 (1934).
[71] L. D. Rampino and F. F. Nord, J. Am. Chem. Soc., **63**, 2745 (1941).
[72] E. Amouyal, D. Grand, A. Moradpour and P. Keller, Nouv. J. Chim., **6**, 241 (1982).
[73] J. Kiwi and M. Grätzel, J. Am. Chem. Soc., **101**, 7214 (1979).
[74] (a) O. Johansen, A. Launikonis, J. W. Loder, A. W. H. Mau, W. H. F. Sasse, J. D. Swift and D. Wells, Aus. J. Chem., **34**, 981 (1981); (b) *ibid.*, **34**, 2347 (1981).
[75] A. Harriman and A. Mills, J. Chem. Soc. Faraday Trans. 2, **77**, 2111 (1981).
[76] K. Mandal and M. Z. Hoffman, J. Phys. Chem., **88**, 185 (1984).
[77] R. Rafaeloff, Y. Haruvy, J. Binenboym, G. Baruch and L. A. Rajbenbach, J. Mol. Catal., **22**, 219 (1983).
[78] M. T. Nenadovic, O. I. Micic, T. Rajh and D. Savic, J. Photochem., **21**, 35 (1983).
[79] T. Matsuo, T. Sakamoto, K. Takuma, K. Sakura and T. Ohsako, J. Phys. Chem., **85**, 1277 (1981).
[80] D. Miller and G. Mc Lendon, Inorg. Chem., **20**, 950 (1981).
[81] A. M. Braun in *Photochemical Conversion and Storage of Solar Energy*, edited by E. Pelizzetti and M. Schiavello, (Kluwer, Dordrecht,1991) p. 551 and references therein.
[82] E. Amouyal, Sci. Pap. Inst. Phys. Chem. Res., (Jpn), **78**, 220 (1984).
[83] G. L. Gaines, Jr., J. Phys. Chem., **83**, 3088 (1979).
[84] P. Koffi, Thesis, (Université Paris-Sud, Orsay, 1986).
[85] J. A. Farrington, M. Ebert and E. J. Land, J. Chem. Soc. Faraday Trans. 1, **74**, 665 (1978).
[86] J. A. Farrington, M. Ebert, E. J. Land and K. Fletcher, Biochim. Biophys. Acta, **314**, 372 (1973).
[87] E. J. Nanni, Jr., C. T. Angelis, J. Dickson and D. T. Sawyer, J. Am. Chem. Soc. **103**, 4268 (1981).
[88] M. Z. Hoffman, J.Phys.Chem., **92**, 3458 (1988).
[89] P. Keller, A. Moradpour, E. Amouyal and H. Kagan, J. Mol. Catal., **7**, 539 (1980).
[90] P. Keller and A. Moradpour, J. Am. Chem. Soc., **102**, 7193 (1980)
[91] A. Launikonis, J. W. Loder, A. W. H. Mau, W. H. F. Sasse, L. A. Summers and D. Wells, Aus. J. Chem., **35**, 1341 (1982)
[92] A. Launikonis, J. W. Loder, A. W. H. Mau, W. H. F. Sasse and D. Wells, Isr. J. Chem., **22**, 158 (1982).
[93] J. M. Lehn, J. P. Sauvage and R. Ziessel, Nouv. J. Chim., **3**, 423 (1979).
[94] K. Kalyanasundaram and M. Grätzel, Angew. Chem. Int. Ed. Engl., **18**, 701 (1979).
[95] A. Mills, E. Dodsworth and G. Williams, Inorg. Chim. Acta., **150**, 101 (1988).
[96] E. Borgarello, K. Kalyanasundaram, Y. Okuno and M. Grätzel, Helv. Chim. Acta, **64**, 1937 (1981).
[97] V. Ya. Shafirovich, N. K. Khannanov and V. V. Strelets, Nouv. J. Chim., **4**, 81 (1980).
[98] V. Ya. Shafirovich and V. V. Strelets, Nouv. J. Chim., **6**, 183 (1982).
[99] A. Harriman, G. Porter and P. Walters, J. Chem. Soc. Faraday Trans. 2, **77**, 2373 (1981).

[100] M. Neumann-Spallart, K. Kalyanasundaram, C. Grätzel and M. Grätzel, Helv. Chim. Acta., **62**, 1111 (1980).
[101] K. Chandrasekaran, T. K. Foreman and D. G. Whitten, Nouv. J. Chim., **5**, 275 (1981).
[102] P. A. Christensen, W. Erbs and A. Harriman, J. Chem. Soc. Faraday Trans. 2, **81**, 575 (1985).
[103] V. Ya. Shafirovich, N. K. Khannanov and A. E. Shilov, J. Inorg. Biochem., **15**, 113 (1981).
[104] M. Kaneko, N. Takabayashi and A. Yamada, Chem. Lett., 1647 (1982).
[105] M. Kaneko, N. Takabayashi, Y. Yamauchi and A. Yamada, Bull. Chem. Soc. Jpn., **52**, 156 (1984).
[106] P. A. Christensen, A. Harriman, P. Neta and M. C. Richoux, J. Chem. Soc. Faraday Trans. 1, **81**, 2461 (1985).
[107] J. M. Lehn, J. P. Sauvage and R. Ziessel, Nouv. J. Chim. **4**, 355 (1980).
[108] A. Harriman, I. J. Pickering, J. M. Thomas and P. A. Christensen, J. Chem. Soc. Faraday Trans. 1, **84**, 2795 (1988).
[109] A. W. Rutherford, J. L. Zimmerman, A. Boussac and J. Barber, (editors), *The Photosystems: Structure, Function and Molecular Biology* (Elsevier, New York, 1992) p. 179.
[110] R. J. Debus, Biochim. Biophys. Acta., **1102**, 269 (1992).
[111] C. Philouze, G. Blondin, J. J. Girerd, J. Guilhem, C. Pascard and D. Lexa, J. Am. Chem. Soc., **116**, 8557 (1994).
[112] N. P. Luneva, E. I. Knerelman, V. Ya. Shafirovich and A. E. Shilov, J. Chem. Soc. Chem. Commun. 1504 (1987).
[113] J. P. Collin, J. M. Lehn and R. Ziessel, Nouv. J. Chim., **6**, 405 (1982).
[114] V. Ya. Shafirovich, A. E. Shilov, J. R. Norris, Jr, and D. Meisel, (editors), *Photochemical Energy Conversion* (Elsevier, New York, 1989) p. 173.
[115] M. Calvin, Science, **184**, 375 (1974).
[116] S. R. Cooper and M. Calvin, Science, **185**, 376 (1974).
[117] S. W. Gersten, G. J. Samuels and T. J. Meyer, J. Am. Chem. Soc., **104**, 4029 (1982).
[118] J. P. Collin and J. P. Sauvage, Inorg. Chem., **25**, 135 (1986).
[119] R. Ramaraj, A. Kira and M. Kaneko, Angew. Chem. Int. Ed. Engl., **25**, 825 (1986).
[120] F. P. Rotzinger, S. Munavalli, P. Comte, J. K. Hurst, M. Grätzel, F. J. Pern and A. J. Frank, J. Am. Chem. Soc., **109**, 6619 (1987).
[121] P. Comte, M. K. Nazeeruddin, F. P. Rotzinger, A. J. Frank and M. Grätzel, J. Mol. Catal., **52**, 63 (1989).
[122] P. A. Jacobs, J. B. Uytterhoeven and H. K. Beyer, J. Chem. Soc. Chem. Commun., 128 (1977).
[123] S. Leutwyler and E. Schumacher, Chimia, **31**, 475 (1977).
[124] B. Sulzberger and G. Calzaferri, J. Photochem., **19**, 321 (1982).
[125] R. Beer, F. Binder and G. Calzaferri, J. Photochem., Photobiol. A: Chem., **69**, 67 (1992).
[126] (a) G. Calzaferri in reference [3] p. 141 and references therein, (b) G. Calzaferri, N. Gfeller and K. Pfanner, J. Photochem. Photobiol. A: Chem., **87**, 81 (1995); (c) K. Pfanner, N. Gfeller and G. Calzaferri, J. Photochem. Photobiol. A: Chem., 95, 175 (1996).
[127] J. S. Connolly, J. R. Bolton, M. A. Fox and M. Chanon (editors), *Photoinduced Electron Transfer* (Elsevier, Amsterdam, 1988) Part D, p. 303.
[128] T. J. Meyer, Acc. Chem. Res., **22**, 163 (1989).
[129] M. R. Wasielewski, Chem. Rev., **92**, 435 (1992).

[130] D. Gust, T. A. Moore and A. L. Moore, Acc. Chem. Res., **26**, 198 (1993).
[131] K. Maruyama, A. Osuka and N. Mataga, Pure Appl. Chem., **66**, 867 (1994).
[132] E. Amouyal, H. Bouas-Laurent, J. P. Desvergne, R. Lapouyade and B. Valeur, L'Actualité Chimique, Suppl. **7**, 194 (1994).
[133] C. A. Bignozzi, R. Argazzi, M. T. Indelli and F. Scandola, Sol. Energy Mater. Sol. Cells, **32**, 229 (1994).
[134] J. P. Sauvage, J. P. Colin, J. C. Chambron, S. Guillerez, C. Coudret, V. Balzani, F. Barigelletti, L. De Cola and L. Flamigni, Chem. Rev., **94**, 993 (1994).
[135] H. Dürr, S. Bossmann, R. Schwarz, M. Kropf, R. Hayo and N. J. Turro, J. Photochem. Photobiol. A: Chem. **80**, 341 (1994).
[136] M. N. Paddon-Row, Acc. Chem. Res., **27**, 18 (1994).
[137] (a) J. L. Y. Kong and P. A. Loach in *Frontiers of Biological Energetics: from Electrons to Tissues*, edited by P. L. Dutton, J. S. Leigh and A. Scarpa, (Academic Press, New York, 1978) Vol.1, p. 73; (b) J. L. Y. Kong, K. G. Spears and P. A. Loach, Photochem. Photobiol., **35**, 545 (1982).
[138] I. Tabushi, N. Koga and M. Yanagita, Tetrahedron Lett., 257 (1979).
[139] E. Amouyal and M. Mouallem-Bahout, J. Chem. Soc. Dalton Trans., 509 (1992).
[140] V. Ya. Shafirovich, E. Amouyal and J. A. Delaire, Chem. Phys. Lett., **178**, 24 (1991).
[141] (a) D. Gust, P. Mathis, A. L. Moore, P. A. Liddell, G. A. Nemeth, W. R. Lehman, T. A. Moore, R. V. Bensasson, E. J. Land and C. Chachaty Photochem. Photobiol., **37S**, S46 (1983); (b) T. A. Moore, D. Gust, P. Mathis, J. C. Mialocq, C. Chachaty, R. V. Bensasson, E. J. Land, D. Doizi, P. A. Liddell, W. R. Lehman, G. A. Nemeth and A. L. Moore, Nature **307**, 630 (1984).
[142] S. Nishitani, N. Kurata, Y. Sakata, S. Mizumi, A. Karen, T. Okada and N. Mataga, J. Am. Chem. Soc., **105**, 7771 (1983).
[143] N. Mataga, A. Karen, T. Okada, S. Nishitani, N. Kurata, Y. Sakata and S. Misumi, J. Phys. Chem. **88**, 5138 (1984).
[144] M. R. Wasielewski, M. P. Niemczyk, W. A. Svec and E. B. Pewitt, J. Am. Chem. Soc., **107**, 5562 (1985).
[145] Y. Kanda, H. Sato, T. Okada and N. Mataga, Chem. Phys. Lett., **129**, 306 (1986).
[146] J. D. Batteas, A. Harriman, Y. Kanda, N. Mataga and A. K. Novak, J. Am. Chem. Soc., **112**, 126 (1990).
[147] (a) E. Danielson, C. M. Elliott, J. W. Merkert and T. J. Meyer, J. Am. Chem. Soc., **109**, 2519 (1987) (b) L. F. Cooley, S. L. Larson, C. M. Elliott and D. F. Kelley, J. Phys. Chem., **95**, 10694 (1991).
[148] S. L. Mecklenburg, D. G. Mc Cafferty, J. R. Schoonover, B. M. Peek, B. W. Erickson and T. J. Meyer, Inorg. Chem., **33**, 2974 (1994).
[149] S. C. Hung, S. Lin, A. N. Macpherson, J. M. De Graziano, P. K. Kerrigan, P. A. Liddell, A. L. Moore, T. A. Moore and D. Gust, J. Photochem. Photobiol. A. Chem., **77**, 207 (1994).
[150] T. A. Moore, D. Gust, S. Hatlevig, A. L. Moore, L. R. Makings, P. J. Pessiki, F. C. De Schryver, M. Van Der Auweraer, D. Lexa, R. V. Bensasson and M. Rougee, Isr. J. Chem., **28**, 87 (1988).
[151] D. Gust, T. A. Moore, A. L. Moore, D. Barrett, L. O. Harding, L. R. Makings, P. A. Liddell, F. C. De Schryver, M. Van der Auweraer, R. V. Bensasson and M. Rougee, J. Am. Chem. Soc., **110**, 321 (1988).
[152] D. Gust, T. A. Moore, A. L. Moore, L. R. Makings, G. R. Seely, X. Ma, T. T. Trier and F. Gao, J. Am. Chem. Soc., **110**, 7567 (1988).
[153] D. Gust, T. A. Moore, A. L. Moore, S. J. Lee, E. Bittersmann, D. K. Luttrull, A. A. Rehms, J. M. De Graziano, X. C. Ma, F. Gao, R. E. Belford and T. T. Trier, Science, **248**, 199 (1990).

[154] M. N. Paddon-Row and J. W. Verhoeven, New J. Chem. **15**, 107 (1991).
[155] H. Heitele, P. Finckh, S. Weeren, F. Pollinger and M. E. Michel-Beyerle, J. Phys. Chem., **93**, 5173 (1989).
[156] G. L. Closs and J. R. Miller, Science, **240**, 440 (1988).
[157] S. L. Mecklenburg, B. M. Peek, J. R. Schoonover, D. G. Mc Cafferty, C. G. Wall, B. W. Erickson and T. J. Meyer, J. Am. Chem. Soc. **155**, 5479 (1993).
[158] S. Aono, N. Kaji and I. Okura, J. Chem. Soc. Chem. Commun., 170 (1986).
[159] A. W. H. Mau, W. H. F. Sasse, I. I. Creaser and A. M. Sargeson, Nouv. J. Chim., **10**, 589 (1986).
[160] I. I. Creaser, A. Hammershøi, A. Launikonis, A. W. H. Mau, A. M. Sargeson and W. H. F. Sasse, Photochem. Photobiol., **49**, 19 (1989).
[161] C. Königstein, 1994, Personal communication.
[162] D. Yogev and D. Meisel, in reference [3] p. 75.
[163] M. Fujihira, in reference [3] p. 193.
[164] E. V. Efimova, S. V. Lymar and V. N. Parmon, J. Photochem. Photobiol. A: Chem., **83**, 153 (1994) and references therein.
[165] S. J. Stoessel and J. K. Stille, Macromolecules, **25**, 1832 (1992).
[166] J. Rabani in *Photochemical Conversion: Storage of Solar Energy*, edited by E. Pelizzetti and M. Schiavello, (Kluwer, Dordrecht, 1991) p. 103.
[167] V. Ya. Shafirovich and A. Shilov, Isr. J. Chem., **28**, 149 (1988).
[168] J. H. Fendler, J. Phys. Chem., **89**, 2730 (1985).
[169] T. Matsuo, Pure Appl. Chem., **54**, 1693 (1982).
[170] M. Calvin, I. Willner, C. Laane and J. W. Otvos, J. Photochem., **17**, 195 (1981).
[171] I. Rico-Lattes and A. Lattes, this volume., Ch. 9.
[172] A. Slama-Schwok, D. Avnir and M. Ottolenghi, J. Phys. Chem., **93**, 7544 (1989).
[173] D. Avnir, Acc. Chem. Res., **28**, 328 (1995).
[174] A. Slama-Schwok, M. Ottolenghi and D. Avnir, Nature **355**, 240 (1992).
[175] J. S. Krueger, J. E. Mayer and T. E. Mallouk, J. Am. Chem. Soc. **110**, 8232 (1988).
[176] E. Amouyal, Sol. Energy Mater. Sol. Cells, *38*, 249 (1995).
[177] M. Borja and P. K. Dutta Nature *362*, 43 (1993).
[178] P. K. Dutta, M. Borja and M. Ledney, Sol. Energy Mater. Sol. Cells, **38**, 239 (1995).
[179] G. Sprintschnik, H. W. Sprintschnik, P. P. Kirsch and D. G. Whitten, J. Am. Chem. Soc., **98**, 2337 (1976).
[180] G. Sprintschnik, H. W. Sprintschnik, P. P. Kirsch and D. G. Whitten, J. Am. Chem. Soc., **99**, 4947 (1977).
[181] S. J. Valenty and G. L. Gaines, Jr, J. Am. Chem. Soc., **99**, 1285 (1977).
[182] G. L. Gaines, Jr, P. E. Behnken and S. J. Valenty, J. Am. Chem. Soc. **100**, 6549 (1978).
[183] A. Harriman, J. Chem. Soc. Chem. Commun., 777 (1977).
[184] K. P. Seefeld, D. Möbius and H. Kuhn, Helv. Chim. Acta, **60**, 2608 (1977).
[185] L. J. Yellowlees, R. G. Dickinson, C. S. Halliday, J. S. Bonham and L. E. Lyons, Aust. J. Chem., **31**, 431 (1978).
[186] H. Nijs, J. J. Fripiat and H. Van Damme, J. Phys. Chem., **87**, 1279 (1983).
[187] F. Bergaya, D. Challal, J. J. Fripiat and H. Van Damme, Nouv. J. Chim., **9**, 721 (1985).
[188] A. Harriman, J. Photochem., **25**, 33 (1984).
[189] D. Γ. Katakis, C. Mitsopoulou, J. Konstantatos, E. Vrachnou and P. Falaras, J. Photochem. Photobiol. A: Chem., **68**, 375 (1992).
[190] D. F. Katakis, C. Mitsopoulou and E. Vrachnou, J. Photochem. Photobiol. A: Chem., **81**, 103 (1994).

9 Organized Media and Homogeneous Photocatalysis

ISABELLE RICO-LATTES and ARMAND LATTES
Institut de Chimie Moléculaire Paul Sabatier, Laboratoire des IMRCP, Toulouse, France

1	Introduction		310
2	Reactivity in organized media		311
	2.1	Homogeneous and heterogeneous media	311
	2.2	Choice of system for conducting reactions	311
	2.3	Micelles — lyotropic liquid crystals	312
	2.4	Vesicles	314
		2.4.1 Choice of system for conducting reactions	315
	2.5	Microemulsions	315
		2.5.1 Definition of microemulsions	315
		2.5.2 Domains of existence and phase diagram of microemulsions	316
		2.5.3 Structure of microemulsions	317
		2.5.4 Distribution of the different structures in a pseudoternary phase diagram	319
		2.5.5 Phase behavior: boundaries of one-, two- or three-phase systems	319
		2.5.6 Exchange of time	319
		2.5.7 Microemulsions and reactivity	320
	2.6	Surfactant synthesis and generalization of amphiphile concept	321
	2.7	Aqueous solutions of polymers	322
	2.8	Properties of Organized liquid media	322
		2.8.1 Expected properties of organized systems	322
		2.8.2 Particular case of microemulsions	324
		2.8.3 Structure reactivity relationships in organized systems	324
		2.8.4 Control of Stereochemistry	325
		2.8.5 Water and Water Substitutes in Organized Media	325
3	Photocatalysis		326
	3.1	Definitions	326

Homogeneous Photocatalysis. Edited by M. Chanon
© 1997 John Wiley & Sons Ltd

        3.2   Classification of photocatalytic reactions   327
              3.2.1   Photoinduced catalytic reactions or reactions catalytic in
                      photons   327
4   Examples of reactions in organized molecular systems.
    Structure–reactivity relationships   329
    4.1   Influence of medium on concentration of reactive species   330
    4.2   Structure of medium and propagation of chain reactions. Reaction
          in the bicontinuous phase of nonaqueous microemulsions   333
          4.2.1   Amidation of $C_8F_{17}-CH=CH_2$ by γ-radiolysis   335
          4.2.2   Photoamidation of a mixed olefin $R_FCH=CHR_H$ in
                  homogeneous and microscopically heterogeneous
                  media   336
          4.2.3   Choice of surfactant and cosurfactant   336
          4.2.4   Phase diagram of the microemulsions   336
    4.3   Electron transfer and organized molecular systems   339
          4.3.1   Micellar solutions   343
          4.3.2   Reverse micelles and microemulsions   344
          4.3.3   Vesicles   344
          4.3.4   Polymerized surfactant vesicles   346
          4.3.5   Aqueous solutions of polymers   347
    4.4   Preparation of metallic microparticles as catalysts for redox
          reactions   348
    4.5   Exchange of ligands, metallation of porphyrins   348
5   Conclusion   348
References   349

1 INTRODUCTION

There is a growing awareness of the interest in organized molecular systems for the study of chemical and photochemical reactions. These systems belong to the new discipline of supramolecular chemistry, which may be split into two levels of enquiry:

1. Level 1: host–guest interactions;
2. Level 2: study of molecular aggregates.

These both depend on low-energy molecular interactions sharing the following properties: (1) molecular recognition; (2) transport from one phase to another; (3) catalysis.

Organized media, which are employed for facilitating chemical processes, mainly belong to the second level. These media are prepared by dissolving amphiphilic molecules in water or organic solvents, when they self-associate into colloidal entities adopting different forms (micelles, microemulsions, vesicles, etc.). They excel as solvents with the further benefit of transparency, which confers particular value in photochemical applications.

The interest in such systems lies in their utilization as media for conducting reactions, where the selective dissolution of reactants, the stabilization of species and the bringing together of reactants all favor catalytic processes. Such systems clearly show promise for photocatalytic processes, although this does raise the question of the definition of photocatalysis, which has encompassed rather contradictory notions, i.e. photochemistry, the source of photonic energy, and catalysis, which does not supply energy but facilitates reactions by decreasing the activation energy. All combinations of these two processes have been examined. We will discuss a few selected examples out of the wide range of systems investigated so far.

Studies on photocatalytic reactions in organized media extend from investigations on the structures of such media to the gamut of reaction processes that can be carried out within them. We will consider here relationships between photocatalysis and the structure of the medium. After describing the different systems and examining the possible limits of photocatalytic reactions we will discuss, using a few selected examples, the role of interfaces and the influence of structural modifications on reactions carried out in such systems.

2 REACTIVITY IN ORGANIZED MEDIA

2.1 HOMOGENEOUS AND HETEROGENEOUS MEDIA

Two prime objectives of chemistry, namely comprehension and improvement of chemical reactivity are highly dependent on the medium in which reactions are conducted. The limitations of homogeneous media, which have the most long-standing use, has spurred the development of new solvent systems [1]. Most of the heterogeneous media that have been developed are macroscopically heterogeneous, such as those used in phase-transfer catalytic reactions [2]. Organized media can, however, combine the advantages of homogeneous (which they are macroscopically) and heterogeneous media (which they are microscopically). They can thus be defined as microscopically heterogeneous solvent media.

Such systems have emerged in the natural evolution of chemistry, and are now encompassed in the field of supramolecular chemistry [3]. The implementation of complex systems, which may comprise a large number of components, places the accent on systems rather than species. Compared to the first level, or host–guest system involving one or two components, the molecular aggregates in complex systems may contain 100 or more molecules. These ensembles share the properties of the simpler supramolecular systems namely: recognition, transport and catalysis.

2.2 CHOICE OF SYSTEM FOR CONDUCTING REACTIONS

In general, organized systems are prepared from hydrophobic solutes, which highlights the important role of water, which will be discussed below. The solutes belong to four categories:

1. Surfactants, versatile amphiphilic molecules;
2. Planar cyclic or heterocyclic compounds which organize in stacks;
3. Non-planar condensed ring compounds such as sterols which aid formation of aggregates with a limited number of components;
4. Hydrosoluble polymers, which tend to self-associate or organize themselves.

Apart from the systems formed by stacking, which are mainly employed to mimic interactions of purine and pyrimidine bases, media prepared from any of the other solutes can be employed in studies of chemical reactivity.

2.3 MICELLES — LYOTROPIC LIQUID CRYSTALS

Consideration of surfactants gives an indication of the versatility of the molecular arrangements. Surfactants can promote the formation of well-organized monolayers at the various interfaces and aid formation of aggregates in solution producing suspensions of microscopic objects. This gives rise to heterogeneous phases within a liquid [4,5]. In water, spherical or deformed micelles tend to be obtained (Figure 9.1).

As a function of concentration and temperature, cylinders and then cubic or hexagonal liquid crystals and finally lamellae [6,7] may be observed (Figure 9.2 and Table 9.1).

In practice, each amphiphile only displays a small number of the above-mentioned phases, depending on concentration and temperature. A binary water (or solvent)-surfactant phase diagram as a function of temperature can be drawn up for such systems (Figure 9.2) [6]. In non polar solvents, a different type of aggregate is observed [8] consisting of reverse micelles, which may contain a small amount of water (Figure 9.3).

Modification of the structure of the surfactant will influence these transitions, and new aggregates, vesicles or liposomes may be obtained.

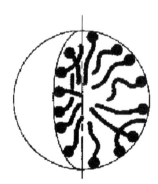

Figure 9.1. Model of direct micelle

ORGANIZED MEDIA AND HOMOGENEOUS PHOTOCATALYSIS 313

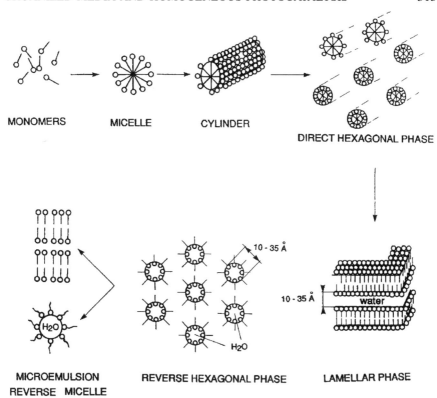

Figure 9.2. Micellar solutions and liquid crystals

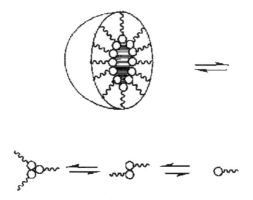

Figure 9.3. Model of reverse micelle

Table 9.1. Stability of solutions of surfactants as a function of concentration and temperature [7]

Micellar solutions $\underset{}{\overset{+H_2O}{\rightleftharpoons}}$ cubic phase I $\underset{}{\overset{+H_2O}{\rightleftharpoons}}$ hexagonal phase $\underset{}{\overset{H_2O}{\rightleftharpoons}}$ cubic phase II $\underset{}{\overset{+H_2O}{\rightleftharpoons}}$ lamellar phase $\underset{}{\overset{+H_2O}{\rightleftharpoons}}$ reverse cubic phase $\underset{}{\overset{+H_2O}{\rightleftharpoons}}$ reverse hexagonal phase $\underset{}{\overset{+H_2O}{\rightleftharpoons}}$ reverse micellar phase

2.4 VESICLES

The media described above consist of a suspension of highly mobile colloidal objects which form and then quickly vanish. Their lifetimes rarely exceed 10^{-3} s. The use of rigid amphiphilic molecules comprising two and even three hydrophobic chains can give rise to more stable systems. Furthermore, vesicles may be produced by mixing micellar solutions of anionic and cationic surfactants at near-equimolecular ratios (cationic micellar mixtures). Vesicles are enclosed bilayer structures (Figure 9.4). The term "liposome" is restricted to vesicles formed from natural phospholipids (with a mean diameter ranging from 200 to 400 Å).

Three types of vesicle can be distinguished [9]:

1. Multilamellar vesicles (MLV) consisting of several concentric bilayers in an onion skin arrangement. They range in size from 100 to 1000 nm.
2. Small unilamellar vesicles (SUV) formed from a single bilayer with diameters ranging from 20 to 100 nm.

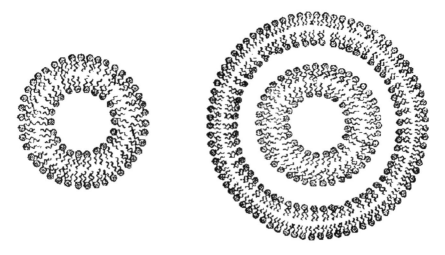

Figure 9.4. Model of vesicle

ORGANIZED MEDIA AND HOMOGENEOUS PHOTOCATALYSIS 315

3. Large unilamellar vesicles (LUV), also formed from a single bilayer but with diameters from 100 nm to 1 µm.

MLV are prepared as follows:

1. Stirring an aqueous mixture containing swollen surfactant in excess water; or
2. From a film of surfactant placed on a surface by evaporation of solvent and addition of water.

An external source of energy is often required to form monolamellar vesicles. SUV can be prepared from MLV as follows:

1. Ultrasonic dispersion of surfactants in water at temperatures above the phase transition temperature [10]; or
2. By shearing (French press exclusion); or
3. By dialysis of detergent after forming mixed micelles with an appropriate monocatenary surfactant, the surfactant is removed by dialysis.

LUV are prepared: (1) from water in oil microemulsions by removal of the organic solvent under reduced pressure, (2) fusion of SUV by addition of calcium salts.

Vesicles may last from several days to months and in some cases even years depending on the composition, purity, salt concentration of the medium, temperature, etc. [11].

2.4.1 Choice of System for Conducting Reactions

The choice of system depends on the objective. Monolayers, direct and reverse micelles and more complex aggregates may be employed for mechanistic studies to gain further understanding of the course of a chemical reaction. Synthesis for preparative purposes necessitates larger quantities, and a microemulsion is the medium of choice.

2.5 MICROEMULSIONS

2.5.1 Definition of Microemulsions

Addition of hydrocarbon and alcohol to micellar solutions consisting generally of water and surfactant can give rise to a variety of microheterogeneous phases. Most of these media, albeit of different structures, are encompassed by the term microemulsion (Table 9.2).
2). The following types of system may be observed:

1. Four-component microemulsions: surfactant, cosurfactant (alcohol, short-chain amine, etc.), water and hydrocarbon. The microemulsions initially described by Schulman belong to this type of system. The salt content of the water, which represents an additional parameter, has been extensively

Table 9.2. Characteristics of emulsions and microemulsions

Properties	Emulsions	Microemulsions
Composition	Water, surfactant, oil	Water, surfactant Oil, *cosurfactant*
Aspect	Turbid, milky	*Clear*
Formation	With energy	*Spontaneous*
Aggregate size	$d \approx 1\mu m$	$d < 500$ Å

investigated by Winsor [12] and is of widespread application in tertiary oil recovery.

2. Microemulsions also include three component systems: (a) minus cosurfactant in systems with non-ionic surfactants [13] (b) in certain systems devoid of surfactant [14]. However, these systems share many of the physical characteristics of four-component systems.

3. It has recently been shown that the water in four-component systems can be replaced by other structured polar solvents such as formamide [15]. This is due to the high dielectric constant of formamide. This is an essential factor for formation of direct ionic micelles as the electrolyte is completely dissociated in dilute solution. This solvent favors formation of strong hydrogen bonds thereby stabilizing the micelle.

Friberg [16] and Friberg and Lapeczynska [17] have also described waterless systems based on ethylene glycol or glycerol. However, these systems are highly viscous and their properties are rather different from those of the aqueous and formamide-based systems.

We will focus our discussion on the three- and four-component systems consisting of water or formamide, oil, surfactant and cosurfactant.

2.5.2 Domains of Existence and Phase Diagram of Microemulsions

The domain of existence of a microemulsion is defined by the spontaneous and abrupt transition from a cloudy emulsion to a transparent microemulsion [18]. The thermodynamic stability of such systems is a matter of debate at the present time and has been the subject of numerous studies [19]. They are, however, stable in time and resist separation for months and even years.

Numerous authors have described the phase diagrams of microemulsions as they have particular application for the tertiary recovery of oil, although more recently the results of structural studies have been reported [20]. Similar diagrams are obtained on replacing water with formamide [21]. The domains of existence of these systems are determined from the phase diagram. The boundaries are defined by the appearance of multiphase systems.

In the case of four-component systems, three-dimensional diagrams are produced, which are hard to interpret. Pseudo-ternary phase diagrams are thus

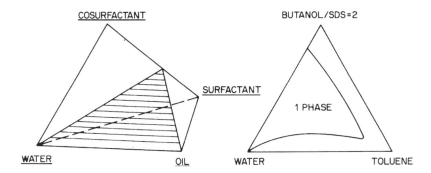

Figure 9.5. Pseudoternary phase diagram of microemulsion

preferred in which two components are lumped together in a given concentration ratio to form a pseudo-constituent. A typical example is shown in Figure 9.5.

2.5.3 Structure of Microemulsions

A wide variety of methods have been employed to investigate the structure of microemulsions. They include scattering of light, neutrons or X-rays, nuclear magnetic resonance (NMR), fluorescence spectroscopy, measurement of electrical conductivity or interfacial tension for polyphasic systems of the Winsor I, II and III types [22-25].

The principal structural characteristics of microemulsions are as follows:

1. Water in oil microemulsions (reverse micelles: W/O or L2), in which the dispersed phase consists of approximately spherical micelles (Figure 9.6b) comprising an aqueous core surrounded by a monomolecular layer of surfactant and cosurfactant. These micelles are dispersed in a continuous phase of oil and alcohol (with traces of water). The radii of the droplets range from 50 to 100 Å and increase with increase in water/surfactant ratio independently of the amounts of alcohol and oil [20].

2. Oil in water microemulsions (direct micelles: O/W or L1). Fewer studies have been devoted to water-rich microemulsions than to the reverse micelles described above. The main reason is their polydispersed nature. The formation of spherical droplets of a variety of sizes in a given system hampers precise measurement [26,27]. Although there is some lack of agreement [28,29], it is generally recognized that the L1 micelles are larger than the L2 micelles, and have diameters ranging from 100 to 500 Å (Figure 9.6a).

3. Phase inversion (transition of L1 to L2 micelles) and structure. The transition of W/O to O/W microemulsions has been studied by NMR [31] and by

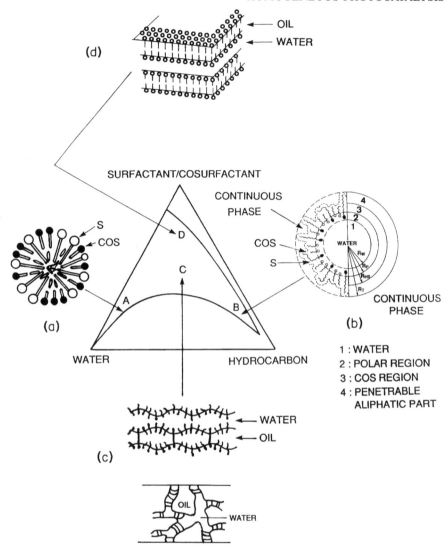

Figure 9.6. Structure of microemulsions: (A) direct micelles; (B) reverse micelles; (C) bicontinuous zone; (D) lamellar phase

measuring viscosity [30] or electrical conductivity [32]. Cazabat et al. have investigated this process in systems based on ionic surfactants by measuring electrical conductivity [33].

— for a rigid interfacial film, the droplets may be concentrated up to a volumic fraction θ of around 0.3. Phase inversion is relatively abrupt.

— for a fluid interfacial film, the electrical conductivity rises abruptly when θ reaches a value of around 0.1. This is a reflection of percolation or an interconnection of droplets over macroscopic distances [34]. The structure of the droplet is then progressively lost above the threshold of percolation. The zone of phase inversion is thus wide: the microemulsions are said to have a bicontinuous structure (Figure 9.6c).

2.5.4 Distribution of the Different Structures in a Pseudoternary Phase Diagram

Within the monophasic domain of a pseudoternary diagram, several zones can be defined in the system described above (Figure 9.6) as follows:

1. In the water-rich zone A L1 micelles are found;
2. In the oil-rich zone B, L2 micelles are found;
3. The bicontinuous phase lies in the intermediate zone C;
4. Lamellar structures may be observed in the surfactant-rich zone D. This zone has been little studied, and it corresponds to the limiting case of this type of organized medium.

2.5.5 Phase Behavior: Boundaries of One- Two- or Three-phase Systems

Phase separations departing from the monophase domain can be observed on altering the composition of the system. Multiphase systems may exist in various types of equilibrium: microemulsions in equilibrium with excess oil, water or both. These are referred to as Winsor I, II or III systems [12], and by extension, the monophase region is commonly referred to as Winsor IV. Figure 9.7 represents a phase diagram for the simplest microemulsion system (three components: water, oil and surfactant) as a function of salt concentration.

2.5.6 Exchange Time

The diagrams illustrated above provide a snapshot of the system at a given moment. Renewal of the structure of such systems, however, must be taken into account, especially for comprehension of chemical reactions in such media [35]. This process is poorly understood, and the current theories, essentially of a thermodynamic nature, do not satisfactorily account for the overall behavior. Nevertheless, a number of properties of the system clearly depend on structure. Viscosity is a good example. The fact that it is not significantly higher than in water is indicative of a continuous renewal of structure. The renewal time has been estimated by Zana [36] to be around 10^{-6}–10^{-7} s^{-1}. It undoubtedly stems from the exchange of molecules of surfactant and cosurfactant between the two separated interfaces. Furthermore, the lifetime of the structure depends on the composition of the microemulsion.

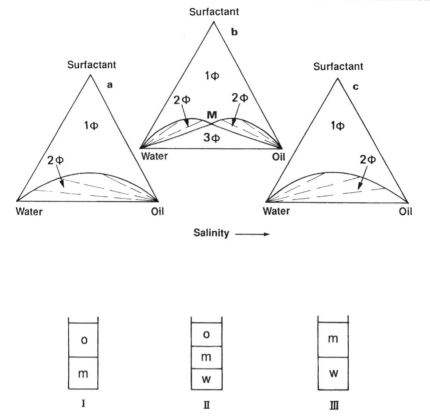

Figure 9.7. Multiphase systems: Winsor I, excess oil in equilibrium with microemulsion; Winsor II, excess water in equilibrium with microemulsion; Winsor III, coexistence of three phases — water, microemulsion, oil

2.5.7 Microemulsions and Reactivity

These mixtures, which commonly consist of four components including one or two reactants, are complex and can give rise to difficulties in product separation. This can be partially resolved by turning one of the components of the system into a reactant. This is referred to as the principle of molecular economy [37]. Transition into polyphasic domains also aids separation of products, while the use of polymer surfactant mixtures reduces the amount of surfactant and facilitates removal at the end of the process.

All the above-mentioned systems can be distinguished on the basis of certain properties. They all include interfaces of varying degrees of permeability. The permeability can be modified by using polymerizable surfactants.

Polymerized systems can be obtained from micelles, microemulsions or vesicles [38–43]. In attempts to stabilize the objects and control interface

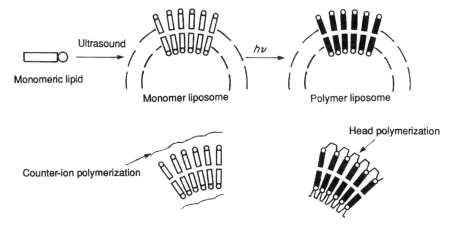

Figure 9.8. Polymerized organized systems

permeability and object size, surfactants functionalized with polymerizable groups (vinyl, methacrylate, diacetylene, isocyano, styrene, etc.) have been employed. Such groups can also be introduced in the polar moiety, lipophilic chains or even counter-ions (Figure 9.8). After formation of colloidal objects, polymerization is carried out by conventional methods such as irradiation, use of initiators, etc. Along with a reduction in their permeability, such suspensions are both more stable and less sensitive to the presence of alcohol and detergents.

2.6 SURFACTANT SYNTHESIS AND GENERALIZATION OF AMPHIPHILE CONCEPT

The surfactant is a key component of the organized systems used in studies of chemical reactivity, and it is important to use an appropriate surfactant for the process under consideration. This requires a range of suitable surfactants at one's disposal or a method for modular synthesis of these vital components [44]. The aim of modular synthesis is to transform suitable molecules into amphiphiles for production of organized systems. These systems are designed to accept certain types of substrate or produce automicelles where the reaction is carried out. Isolated molecules and aggregates exhibit marked differences in behavior. Mention should also be made of polymerizable surfactants, surfactants immobilized on gels, and destructible surfactants which can be readily removed after the reaction.

The structural diversity of these surfactants leads to a more general concept of an amphiphile [45]. If it is accepted that aggregates arise from competition between opposing interactions (hydrophile, hydrophobe), it follows that molecules could be constructed bearing structural features with individual affinities for incompatible solvents. A typical example is that of the mixed hydrogenated and fluorinated molecules, which spontaneously segregate. When dissolved in a structurally homogeneous solvent, whether hydrocarbon or fluorinated compound, aggregates bearing microscopic interfaces are formed. This shows the interest of highly fluorinated compounds, and perfluorinated surfactants in particular [46].

Research is currently focussed on the development of surfactants producing aggregates with particular affinities for certain substances on either physicochemical or geometrical grounds (selectivity and stereoselectivity).

2.7 AQUEOUS SOLUTIONS OF POLYMERS

Hydrophobic interactions within polymers can aid formation of organized structures. In aqueous solution, these polymers exist either as globules or an extended coil or both [47]. In the globular conformation, macromolecules form a microemulsion-like hydrophobic microphase. In the same way, reverse micelles may be produced with polymeric surfactants in non-polar organic solutions [48]. All these media can solubilize substrates at a variety of concentrations, giving rise to a stable solution, which remains stable even on extensive dilution.

It is also possible to use mixtures of polymers and surfactants. A globular chain can be solubilized using an amphiphilic surfactant, and the exact effect will depend on the properties of the particular surfactant.

2.8 PROPERTIES OF ORGANIZED LIQUID MEDIA

2.8.1 Expected Properties of Organized Systems

Liquid organized media can both solubilize and localize reactants. Substances dissolved in these solutions will distribute throughout the continuous phase as well as in the objects (Figure 9.9). This compartmentation of reactants and products can be exploited to:

1. Locally modify concentrations or maintain concentration gradients at the reaction site,
2. Prevent approach of unwanted reactants;
3. Favor charge separation.

Another property is the ordering of molecules, which may make them react or induce them to react in a selective manner. This can be accomplished by the preferential localization of sufficiently polar organic compounds at interfaces. Albeit to a lower degree than in the crystalline state, the molecules in monolayers are highly ordered. Micellar solutions and microemulsions also possess a degree of molecular ordering, and solutions of polymers are sufficiently organized to give rise to behavior that is significantly different from that of a disordered molecular solution (Table 9.3).

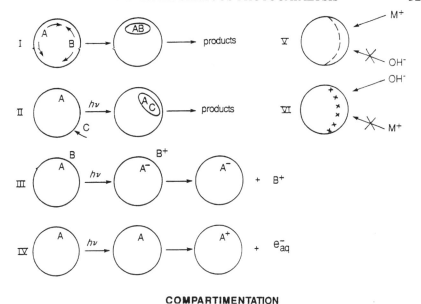

Figure 9.9. Organized molecular systems and compartmentation

Table 9.3. Molecular order

Crystal > Monolayer > Micelles microemulsions
> Polymer solutions
> Molecular solutions

Table 9.4. Properties of organized liquid media

Solubilization, localization, organization of the reactants

Alteration of the concentration at the microscopic level and maintenance of reactant concentration gradients

Modification of ionization potential, oxidation–reduction potentials

Modification of reaction mechanism and velocity, by stabilization of reactants, intermediates or products

Separation of the products and /or the charges

Thus it is possible to perform reactions in microemulsions that could only be carried out in the solid phase [49]. The localization at interfaces and compartmentation accompanied by a change of site after transformation will affect equilibria, and can bring about changes in ionization and redox potentials. (Table 9.4)

Stabilization of reactants, intermediates and products may also modify the mechanism and rate of reaction. This will depend on the order of the reaction.

The rate of unimolecular reactions will be affected by the nature of the medium, while bimolecular reactions will also be influenced by concentration.

2.8.2 Particular Case of Microemulsions

These represent media of choice for both mechanistic studies and preparative reactions since:

1. The droplets are larger than in a simple micellar solution due to the high concentration of amphiphiles. They thus possess higher dissolving power. The problem of elimination of surfactant can be overcome by optimization of the phase behavior and by the transition to a polyphasic system for product separation,
2. The high degree of dispersion makes microemulsions transparent, which confers a particular advantage for photochemical reactions.
3. They are fluid and stable over time.
4. A direct micellar system (O/W) can be transformed into a reverse micellar system (W/O). This may favor consumption of a large amount of one of the constituents (if it is also one of the reactants), while remaining in the macroscopically homogeneous domain.

2.8.3 Structure Reactivity Relationships in Organized Systems

As a first approximation, such media may be described as liquid suspensions of organized objects distributed in an ordered fashion. The characteristics of these objects include an interface in contact with the first liquid phase, an interphase (tail of amphiphiles) and a second liquid phase.

There is a relationship between the structure of such systems and reactivity. Certain reactions only take place in direct micelles while others only occur in reverse micelles. Localization of one molecule in each aggregate is adapted for transfer of photochemical energy. In this case, direct micelles are most suitable, whereas microemulsions and vesicles are more appropriate for organizing high concentrations of polar and a polar molecules in each aggregate. These properties have been exploited for conversion of solar energy.

Other processes may only take place in bicontinuous media. These media, of which microemulsion are typical examples, have particular properties including a mobile interface and very low interfacial tension, which facilitate exchange of matter.

From the standpoint of molecular economy, *this shows the importance of methods for the synthesis of amphiphiles of different structures, and the need for comprehension of the relationship between the behavior of such compounds in solution and their molecular characteristics.*

Conversely, the study of such reactions can provide information on the nature of the media. Chemical reactions can thus be employed *as molecular probes for structural studies.* This requires information on *the influence of solutes on the stability of such structures.* It should also be noted that spontaneous formation of aggregates can occur even in the absence of amphiphiles [50]. Abrupt alteration of reaction behavior may reflect a change in structure of the medium.

Molecular segregation can give rise to systems in which different types of object coexist (namely cellular organelles) each with its specific properties. This can facilitate reactions in stages.

2.8.4 Control of Stereochemistry

Microheterogeneities of organized systems can produce reactant combinations with applications in most domains of chemistry (see Table 9.5). Such systems confer considerable benefits in stereoselective synthesis. It is recognized that stereoselectivity is greatest in systems possessing specific and relatively rigid binding sites for a given substrate and transition state. For chiral resolution, the binding energies and/or reactivities of the enantiomers should differ. Vesicles and polymerized micelles show considerable promise for this type of application. It is of interest in this respect that the permeability of H^+ and OH^- ions is much lower in polymerized vesicles than in non-polymerized analogs.

2.8.5 Water and Water Substitutes in Organized Media

The prime role of water has been mentioned, and for the organic chemist this raises the possibility of conducting reactions in water. This clearly has far-reaching practical applications.

Generalization of the amphiphile concept suggests that molecular aggregation could be observed in other solvents. The degree of organization, however, depends on the cohesion properties (structure) of the liquid. A number of solvents may be employed instead of water. They include, formamide, glycerol, ethylene glycol and tetraethylammonium nitrate. These substitutes may confer benefits from the modification of properties such as polarity and solubility. The number of reactions carried out under such conditions is growing steadily as they have applications in areas of economic importance (Table 9.5 p. 326).

A higher degree of control of chemical reactivity is generally observed, and further progress in the physicochemical comprehension of such systems will undoubtedly aid matching of structure to reactivity. Attention is currently being devoted to the following issues:

1. Generalization of amphiphile concept and modular synthesis of surfactants;
2. Furthering understanding of structure of stability of these systems in the presence of reactants, with application of molecular economy to microemulsions;
3. Development of complex systems where objects with defined affinities and reactivities coexist;
4. Control of regioselectivity and stereochemistry;
5. Development of artificial enzymes.

Table 9.5. Examples of chemical and photochemical reactions carried out in organized systems

Organic reactions
Basic hydrolysis of long-chain esters
Synthesis of macrocyclic lactones (inter- and intramolecular control)
Mercuration of dienes (inter-intramolecular competition)
Catalytic formation of ketones
Nucleophilic substitutions (aliphatic and aromatic)
Variation of constants of acidity, proton transfer
Benzidinic rearrangement
Redox reactions
Nucleophilic additions
Esterolysis (stereoselective)
Cycloaddition (endo/exo)
Eliminations
Wacker process for oxidation of olefins
Oxidation of aldehydes to acids
Mutarotation of oses
Polymerization
Enzymatic reactions
Preparations of materials and their precursors

Inorganic reactions
Metallation of porphyrins
Electrochemical reactions (Cd(II), The(I), ferri and ferrocyanides)
Redox reactions
Ligand transfer
Hydration of complexes
Preparation of new catalysts

Photochemical reactions
Photoinduced removal of hydrogen by excited carbonyls
Fragmentation reactions as applied to isotopic enrichment
Photocycloadditions
Photodimerization
Photoisomerization
Photoinduced substitution of aromatic compounds
Photosensitized oxidation
Electron transfer
Photopolymerization
Solar energy conversion

An important application for organized molecular systems is in catalytic, particularly photocatalytic, reactions.

3 PHOTOCATALYSIS

3.1 DEFINITIONS

The definition of photocatalysis is still a matter of debate, and has led to a variety of interpretations. Gerassimov and Parmon [51] in a recent review have concluded that there is as yet no rigorous and generally accepted definition of the concept. The commission on photochemistry of the International Union of Pure and Applied Chemistry has recommended two definitions concerning two different processes [52].

1. *Photocatalysis* — a catalytic reaction involving light absorption by a catalyst or by a substrate.
2. *Photoassisted catalysis* — a catalytic reaction involving production of a catalyst by absorption of light. These recommendations have not been accepted by all photochemists [53–57], and the lack of agreement hinges essentially on the resemblance of two techniques of different modes of action namely:
   (a) Photochemistry, which supplies energy by irradiation;
   (b) Catalysis, which is not involved energetically, but has an influence is on the rate of reaction rather than on the position of an equilibrium.

A definition of photocatalysis in terms of the kinetic expression of photochemical processes, namely from quantum yield, emerges from an examination of these two techniques. Two types of reaction can thus be distinguished:

1. Those that are catalytic in photons, where, in theory, one photon enables transformation of an infinite number of molecules.
2. Those requiring one photon for each molecule transformed, but where the catalytic species may give rise to a high turnover. Even with this definition, there is a lack of distinction between true photochemical reactions and chain reactions that are initiated photochemically. Consideration of the mechanism of this type of transformation can simplify the approach by restricting photocatalysis to two types of process:
   (a) photochemical production of a catalyst;
   (b) involvement of photochemical activation in a stage of the catalytic cycle.

This produces a broad definition of photocatalysis encompassing all reactions involving both light and a catalyst. Hennig [58] has proposed that photocatalysis should be defined as broadly as possible in order to limit the number of categories of photocatalytic reaction. By enlarging the scope to: (i) precursors and initiators activated by light, and (ii) by recognizing that an excited species may be involved energetically, a representative list of reactions can be drawn up based on the classifications of Hennig et al. [59] and Gerassimov and Parmon [51].

3.2 CLASSIFICATION OF PHOTOCATALYTIC REACTIONS

3.2.1 Photoinduced catalytic reactions or reactions catalytic in photons:

1. Photogenerated catalyst

$$\text{Cat} \xrightarrow{h\nu} \text{Cat}^* \quad (\text{Cat}^* = \text{excited form of the catalyst})$$

$$\text{Cat}* \longrightarrow \text{Cat}' \, (\text{Cat}' = \text{active form of the catalyst})$$
$$\text{Cat}' + S \longrightarrow P + \text{Cat}' \quad (1)$$

2. Photoinduced chain reactions:

$$I \xrightarrow{h\nu} I*$$
$$I* + S \longrightarrow S' + I \, (I = \text{photoinitiator}) \quad (2)$$
$$S' + S \longrightarrow S' + P$$

3. Photoinduced chain reaction initiated by a sacrificial agent:

$$S \xrightarrow[I]{h\nu} S*$$
$$S \longrightarrow S' + I \, (I = \text{photoinitiator}) \quad (3)$$
$$S' + S \longrightarrow S' + P$$

3.2.2 Photoassisted catalyst — catalysis with assistance of light or reactions requiring continuous irradiations

1. Photoassisted reaction with a photoassistor:

$$\text{Cat} \xrightarrow{h\nu} \text{Cat}*$$
$$\text{Cat}* + S \longrightarrow \text{Cat} + P \quad (4)$$

2. Photoassisted reaction with a short-lived intermediate in ground state as catalyst:

$$\text{Cat} \xrightarrow{h\nu} \text{Cat}*$$
$$\text{Cat}* \longrightarrow \text{Cat}' \, (\text{Cat}' = \text{short-lived intermediate in ground state}) \quad (5)$$
$$\text{Cat}' + S \longrightarrow \text{Cat} + P$$

3. Photosensitized reactions:

$$\text{Cat} \xrightarrow{h\nu} \text{Cat}*$$
$$\text{Cat}* + S \longrightarrow \text{Cat} + S* \, (\text{Cat} = \text{sensitizer}) \quad (6)$$
$$S* \longrightarrow P$$

4. Catalyzed photoreactions:

a) $$S \xrightarrow{h\nu} S^*$$
$$S^* + Cat \longrightarrow Cat + P \tag{7a}$$

b) $$S \xrightarrow{h\nu} P'$$
$$P' + Cat \longrightarrow P + Cat \tag{7b}$$

5. Phototransfer of electrons. Reactions in accordance with photosensitized process (reaction 6) and with photoreaction (7). In some cases, energy and electron transfer may not be readily discriminated.

$$Cat \xrightarrow{h\nu} Cat^*$$
$$Cat^* + S \longrightarrow Cat^+ + S^-$$
$$S \longrightarrow P \tag{8}$$
$$R + Cat^+ \longrightarrow Cat + R^+$$
$$R^+ \longrightarrow Q$$

All these reactions present favorable characteristics for organized systems employed as reaction media. They all involve a strong interaction between species: substrate and catalyst. This interaction can occur in the following ways:

1. From the ground state, for example in photochemical reactions of substrates which only absorb light after complexation with a catalyst:

$$Cat + S \rightleftharpoons [Cat - S]$$
$$[Cat - S] \xrightarrow{h\nu} [Cat - S]^* \longrightarrow Cat + P \tag{9}$$

2. During formation of exciplexes

a) $$Cat \longrightarrow Cat^*$$
$$Cat^* + S \longrightarrow [Cat - S]^* \tag{10a}$$

b) $$S \longrightarrow S^*$$
$$S^* + Cat \longrightarrow [Cat - S]^* \tag{10b}$$
$$[Cat - S]^* - \text{exciplex}$$

processes which represent alternatives to some of the photocatalytic reactions described above.

4 EXAMPLES OF REACTIONS IN ORGANIZED MOLECULAR SYSTEMS. STRUCTURE–REACTIVITY RELATIONSHIPS

We have already seen that the structure of the medium has an important influence on chemical reactivity. As illustrated in Figure 9.8, this may be exerted on various behaviors, such as:

1. Proximity of reactants, which can give rise to exciplexes in photochemical reactions;
2. Propagation of chain reactions, only possible in bicontinuous media;
3. Charge separation.

4.1 INFLUENCE OF MEDIUM ON CONCENTRATION OF REACTIVE SPECIES

To illustrate this behavior, we will consider the probe reaction of the photoisomerization of 2-acetonaphthone oxime O-methyl ether studied by Padwa and Albrecht [60]:

anti (E)   anti (Z)   (11)

Irradiation led to a photostationary state whose composition is dependent on:

1. Temperature: high temperature reduced yield of the *anti* form;
2. Solvent: higher polarity increased the fraction of the *syn* form;
3. Concentration: high oxime ether concentration enhanced the fraction of the *anti* isomer in the photostationary state.

On spectroscopic analysis, there was no evidence for an association of ground-state molecules, and the observed fractions were consistent with the involvement of an excimer inducing *syn–anti* isomerization with a decay ratio differing from that of the excited monomer.

The mechanism can be summarized as follows:

$$\text{Monomer} \longrightarrow \text{monomer}^*(^1S) \longrightarrow \text{predominantly Z } syn$$

$$\text{Monomer} \longrightarrow [\text{monomer}^*(^1S)]^* + \text{monomer}] \longrightarrow \text{excimer}$$

$$\longrightarrow \text{predominantly E } anti$$

The investigations of Padwa on the photoisomerization in benzene and pentane were extended by:

1. Using different wavelengths of irradiation. It was found that the photostationary state composition was not wavelength-dependent.
2. Examining the effect of concentration in different solvents [61] including cyclohexane and more polar protic solvents such as acetonitrile and various alcohols. A linear change with concentration was observed. The relationship was comparable in cyclohexane, benzene and acetonitrile, although a higher proportion of the *syn* isomer was observed in the solvents of higher polarity. The main finding was that the concentration effect was more marked in protic than in aprotic solvents (Figure 9.10).

The percentage of the *syn* isomer was consistent with a lower polarity of the excimer compared to the excited monomer (singlet). The influence of protic solvents was not mediated by this process as methanol and acetonitrile have closely similar polarities, although the excimer fluorescence emission was more readily observed in methanol than in the other solvents.

The E $\rightleftarrows$ Z photoisomerization was then investigated in various microemulsion zones of the phase diagram using different surfactants and cosurfactants [62]. Since the results were similar in both anionic and cationic microemulsions, we will discuss the results obtained with water/benzene/SDS/butanol ($\frac{1}{2}$). Quite different results were observed in microemulsions than in the pure solvents, and

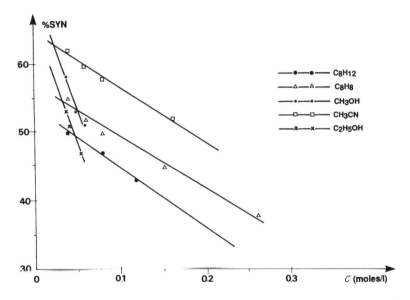

Figure 9.10. Photoisomerization of 2-acetonaphthone oxime O-methyl ether in different solvents

were found to depend on the nature of the continuous phase. We will describe the behavior of the compound over the whole phase diagram (Figure 9.11):

1. In the benzene-rich area (C) (W/O with benzene as continuous phase), the change was linear with concentration, like that observed in non-polar solvents such as pure benzene. The oxime ether was localized in the bulk benzene phase and the presence of micelles did not affect the photoisomerization reaction,
2. In contrast in the microemulsions of areas (A), (B) and (D), there was a discontinuous change with concentration. The plot of the proportion of Z isomer against concentration exhibited a discontinuity represented by two lines. The concentration at the intersection of these two lines was referred to as the saturation concentration Cs. The results were consistent with a localization of the oxime ether at the interface of benzene-rich micelles as this compound is insoluble in both water and mixtures of water and butanol. At *low concentration*, the compound lies at the interface, the reaction medium is protic and the effect of concentration resembles that observed in methanol. With *rise in concentration* the excess oxime either passes into the micelles

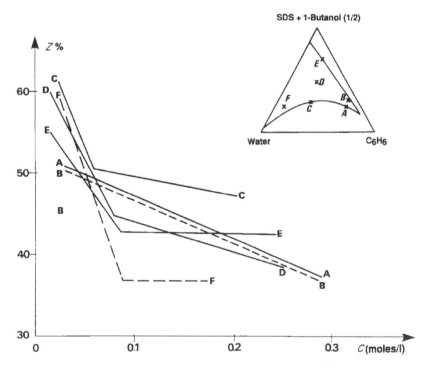

Figure 9.11. Photoisomerization of 2-acetonaphthone oxime O-methyl ether in microemulsions

and thus into benzene. The effect observed is the result of two processes:
(a) photoisomerization in protic medium at the interface (at constant concentration);
(b) photoisomerization in oil at high oxime ether concentration.

Thus the proportion of Z at the photostationary equilibrium is not greatly affected by the total concentration of oxime ether.

For this reaction, *processes at the interface* are significant, especially if there is a relation between the saturation concentration and the nature of the interface for the microemulsions situated in the middle of the diagram (D).

1. Two microemulsions with the same percentage of surfactant led to the same saturation concentration;
2. The saturation concentration was a linear function of percentage of surfactant.

In this example, interfacial processes appear to play an important role, and there was a close relationship between the saturation concentration and the nature of the interface. To provide further evidence, the nature of the interface was altered by addition of salts (sodium and potassium chloride). In all cases, addition of 1% salt reduced the saturation concentration. With increase in salt concentration to 2%, there was no further change in saturation concentration. These results indicate that the interface is only able to take up a limited amount of substrate, which depends on the availability of a number of sites. This number increases with increasing proportion of SDS and is reduced by addition of salts. Taken together these observations show the important role played by the interface in this photoisomerization reaction.

4.2 STRUCTURE OF MEDIUM AND PROPAGATION OF CHAIN REACTIONS. REACTION IN THE BICONTINUOUS PHASE OF NONAQUEOUS MICROEMULSIONS

We have seen above that a microemulsion is defined in a general way as a transparent medium consisting of water, oil (saturated or unsaturated hydrocarbon), a surfactant and a cosurfactant (a short-chain amphiphile such as an alcohol or amine). Depending on the proportions of the constituents, three main types of structure can be distinguished: reverse micelles (W/O), direct micelles (O/W) or bicontinuous structures (Figure 9.6). Reactions carried out in microemulsions usually take place either in the direct micellar zone or more commonly in the reverse micellar zone [11]. There are few reports of reactions taking place in the bicontinuous phases [63].

It was shown for the first time that microemulsions based on formamide instead of water could be employed for carrying out chemical reactions. Formamide was employed in view of the low solubility of many organic compounds in water. Moreover, formamide can be utilized as a reactant in microemulsions in which the two main components (oil and formamide) are reactants. We succeeded in amidating the olefin $C_8H_{17}CH=CH_2$ by γ-radiolysis in a formamide (F) microemulsion containing 1,1,2-trihydroperfluoro-1-decene ($C_8F_{17}CH=CH_2$) as oil (O), potassium 2,2,3,3-tetrahydroperfluoroundecanoate ($C_8F_{17}C_2H_4,CO_2K$) as surfactant (S) and 1,1,2,2-tetrahydroperfluorohexanol ($C_4F_9C_2H_4OH$) as cosurfactant (CoS) [64,65]. The phase diagram of this system [66] was explored after γ-radiolysis. The structures of the various microemulsions were determined by measuring self-diffusion by ^1H NMR [67]. The results are shown in Figure 9.12 and Table 9.6.

The self-diffusion coefficients (cf. Table 9.6) defined three microemulsion

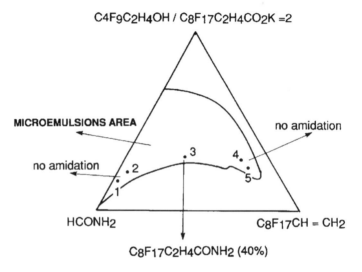

Figure 9.12. γ-radiolysis at 25 °C of microemulsion system (HCONH$_2$, $C_8F_{17}CH=CH_2$, $C_4H_9C_2H_4OH/C_8F_{17}CH_2CO_2K = 2$)

Table 9.6. Self-diffusion coefficients of HCONH$_2$ ($D1$) and C_8F_{17} CH=CH$_2$ ($D2$) in microemulsions at 25 °C

	Microemulsions (% wt)					
No.	F	O	S	CoS	$D1$	$D2$
1	87	7	2	4	3.62	0.54
2	78	10	4	8	3.37	0.60
3	44	44	4	8	0.85	1.21
4	12	61	9	18	0.60	2.89
5	10	66	8	16	0.60	2.98
Pure HCONH$_2$					5.21	
Pure C$_8$F$_{17}$CH=CH$_2$						5.32

F=HCONH$_2$, O=C$_8$F$_{17}$CH=CH$_2$, S=C$_8$F$_{17}$C$_2$H$_4$CO$_2$K, CoS=C$_4$F$_9$C$_2$H$_4$OH
Unit for $D = 10^{-10}$ m^2 s^{-1} $\pm$ 0.08 $\times$ 10^{-10}m^2 s^{-1}

zones closely resembling those of the corresponding aqueous systems:

1. Microemulsions 1 and 2 have an O/F structure similar to that of direct micelles (O/W); the olefin in droplets has a much lower self-diffusion coefficient than that of the pure liquid, while formamide, making up the continuous phase, diffuses readily.
2. In contrast, microemulsions 4 and 5 have a F/O structure similar to reverse micelles (W/O); the formamide in droplets has a low self-diffusion coefficient unlike that of the olefin.
3. In microemulsion 3, the coefficients of self-diffusion of formamide and oil are similar, indicating the presence of a bicontinuous phase, not organized in micelles. This type of structure, well known in aqueous media [6] was the first example reported in nonaqueous microemulsions.

These results show that by judicious choice of surfactant and cosurfactant, it is possible to obtain formamide-based microemulsion systems analogous to those produced in water.

4.2.1 Amidation of $C_8F_{17}-CH=CH_2$ by γ-radiolysis

The following results were obtained (cf. Figure 9.12):

1. In microemulsions 1, 2, 4 and 5, the olefin did not react, and only the oxamine $CONH_2-CONH_2$ was obtained in low yield ($\approx 5\%$ with respect to the starting formamide) [8].
2. The amidation reaction only took place in microemulsion 3. The terminal amide $C_8F_{17}CH_2CH_2CONH_2$ was isolated in a 40% yield with respect to the starting olefin, which is the limiting reactant [8]. The oxamide $CONH_2-CONH_2$ was also obtained in low yield ($\approx 5\%$ with respect to the starting formamide).

These results were interpreted in terms of the general mechanism of an amidation reaction [68]:

Initiation: $\qquad HCONH_2 \xrightarrow{\gamma} \cdot H + \cdot CONH_2$

Propagation:

$$C_8F_{17}-CH=CH_2 + \cdot CONH_2 \longrightarrow C_8F_{17}-CH^\bullet-CH_2CONH_2$$
$$C_8F_{17}-CH^\bullet-CH_2CONH_2 + HCONH_2 \longrightarrow C_8F_{17}-CH_2CH_2CONH_2 + \cdot CONH_2 \quad (12)$$

Termination: $\qquad 2 \cdot CONH_2 \longrightarrow CONH_2-CONH_2$

In the reverse microemulsions 4 and 5, formamide was confined to the droplets, as demonstrated by the self-diffusion measurements, the carbamoyl radicals diffuse little, being restrained in a micellar cage, and only the oxamide was formed.

On the other hand, in the direct microemulsions 1 and 2, the olefin is confined in the micelles. The self-diffusion measurements showed that the olefin in fact diffused little into the medium, and so did not interact greatly with the carbamoyl radicals. However, in microemulsion 3 with a bicontinuous structure, the constituents (also reactants) diffuse together (Table 9.6), and suitable contact between formamide and the olefin can take place hence enabling a reaction to take place.

These results show the potential of the relatively orderless bicontinuous phase as a medium for chemical reactions. These structures which favor simultaneous diffusion of reactants enable reactions to be carried out that would be impossible in a strictly micellar medium.

4.2.2 Photoamidation of a Mixed Olefin $R_FCH=CHR_H$ in Homogeneous and Microscopically Heterogeneous Media [69].

Photoamidation by formamide of the olefin $C_8F_{17}CH=CHC_6H_{21}$ was first carried out in a dilute medium of *tert*-butyl alcohol leading to a preferential attack of the carbamoyl radical on the carbon bearing the alkyl chain [70].

$$C_8F_{17}CH=CH-C_{10}H_{21} + HCONH_2 \longrightarrow C_8F_{17}CH_2-CH(CONH_2)-C_{10}H_{21} (a)$$
$$+ C_8F_{17}-CH(CONH_2)-CH_2-C_{10}H_{21} (b)$$
(13)

Since formamide microemulsions are transparent in the visible range, they could be utilized for photoamidation.

4.2.3 Choice of Surfactant and Cosurfactant

In order to obtain microemulsions using the olefin $C_8F_{17}-CH=CH-C_{10}H_{21}$, a hydrogenated surfactant and a fluorinated surfactant or vice versa are required. This is due to the segregation between hydrogenated and fluorinated chains [71,72].

We selected as surfactants the ethoxylated nonylphenols of general formula

$$C_9H_{19}C_6H_4(OCH_2CH_2)_n-OH \qquad (14)$$

for which it is possible to vary the number of ethoxy units n. We chose the fluorinated alcohol $C_4F_9C_2H_4OH$ as cosurfactant.

4.2.4 Phase Diagram of the Microemulsions

In the Winsor nomenclature, only two- and three-phase states are distinguished in the phase diagrams of microemulsions:

Winsor I (W I): two-phase state for a microemulsion in equilibrium with an organic phase;

Winsor II (W II): two-phase state for a microemulsion in equilibrium with an aqueous phase (or formamide);

Winsor III (W III): three-phase state for a microemulsion in equilibrium with both an organic and an aqueous phase (or formamide)

Winsor IV (W IV): one-phase state for microemulsions.

The Winsor systems were determined and are indicated on the phase diagram (Figure 9.13).

Amidation of the olefin was carried out in microemulsions A, B and C (Figure 9.13). These microemulsions with a high surfactant and cosurfactant content (65%) maintained a homogeneous reaction medium throughout the course of the reaction. The results obtained along with those obtained in *tert*-butanol are listed in Table 9.7.

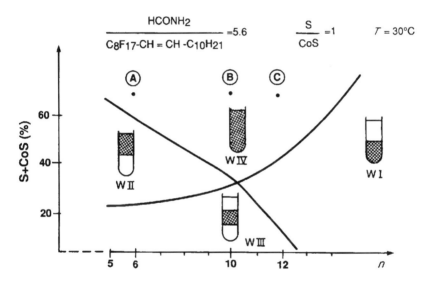

Figure 9.13. Phase diagram of system ($HCONH_2$, $C_8F_{17}CH=CHC_{10}H_{21}$, $C_9H_{19}C_6H_4(OCH_2CH_2)_nOH$ and $C_4H_9H_4OH$. WI = Winsor I system (oil/microemulsion), WII = Winsor II system (microemulsion/formamide), WIII = Winsor III system (oil/microemulsion/formamide), and WIV = Winsor IV system (microemulsion); S = surfactant = $C_9H_{19}C_6H_4(OCH_2CH_2)_nOH$, and CoS = cosurfactant = $C_4F_9C_2H_4OH$

Table 9.7. Photoamidation of $C_8F_{17}CH=CHC_{10}H_{21}$ in microemulsions

Microemulsions	Regioselectivity	Yield (%)
A	0.42	71
B	0.40	89
C	0.44	94
t-butanol	7.71	64

The following comments can be made on these results:

1. The regioselectivities observed in the microemulsions were the opposite of those observed in *tert*-butanol.
2. The yields were considerably higher in the microemulsions than in *tert*-butanol. It can be seen that increasing the number of ethoxy groups n also increased the yield.

The results may be interpreted in terms of the reaction mechanism described in a previous study [69]. This involves a free radical chain reaction initiated with the photoreduction of acetone (excited to the triplet state) by formamide:

Initiation: $CH_3COCH_3 \xrightarrow{h\nu} {}^1CH_3COCH_3^* \longrightarrow {}^3CH_3COCH_3^*$

$${}^3CH_3COCH_3^* + HCONH_2 \longrightarrow (CH_3)_2C^{\bullet}OH + {}^{\bullet}CONH_2$$

Propagation: $R_F-CH=CH-R_H + {}^{\bullet}CONH_2 \longrightarrow R_F-CH^{\bullet}-\underset{\underset{CONH_2}{|}}{CH}-R_H$

$$+ \ R_F-\underset{\underset{CONH_2}{|}}{CH}-CH^{\bullet}-R_H$$

$R_F-CH^{\bullet}-\underset{\underset{CONH_2}{|}}{CH}-R_H + HCONH_2 \longrightarrow R_F-CH_2-\underset{\underset{CONH_2}{|}}{CH}-R_H + {}^{\bullet}CONH_2$

a

$R_F-\underset{\underset{CONH_2}{|}}{CH}-CH^{\bullet}-R_H + HCONH_2 \longrightarrow R_F-\underset{\underset{CONH_2}{|}}{CH}-CH_2-R_H + {}^{\bullet}CONH_2$

b

Termination:
$$2 \ {}^{\bullet}CONH_2 \longrightarrow (CONH_2)_2 \text{ ethanediamide or oxamide}$$

(15)

The oxamide was the only by-product obtained with a yield <5% with respect to the starting olefin.

In tert-butanol, the results obtained for photoamidation of $C_8F_{17}-CH=CH-C_{10}H_{21}$, indicated a predominant role of steric effects accounting for the preponderance of isomer *a*. *In microemulsions*, the results were less readily interpretable, but could be accounted for in terms of the microstructure of the reaction medium. Light-scattering studies indicated the presence of polydispersed aggregates with sizes ranging from 500 to 800 Å. The aggregates were thus predominantly lamellar rather than spherical. The olefin was thus situated in one channel, while the formamide was in another, and the reaction took place mainly at the interface. Two possibilities suggested themselves to account for the observed regioselectivities.

The olefin was partially anchored in the interfacial film by the hydrogenated chain, or *the olefin was completely anchored in the film* along with molecules of surfactant. In this latter situation, the α and β carbons of the olefin are equally sterically hindered. Thus polar effects would predominate, with amidation taking place at the α carbon on the fluorinated chain.

In order to try and differentiate between these two possibilities, a further set of experiments were carried in microemulsion D in which the surfactant was fluorinated ($C_6F_{13}C_2H_4$ $(OC_2H_4)_{12}OH$) and the cosurfactant hydrogenated ($C_6H_{13}OH$). Photoamidation was performed under the same conditions as for the other microemulsions. There was no change in regioselectivity, which tended to rule out the possibility of partial anchorage of the olefin in the interfacial film. If this had been the case, the opposite regioselectivity would have been observed. The olefin would have been retained in the interfacial film by the fluorinated chain along with fluorinated surfactant, and amidation would have occurred preferentially on the β carbon of the fluorinated chain. This was not, however, the case. It would thus appear that the olefin was totally anchored in the interfacial film, and the regioselectivity was essentially governed by polar factors.

The first experiment designed to model photosynthesis was reported in 1930 by the Russian chemist, Oparine. In an attempt to demonstrate the role of compartmentation in a putative origin of life, he investigated the formation of colloidal objects by dispersing mixtures of a protein (histone) and a polysaccharide (gum arabic) in water. He obtained a suspension of small droplets which he referred to as coacervats. Since these droplet sedimented rapidly, he sought a way of stabilizing them by carrying out simple chemical reactions such as:

1. Enzymatic formation of starch from glucose-1-phosphate using a phosphorylase;
2. Double enzymatic reaction using a second enzyme, an amylase which together with the phosphorylase produce a maltose from glucose-1-phosphate;
3. Redox reaction catalyzed by a dehydrogenase for reduction of methyl red.

Encouraged by these preliminary experiments, Oparine attempted to construct a model of photosynthesis in the coacervats. He succeeded in dissolving chlorophyll and by compartmentation carried out a series of processes resembling those of photosynthesis. Chlorophyll excited photochemically can donate an electron to methyl red enabling transfer of hydrogen or reduction. Oxidized chlorophyll was reduced by ascorbic acid thereby returning to its ground state, while the oxidized ascorbic acid was liberated. This experiment is illustrated in Figure 9.14.

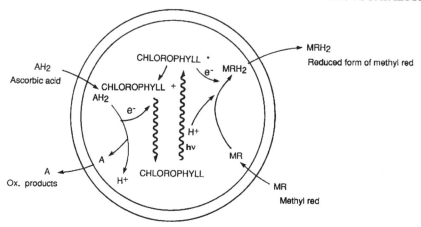

Figure 9.14. Simulation of photosynthesis in coacervat

4.3 ELECTRON TRANSFER AND ORGANIZED MOLECULAR SYSTEMS

Conversion of solar energy by means of artificial photosynthetic systems will generally require a series of redox reactions. These photoredox systems require a donor and acceptor species, one of which is induced into the excited state (either directly or by photosensitization):

$$A \xrightarrow{h\nu} A^* \quad \text{or} \quad D \xrightarrow{h\nu} D^* \tag{16}$$

An electron transfer takes place between one entity in the excited state and the other in the ground state:

$$A^* + D \longrightarrow A^- + D^+ \quad A + D^* \longrightarrow A^- + D^+ \tag{17}$$

A sequence of secondary reactions (dark reactions) is required to generate the energy-storing products. Thus water may be transformed into oxygen and hydrogen, the latter being the storage product as its recombination with oxygen is highly energetic.

One of the problems is avoidance of transfer of an electron between A^- and D^+, which will nullify the effect of the initial reaction. A source of inspiration for resolution of this problem and the development of efficient artificial systems is the efficacy of natural photosynthesis [73–75]. It should be borne in mind that in biological systems the photons captured by chlorophyll during the light reaction supply the energy required to oxidize water:

$$2H_2O \longrightarrow O_2 + 4H^+ + 4e^- \quad E = +0.82 \text{ V} \tag{18}$$

The four electrons and two of the four protons are used to reduce $NADP^+$ into NADPH. The energy stored in NADPH along with that supplied by ATP is used to convert CO_2 and hydrogen into glucose and water:

$$CO_2 + 2H^+ + 2NADPH \longrightarrow \tfrac{1}{6}(C_6H_{12}O_6) + H_2O \quad (19)$$

Around 43% of the energy in the solar spectrum lies in the visible region. Since chlorophyll absorbs photons at 700 nm, the photosynthetic reaction can only be performed by a system involving a series of photochemical reactions.

In biological systems, two photochemical reactions occur in series in order to extract electrons from water and provide them with enough energy to reduce $NADP^+$ (Figure 9.15).

Since the production of O_2 requires 4e, and two photochemical reactions are involved, a minimum of 8e is required for each molecule of oxygen liberated. In artificial systems, the most interesting redox reactions are those involving more than one electron. A method of storing electrons is thus required in order to generate the electrochemical equivalents needed for the overall reaction:

$$2H_2O \xrightarrow{h\nu} 2H_2 + O_2 \quad (20)$$

To the half-reaction of oxidation of water described above, the half-reaction of reduction should be added:

$$2H^+ + 2e \longrightarrow H_2 \quad E = -0.41 \text{ V} \quad (21)$$

a process which involves two electrons and corresponds to the best thermodynamic solution.

In summary, the problems that have to be overcome for an abiotic photosynthetic system are:

1. *Finding a dye (chlorophyll substitute) that absorbs as much solar radiation as possible.* We will discuss here the use of the $Ru(bipy)_3^{2+}$ complex,

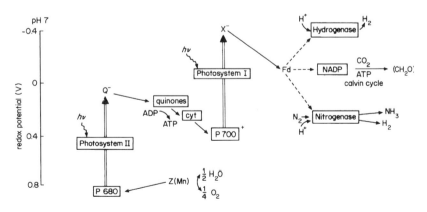

Figure 9.15. The Z scheme for photosynthesis

which also has suitable redox properties [76]. In the excited state the redox potentials of this complex differ from those in the ground state leading to reducing or oxidizing entities that are thermodynamically capable of reducing or oxidizing water.

$$\begin{aligned} \text{Oxidizing pair: } Ru(bipy)_3^{3+/2+} \quad & E = 1.27 \text{ V} \\ \text{Reducing pair: } Ru(bipy)_3^{3+/2+*} \quad & E = -0.84 \text{ V} \end{aligned} \quad (22)$$

2. *A system for storing electrons like $Rh(bipy)_3^{3+}$ which can be reduced in stages:*

$$+1e^- \quad +1e^-$$

or methyl viologen MV^{2+} [78]:

$$CH_3 - \overset{+}{N} \underset{}{\bigcirc} - \underset{}{\bigcirc} \overset{+}{N} - CH_3 \quad (23)$$

homologs of MV^{2+} such as HMV [79]:

$$C_7H_{15} - \overset{+}{N} \underset{}{\bigcirc} - \underset{}{\bigcirc} \overset{+}{N} - CH_3 \quad (24)$$

or the general series C_n MV.

3. *A membrane to prevent recombination of ions*. The multielectronic nature of the splitting of water shows that the oxidation and reduction processes need to be separated in order to prevent loss of energy by remission of light or recombination of ions.

4. *Catalysts to accelerate the redox reactions (Pt or other colloidal metal)*. It should be borne in mind that biological systems use the water molecule itself as electron donor and acceptor. Artificial photochemical splitting can be represented by two half-reactions of which one, reduction with formation of hydrogen (of interest for energy storage), requires an electron donor. These sacrificial compounds (amine, alcohol, EDTA) enable high turnover regeneration of the chemical reducing agent in a cyclic process [80].

An organized molecular system may be utilized at each stage of such processes, by first separating the compounds entering the redox processes and subsequently by preventing recombination of ions. Their ability to concentrate reactants in a small volume can also be exploited for the preparation of highly dispersed catalysts. At the macroscopic level these effects will largely influence the photoinduced charge separation process by:

1. Quenching the excited state;

2. Blocking the thermal back electron-transfer reaction.

The electrical double layer and high surface potential of the interface impinge on both processes. Important factors are the association between sensitizer and quencher along with the exchange of compounds and ions between the organized molecular systems and the bulk solvent.

4.3.1 Micellar Solutions

The quenching of Ru(bpy)$_3^{2+}$ by MV was found to be markedly enhanced in SDS micelles [81–84]. In fact, the intensity of the transfer is effectively determined by the fraction of collisions between the excited ruthenium complex and MV molecules. Since MV rapidly leaves the micelle, more hydrophobic compounds such as benzviologen [85] which bind more strongly to SDS or heptylviologen can be employed [86]. There is thus an important effect of proximity in these organized molecular systems.

In an intensive study of these processes, Grätzel analyzed the role of cationic micelles of CTAC using an amphiphile, N-tetradecyl, N-methylviologen and a water-soluble sensitizer. Rapid solubilization of the viologen in the micelle and the obstruction of entry of the oxidized sensitizer was found to lead to a 400-fold inhibition of the back reaction [87,88].

By using the properties of exchange of counter ions, Grätzel's group examined the system of anionic micelles formed with copper lauryl sulfate (Cu(LS)$_2$). In this case the sensitizer was solubilized within the micelles, and electron transfer was carried out by the reduction of Cu^{2+} to Cu^+ and the oxidation of sensitizer. This transfer is fast (around 1 ns) and is followed by a Cu^+/Cu^{2+} exchange at the water–micelle interface. The Cu^+ ion may then reduce an anionic electron acceptor (e.g. the pair Fe(CN)$_6^{3-}$/Fe(CN)$_6^{4-}$), which cannot penetrate the micelle due to electrostatic repulsions. This effectively separates the charges [89,90].

Matsuo has described a system consisting of N-dodecyl, N-methylviologen, CTAC and EDTA as reducing agent [91]. In this system, the quantum yield for reduction of viologen was 21-fold that of the same system with MC. This is an excellent example of the alteration of the hydrophobicity of species solubilized in water where the effect of quenching is predominant:

$$\text{Ru(bpy)}_3^{2+} + \text{C}_{12}\text{MV}^{2+} \longrightarrow \text{Ru(bpy)}_3^{3+} + \text{C}_{12}\text{MV}^+ \quad (25)$$

The exchange takes place in water, but $C_{12}MV^+$ being less hydrophilic than $C_{12}MV2^+$ enters the micelle and prevents the back reaction.

Micellar systems comprising a chromophore and a quencher with opposite charges may also be envisioned such as Pd(II)[tetra(p-sulfonatophenyl)]porphyrin (Pd(TPPS)$^{4-}$) as chromophore and MV^{2+} or N,N' dibenzyl-2,2'-bipyridine (BV^{2+}) as quencher. Irrespective of the type of surfactant (SDS or CTAC) the reduced viologens have long lifetimes, and the higher the hydrophobicity the greater the stabilization [85].

4.3.2 Reverse Micelles and Microemulsions

Experiments on electron transfer have been carried out using the pyrene-dimethylaniline (DMA) system in reverse CTAC micelles [92]. In this system, charges are only weakly separated, probably as a result of the low polarity of the environment, and a higher yield has been obtained with a more hydrophilic sensitizer (pyrene tetrasulfonate).

Differences between methyl and heptylviologen (HV^{2+}) were evidenced in a laser flash photolysis study in reverse microemulsions. Only HV^+ was detected in significant proportions [86], whereas MV^+ was not detected due to its greater hydrophilicity. On the other hand, since MV^+ is more hydrophobic than HV^{2+} it penetrates deeply in the micelles.

4.3.3 Vesicles

In common with micelles, charged vesicles can also effect charge separation. The differences between the two systems stem from the presence of a water pool within the vesicle. In a simple model of a vesicle, the direct sensitizer–quencher pair may be distributed such that:

1. One lies in the water pool within in vesicle, while the other is bound to the outer surface of the vesicle;
2. Both are situated in the water pool;
3. Both are distributed outside the vesicle.

Case 2 is clearly the most suitable for quenching. The results obtained can be interpreted in the light of the structure of the surfactants used to produce the vesicles.

(a) Liposomes produced from natural phospholipids

Major studies have been carried out by Calvin's group [93–96]. Liposomes with diameters close to 500 Å produced with natural phospholipids have been employed as model membranes (Figure 9.16) for distribution of reactants such as:

1. The sensitizer $Ru(bpy)_3^{2+}$ or derivative amphiphiles produced by introduction of hydrocarbon chains (C_{16}, C_{18}) as amides onto the carboxylic groups born by the pyridine rings. They may distribute on either side of the membrane. The liposome is prepared in the presence of chromophore.
2. Quencher HV^{2+} or propylviologen sulfonate (PVS) on one side of the membrane (e.g. on the outside).
3. Donors such as EDTA or RuO_2 (which can accumulate four electrons) are situated on the other side of the vesicle (on the inside in this case).
4. Colloidal Pt as catalyst on the same side as HV^{2+}.

Electrons are thus transferred from the inside to the outside:

ORGANIZED MEDIA AND HOMOGENEOUS PHOTOCATALYSIS 345

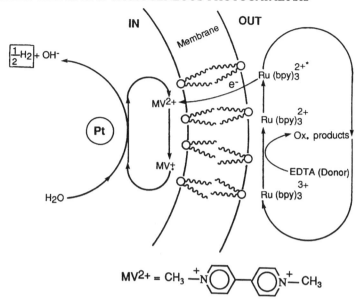

Figure 9.16. Use of liposomes in artificial models of photosynthesis

$$\text{Ru(bpy)}_3^{2+} + \text{HV}^{2+} \longrightarrow \text{Ru(bpy)}_3^{3+} + \text{HV}^+ \qquad (26)$$

and the Ru^{3+} ion is reduced by Ru^{2+} on the internal surface. Although there is doubt about the reality of reaction in view of its low rate [97], on the outside HV^+ is again oxidized by H^+ into HV^{2+}, a reaction catalyzed by colloidal platinum generating hydrogen. On the inside, the donor is oxidized by Ru^{3+} in the presence of RuO_2, efficiently producing oxygen.

Prompted by the example of natural photosynthesis where manganese compounds catalyze the production of oxygen [96], Calvin has proposed a unilamellar vesicular system in which the water-soluble manganese compound is on the inside and the viologene on the outside of the vesicle. This system made using natural (or even synthetic) phospholipids accepts molecules of amphiphilic sensitizer. The chromophore is situated on both the inner and outer surfaces of the membrane. Unfortunately manganese is too strong an oxidizer and tends to destroy the vesicles. In this system, amphiphiles with perfluorinated chains might be more resistant to attack by manganese.

(b) Anionic vesicles

Vesicles formed from dihexadecylphosphate (DHP) have been shown to stabilize the products of a photosensitized electron transfer from diphenylamine to duroquinone. These two substances give rise to radical anions and cations, which are highly stabilized in these vesicles [98]. Highly efficient photoionization of pyrene (P) entrapped in vesicles of DHP has been described, leading to production of P^+ [99].

For artificial photosynthesis, this type of vesicle enables photoinduced diffusion of electrons [100]. In this case, the sensitizer is situated on the outer surface of the vesicles and the quencher MV^{2+} on the inner surface, whereas the donor, triethanolamine (TEOA) is in the bulk solution. On irradiation MV appears on the outer surface.

To anchor the reactant more firmly in the vesicles, Hurst and Thompson [101] utilized N-alkyl-N-methylviologens ($C_n MV^{2+}$) which are bound to both surfaces, and a porphyrin as sensitizer. This derivative, Zn-tetraphenylporphyrin sulfate (ZnTPPS^{4-}), situated in the bulk solution, is oxidized in its excited form by the viologen. The oxidized porphyrin is reduced by TEOA (also in the bulk solution), while the cation $C_n MV^+$ undergoes a redox reaction producing hydrogen. The process is favored for two reasons:

1. The oxidized porphyrin ZnTPPS^{3-} is repelled from the anionic surface of the vesicle and thus remains in the bulk solution;
2. The reduction in charge of the viologen renders it even more lipophilic, further contributing to charge separation.

(c) Cationic vesicles

Most of the studies on these systems have been carried out by Fendler [102] using dodecyl dimethylammonium chloride (DODA) as cationic surfactant to produce vesicles that entrap large molecules. He initially employed as electron donor N-methyl phenothiazine (MPTH), which is situated in the hydrophobic bilayers, combined with the perchlorate of a complex surfactant of ruthenium $RuC_{18}(bpy)2+_3$, as potential electron acceptor, bound to the surface, which enables the following transfer:

$$RuC_{18}(bpy)_2^{3+*} + MPTH \longrightarrow RuC_{18}(bpy)_3^+ + MPTH^+ \qquad (27)$$

The MPTH$^+$ migrates into the bulk solution where it has a long lifetime due to the charge separation induced by the positive surface of DODAC [103].

The lifetime of MPTH$^+$ can be altered by addition of NaCl [103]. Used with chlorophyll as photosensitizer, intermediate species (triplets and radical cations of chlorophyll) with long lifetimes are observed on excitation at 694 mm [102].

4.3.4 Polymerized Surfactant Vesicles

We have already mentioned that polymerization of surfactants bearing vinyl or methacrylate groups can give rise to highly stable vesicles whose permeability can be altered. Such vesicles have been utilized in the photochemical conversion of solar energy [104,105].

4.3.5 Aqueous Solutions of Polymers

Most of the reported examples are based on polyelectrolytes. Such systems can promote charge separation in two ways: (1) under the influence of their electrostatic field; (2) by differential solubility in the aqueous and more hydrophobic phases.

Recently the photoinduced transfer of electrons in two polymers (polyacrylic and polymethacrylic acids, PAH and PMAH) with different solubilities in water has been reported. The photophysical properties of a probe, diphenylanthracene bound covalently to the two polymers was found to be strongly influenced by the polymeric environment. It has also been demonstrated that the hydrophilic ions MV^{2+} quench the first excited singlet state $^1DPA^*$ more efficiently in PAH than in PMAH at low pH, but charges were separated in the PAH–DPA/MV and PAH–DPA/SPV systems at different pH [106]. Studies on a variety of non-ionic polymers have also been reported [107–109].

(a) Anionic polyelectrolytes

Polyvinylsulfates have been employed to investigate the quenching of $Ru(bpy)_3^{2+*}$ by Cu^{2+} ions [110]. The effects were attributed to electrostatic interactions of these ions with the polymer as a function of their charge and consequently of the potential for electron transfer. The ion exchange potential of these polyelectrolytes was thought to play a major part in their activity in this process.

(b) Cationic polyelectrolytes

With a quaternary polyammonium derivative such as polybrene, which is positively charged, Rabani and Sassoon [111] investigated the transfer of electrons between $[Ru(bpy)_2^{2+}(CN^-)_2]^*$ and $Fe(CN)_6^{3-}$. With polybrene the quantum yield is close to unity, while the back reaction is only slightly slowed. This is a result of two opposing actions: an effect on the exchange of species associated at the interface which favors quenching, but also on a poor separation of charges by the weakly charged polymer.

If the donor and acceptor are bound covalently to two different polymers, a third compound is required to relay electrons from the donor to the acceptor. Adapted to the porphyrin/carotenoid/quinone systems, this strategy gives rise to long lifetimes as a result of the charge separation induced by the third compound.

In a recent review Meisel and Matheson [114] cite the example of the technique proposed by Sassoon [112] using 9-methylanthracene (9MeA) as relay. The triplet state of 9MeA reduces MV^{2+} and the resulting oxidized product is reduced in turn by phenothiazine also bound to a polymer.

(c) Non-Ionic polymers

Non-ionic polymers have also been explored. For example, water-soluble porphyrins have been combined with polyvinyl alcohols (PVA) in an attempt to separate charged entities, which circumvents some of the drawbacks of polyelectrolytes. These noncharged polymers can also serve as supports for redox catalysts in the reduction of water in highly reactive microparticles [107,108]. Combinations of porphyrin and PVA in aqueous solution have been shown to favor quenching of methyviologen MV^{2+} and the subsequent separation of charge [109].

4.4 PREPARATION OF METALLIC MICROPARTICLES AS CATALYSTS FOR REDOX REACTIONS

Metallic catalysts in the colloidal state can optimize electron transfer by ordering reactants on their surface. For example, the photoexcitation of colloidal CdS generates electron–hole pairs which may migrate to the surface and promote redox reactions at the water–semiconductor interface. The efficiency of such transfers will depend on the contact area and hence on the size of the particles. For such applications, microparticles may be conveniently prepared in the microreactors constituted by organized molecular systems. Furthermore, the presence of surfactants stabilizes the colloidal suspensions produced.

The particles of SDS may be stabilized by cationic surfactants such as CTAB, which produce positive charges on the surface of CdS, while negative charges can be induced using anionic surfactants such as SDS [115].

The colloidal CdS prepared in polymeric matrices such as Nafion and cellulose confers special properties on such systems [116]. It is also possible to produce particles of colloidal CdS or colloidal CdS in reverse micelles such as AOT–water–isooctane or water in oil CTAB, SDS microemulsions and lecithin vesicles [117]. Vesicles are particularly suited to multielectron transfers. Particles below 5 nm in diameter have been produced in vesicles of DHP, DODAC and polymerized surfactant [118].

4.5 EXCHANGE OF LIGANDS, METALLATION OF PORPHYRINS

An important factor that is often overlooked is the concurrent exchange of ligands in the metallic complexes employed. As an example, the rate constants of aquation of the complexes, $[Cr(C_2O_4)_3]^{3-}$ and $[Co(C_2O_4)_3]^{3-}$ are markedly altered in solutions of dodecylammonium propionate in benzene. Compared to that observed in pure water, the rate of aquation of the cobalt complex is multiplied by a factor of 1500, and that of chromium by a factor of 10^6 [119].

In a similar manner, the metallation of certain complexes is strongly influenced by organized assemblies. The rate of insertion of Cu(II) in tetraphenylporphyrin is multiplied by a factor of 1650.

5 CONCLUSION

We have seen that modular systems can now be constructed for carrying out chemical reactions as suitable preparative methods are now available and the properties of organized molecular systems are increasingly understood. In this non-exhaustive review we have described a number of the molecular assemblies that appear to be suited for photocatalytic reactions.

The key component of these systems is the interface, where many of the reactions take place. Since the number of reaction sites on the interface is a function of its surface area, microemulsions, which should be viewed in their widest sense and not just as reverse micelles, have the advantage of marginally increasing the interfacial area and hence the number of reaction sites. The diversity of structures of microemulsions represents and added advantage, which is offset somewhat by their increased complexity. Nevertheless, the principle of molecular economy described in this review (p. 320) can lead to a significant simplification of these systems by doubling the roles of some of the components in the system. It is often possible to prepare microemulsions in which one of the components is both reactant and constituent of the medium.

The efficiency and orientation of the reaction is strongly influenced by the structure of the microemulsion, and photosensitized chain reactions can be carried out in bicontinuous systems. Such reactions cannot be studied in direct or reverse micellar solutions.

Lastly, the barrier represented by the interface effectively concentrates reactants in certain compartments and restricts exchanges between phases, preventing neutralization of entities of opposite charge. Exploitation of this feature is a promising development as testified by the wide variety of systems used in studies on artificial photosynthesis and the advances made in this ambitious project.

In conclusion we would like to emphasize the innovative nature of the two techniques described, Each is recent and their combination is even more so. This provides much scope for investigations in the two relevant disciplines in attempts to:

1. Synthesize surfactants or polymers which favor assemblies of defined properties;
2. Characterize the physicochemical properties of such systems;
3. Exploit this knowledge for the construction of systems designed to resolve practical problems.

REFERENCES

[1] (a-) E. S. Amis, *Solvent Effects on Reaction Rates and Mechanisms* (Academic Press, New York, 1966); (b-) A. J. Parker, Chem. Rev., **69**, 1, (1969).
[2] M. Makosza and M. Ludwikow, Angew. Chem. Int. Edn., **13**, 665, (1974).
[3] J. M. Lehn, Science, **227**, 849, (1985); Angew. Chem. Int. Ed. Engl. **27**, 89, (1988).
[4] H. Kuhn, Pure and Appl. Chem., **53**, 2105, (1981).
[5] R. Zana, J. Chim. Phys., **83**, 603, (1986).

[6] J. M. Brown, in *Colloïd Science*, edited by D. H. Everett (The Chemical Society, London, 1979) Vol. 3, p. 253.
[7] J. J. Charvolin, Chim. Phys., **80**(1), 15 (1983).
[8] M. P. Pileni, (editor), *Structure and Reactivity in Reverse Micelles* (Elsevier, 1989).
[9] F. Szoka and D. Papahadjopoulos, Ann. Rev. Biophys. Bioeng., **9**, 467 (1980).
[10] L. Sauders, J. Perrin and D. B. Gammack, J. Pharm. Pharmacol., **15**, 155 (1962).
[11] J. H. Fendler and E. J. Fendler, *Catalysis in Micellar and Macromolecular Systems* (Academic Press, New York, 1975).
[12] P. A. Winsor, Solvent Properties of Amphiphilic Compounds, (Butterworths London, 1954).
[13] K. J. Shinoda, Colloid Interface Sci., **24**, 4 (1967).
[14] G. D. Smith, C. E. Donelan and R. E. Barden, J. Colloid Interface Sci., **60**, 488 (1977).
[15] I. Rico and A. Lattes, *Surfactants in Solution*, edited by K. L. Mittal (Plenum Press, New York, 1986).
[16] S. E. Friberg, *Non Aqueous Microemulsions* (CRC Press, 1986).
[17] S. E. Friberg and I. Lapczynska, Prog. Colloid. Polymer. Sci., **56**, 16 (1975).
[18] J. L. Cayias, R. S. Schechter and W. H. J. Wade, J. Colloid Interface Sci., **59**, 31 (1977).
[19] D. Langevin, Mol. Cryst. Liq. Cryst., **138**, 259 (1986).
[20] A. M. Bellocq, J. Biais, P. Bothorel, B. Clin, F. Fourche, P. Lalanne, B. Lemaire and D. Roux, Advances in Colloid and Interface Science, **20**, 167 (1984).
[21] M. Gautier, I. Rico, A. Ahmad-Zadeh Samii, A. de Savignac and A. Lattes, J. Colloid Interface Sci., **112**, 484 (1986).
[22] D. P. Riley and G. Oster, Discuss. Faraday Soc., **11**, 107 (1951).
[23] M. Dvolaitzky, M. Lagues, J. P. Le Pesant, R. Ober, C. Sauterey and C. Taupin, J. Phys. Chem., **84**, 1532 (1980).
[24] M. Dvolaitzky, M. Guyot, M. Lagues, J. P. Le Pesant, R. Ober, C. Sauterey and C. Taupin, J. Phys. Chem., **69**, 3279 (1978).
[25] D. O. Shah, Ed. *Surface Phenomena in Enhanced Oil Recovery* (Plenum Press, 1981).
[26] B. W. Ninham, J. Phys. Chem., **84**, 1423 (1980).
[27] R. Kjellander, J. Chem. Soc. Faraday Trans. **78**, 2025 (1982).
[28] A. Graciaa, J. Lachaise, P. Chabrat, L. Letamendia, J. Rouch and C. Vauchamps, J. Phys. Letters, **39**, 235 (1978).
[29] P. Lianos, Lang, C. Stazielle and R. Zana, J. Phys. Chem., **86**, 1519 (1982).
[30] A. M. Bellocq, J. Biais, B. Clin, P. Lalanne and B. J. Lemanceau, Colloid. Interface Sci. **70**, 524 (1979).
[31] B. Lindman, P. Stilbs and M. E. Moseley, J. Colloid Interface Sci. **83**, 569 (1981).
[32] M. Lagues, R. Ober and C. Taupin, J. Phys. Letters, **35**, 487 (1974).
[33] A. M. Cazabat, D. Chatenay, D. Langevin and A. Pouchelon, J. Phys. Letters, **41**, 441 (1980).
[34] M. Lagues and C. Sauterey, J. Phys. Chem., **84**, 3503 (1980).
[35] R. A. Mackay, Adv. Colloids and Interface Sci. **15**, 131 (1981).
[36] R. Zana, *Surfactants in Solution*, K. L. Mittal and P. Bothorel, Eds. (Plenum Press, New York, 1986).
[37] A. Lattes, A. J. Chim. Phys. **84**(9), 1061 (1987).
[38] J. H. Fendler, Acc. Chem. Res. **13**, 7 (1980).
[39] T. Kunitake and S. Shinkai, Adv. Phys. Org. Chem. **17**, 435 (1980).
[40] L. Gross, H. Ringsdorf and H. Schupp, Angew. Chem. Int. Ed. Engl. **20**, 305 (1987).

[41] J. H. Fendler and P. Tundo, Acc. Chem. Res. **17**, 3 (1984).
[42] J. H. Fendler, in *Surfactants in Solution*, K. L. Mittal; B. Lindman, Eds, (Plenum Press, New York, 1984).
[43] M. Haubs and H. Ringsdorf, New. J. Chem. **11** (2), 151 (1987).
[44] I. Rico, N. Hajjaji-Srhiri, B. Escoula, T. N. De Castro Dantas and A. Lattes, Nouv. J. Chim., **10**, 25 (1986).
[45] E. Perez, J. P. Laval, M. Bon, I. Rico and A. Lattes, J. Fluorine Chem., **39**, 173 (1988).
[46] K. Shinoda, M. Hato and T. Hayashi, J. Phys. Chem., **76**, 909 (1972).
[47] J. P. Couvercelle, J. Huguet and M. Vert, Macromolecules, **24**, 6452 (1991).
[48] Y. L. Khmelnitsky, A. K. Gladilin, V. L. Roubailo, K. Martinek and A. V. Levashov, Eur. J. Biochem., **206**, 737 (1992).
[49] H. Amarouche, C. de Bourayne, M. Riviere and A. C. R. Lattes, Acad. Sci. Paris, **298**, série II, 121 (1984).
[50] E. Perez, N. Alandis, J. P. Laval, I. Rico and A. Lattes, Tetrahedron Lett. **28**, 2343 (1987).
[51] O. V. Gerasimov and V. N. Parmon, Russian Chemical Reviews, **61**(2), 154 (1992).
[52] S. E. Braslovsky and K. N. Houk, Pure and Appl. Chem., **60** (7), 1055 (1988).
[53] H. Kish and H. Hennig, EPA Newslett. **19**, 23 (1983).
[54] R. G. Salomon Tetrahedron, **39**, 485 (1983).
[55] C. Kutal, Coord. Chem. Rev. **64**, 191 (1985).
[56] E. Pelizetti and N. Serpone, Eds. *Photocatalysis* (Wiley, New York, 1989).
[57] F. Chanon and M. Chanon, Photocatalysis, New York, 489 (1989).
[58] H. Hennig, R. Billing and H. Knoll, in *Photosensitization and Photocatalysis Using Inorganic and Organometallic Compounds*, K. Kalyanasundaram, M. Gratzel, M. Eds. (Kluwer, Dordrecht, 1993).
[59] H. Henning, L. Weber and D. Rehorek, *Photosensitive Metal-organic Systems* C. Kutal and N. Serpone, (Eds) Adv. Chem. Ser., vol. 238, Washington (1993).
[60] (a) A. Padwa, F. J. Albrecht, Am. Chem. Soc., **94**(2), 1000 (1972), (b) A. Padwa, F. J. Albrecht, Org. Chem., **39**(16), 2361 (1974).
[61] I. Rico, M. T. Maurette, E. Oliveros, M. Riviere, A. Lattes, Tetrahedron Lett. 4795 (1978).
[62] I. Rico, M. T. Maurette, E. Oliveros, M. Riviere, A. Lattes, Tetrahedron, **36**, 1179 (1980).
[63] F. Candau, I. Zekhnini, J. P. J. Durand, Colloid. Interface Sci. **114**, 398 (1986).
[64] I. Rico, A. Lattes, *Microemulsions Systems*, H. Rosano, M. Clausse, Eds. (Marcel Dekker, New York) **23**, 341 (1987).
[65] I. Rico, A. Lattes, K. P. Das, B. Lindman, J. Am. Chem. Soc., **111**, 7266 (1989).
[66] I. Rico, A. Lattes, J. Colloid Interface Sci., **102**, 1984 (1984).
[67] (a) P. Stilbs, Progr. NMR Spectrosc., **19**, 1 (1987); (b) K. P. Das, A. Cegie, B. Lindman, J. Phys. Chem., **91**, 2938 (1987).
[68] J. Rokach, C. H. Krauch, D. Elad, Tetrahedron Lett. **28**, 3953 (1966).
[69] M. Gautier, I. Rico and A. Lattes, J. Org. Chem. **55**(5), 1500 (1990).
[70] M. Gautier, I. Rico and A. Lattes, J. Fluorine Chem. **44**, 419 (1989).
[71] C. Ceschin, J. Roques, M. Malet-Martino and A. Lattes, J. Chem. Technol. Biotechnol., Chem. Technol. **35A**, 75 (1985).
[72] (a) P. Mukerjee and K. J. Mysels, Colloidal Dispersions, Micellar Behav. **17**, 239 (1970); (b) D. G. Kruger, S. Rajan and C. A. Kingsbury, J. Solution Chem. **11**, 79 (1982).
[73] M. Calvin, Théor. Biol. **1**, 258 (1961).
[74] M. Gibbs, Ed. *Structure and Function of Chloroplast*; Springer-Verlag, Berlin (1971).

[75] E. Govindjee, Ed. *Bioenergetics in Photosynthesis* Academic Press, New York, (1975).
[76] K. Kalyanasundaram, Coord. Chem. Rev. **46**, 159 (1982).
[77] I. Willner and R. J. Maidan, Chem. Soc., Chem. Commun. 876 (1988).
[78] E. Amouyal, B. Zidler, Isr. J. Chem. **22**, 117 (1982). See also Chapter 8 of this book.
[79] M. Meyer, C. Wallberg, B. H. Kurihara and J. H. Fendler, J. Chem. Soc., Chem. Commun., 90 (1984).
[80] P. Schwarz, S. Bossmann, A. Guldner, H. Dur and Langmuir, **10**, 4483 (1994).
[81] R. Schmehl and D. G. Whitten, J. Am. Chem. Soc., **102**, 1938 (1980).
[82] R. Schmehl, G. W. Whitesell and D. G. Whitten, J. Am. Chem. Soc., **103**, 3761 (1981).
[83] M. Tachiya, Chem. Phys. Lett., **69**, 605 (1980).
[84] U. Goselle, U. K. A. Klein and M. Hauser, Chem. Phys. Lett. **68**, 291 (1979).
[85] D. G. Whitten, J. C. Russell and R. H. Schmehl, Tetrahedron, **39**, 2455 (1982).
[86] S. Atik and J. K. Thomas, J. Am. Chem. Soc. **103**, 43647 (1981).
[87] P. E. Brugger and M. Gratzel, J. Am. Chem. Soc. **102**, 2461 (1980).
[88] P. A. Brugger, P. P. Infelta, A. M. Braun and M. Gratzel, J. Am. Chem. Soc. **103**, 320 (1981).
[89] Y. Moroi, A. M. Braun and M. Gratzel, J. Am. Chem. Soc. **101**, 567 (1979).
[90] Y. Moroi, P. P. Infelta and M. Gratzel, J. Am. Chem. Soc. **101**, 573 (1979).
[91] T. Matsuo, K. Takuma, Y. Tsustsoi and T. Nishizima, J. Coord. Chem. **10**, 195 (1980).
[92] (a) M. Gratzel, *Micellization and Microeulsions*, K. L. Mittal, Ed., Plenum Press, New York, **2**, 531 (1977); (b) J. K. Thomas, Acc. Chem. Res., **133**, 10 (1977).
[93] W. E. Ford, J. W. Otvos and M. Calvin, Nature, **274** 507 (1978).
[94] I. Willner, W. E. Ford, J. W. Otvos and M. Calvin, Nature **280** 828 (1979).
[95] W. E. Ford, J. N. Otyos and M. Calvin, Proc. Natl. Acad. Sci. U.S.A., **76**, 3590 (1979).
[96] M. Calvin, Photochem. Photobiol. **23** 425 (1976).
[97] A. W. H. Mau, C. B. Huang, N. Kakuta, A. J. Bard, A. Campion, M. A. Fox, J. M. White and S. E. Webber, J. Am. Chem. Soc., **106**, 6537 (1984).
[98] M. S. Tunuli and J. H. Fendler, J. Am. Chem. Soc. **101**, 4030 (1979).
[99] J. Escabi-Perez, A. Romero, S. Lukac and J. H. Fendler, J. Am. Chem. Soc., **101**, 2231 (1979).
[100] L. Y. Lee, J. K. Hurst, M. Politi, K. Kurihara and J. H. Fendler, J. Am. Chem. Soc., **105**, 370 (1983).
[101] J. K. Hurst and H. P. Thompson, J. Membr. Sci., **28**, 3 (1986).
[102] J. H. Fendler, Acc. Chem. Res., **13**, 7B (1980).
[103] P. P. Infelta, M. Gratzel and J. H. Fendler, J. Am. Chem. Soc., **102**, 1479 (1980).
[104] J. H. Fendler, Science, **223**, 888 (1984).
[105] K. Kurihara, P. Tundo and J. H. Fendler, J. Phys. Chem. **87**, 3777 (1983).
[106] M. Sanquer-Barrie and J. A. Delaire, New. J. Chem., **16**, 801 (1992).
[107] J. Kiwi, *Energy Sources Through Photochemistry and Catalysis*, Ed. M. Gratzel, Academic Press, New York, p. 297 (1983).
[108] P. Keller and A. Moradpour, J. Am. Chem. Soc., **102**, 2193 (1980).
[109] M. D. Yanuck and R. H. Schmehl, Chem. Phys. Lett. **122**(1,2), 133 (1985).
[110] (a) I. A. Taha and H. Morawetz, J. Am. Chem. Soc., **93**, 829 (1971); (b) D. Meisel and M. S. Matheson, J. Am. Chem. Soc., **99**, 6577 (1977).
[111] J. Rabani and R. F. Sassoon, J. Photochem., **29**, 7 (1985).
[112] R. F. Sassoon, J. Am. Chem. Soc., **107**, 6133 (1985).

[113] J. Olmsted, S. F. McClanahan, E. Danielson, J. N. Younathan and T. J. Meyer, J. Am. Chem. Soc., **109**, 3297 (1987).
[114] D. Meisel, M. S. Matheson, Photocatalysis and N. Serpone, E. Pelizettti, Eds. Wiley, (1989).
[115] J. K. Thomas and S. Hashimoto, New J. Chem., **11**(2) 145 (1987).
[116] (a) J. Kuczynski, B. Milosavljevic and J. K. Thomas, J. Phys. Chem. **87**, 3368 (1983); (b) D. Meissner, R. Memming and B. Kastening, Chem. Phys. Lett. **96**, 34 (1983).
[117] P. Lianos and J.K. Thomas. J. Colloid Interface Sci. **117**, (2) 505 (1987).
[118] (a) Y. M. Tricot, A. Emeren and J. H. Fendler. J. Phys. Chem. **89**, 4721, (1985); (b) J. H. Fendler, Isr. J. Chem., **25**, 3 (1985).
[119] C. J. O'Connor, E. J. Fendler and J. H. Fendler. J. Am. Chem. Soc., **95**, 600 (1973), ibid. J. Chem. Soc. Dalton trans, 1974, 625; ibid. J. Am. Chem. Soc., **96**, 370 (1974).
[120] K. Letts and R. A. Mac Kay, Inorg. Chem. **14**, 2990, (1975); ibid., **14** 2993 (1975).

10 Photosynthesis, a Natural Model for Photocatalysis

PAUL MATHIS
Section de Bioénergétique Bât. 532, CEA-Saclay, 91191 Gif-sur-Yvette, France

1	Introduction	355
2	Molecules and concepts	358
	2.1 Proteins, membranes	358
	2.2 Antenna, reaction centers	358
	2.3 Pigments: chlorophylls, carotenoids, linear tetrapyrroles	359
	2.4 Transfer of energy, electron or proton	361
	2.5 Energetics and yields	364
	2.6 Excited states: singlet, triplet	366
	2.7 Summary of reaction centers	368
3	The reaction center of purple bacteria	369
	3.1 Functional properties	369
	3.2 Isolation — composition	373
	3.3 Structure	375
	3.4 Relation between structure and function: overall features	377
	3.5 Relations between structure and function: detailed considerations	377
4	Other types of reaction center	386
	4.1 PS2 reaction center of oxygenic photosynthesis	386
	4.2 PS1 and bacterial analogs	391
	4.2.1 Sequence of electron carriers	391
	4.2.2 Large core antenna	393
	4.2.3 Electron donor and electron acceptor to PS-1	393
5	Conclusions	393
	References	395

1 INTRODUCTION

Photosynthesis is the process by which biological organisms make use of solar energy to satisfy the energy requirements which are needed for the synthesis of their biomolecules. These molecules, which are very diverse, permit the

Homogeneous Photocatalysis. Edited by M. Chanon
© 1997 John Wiley & Sons Ltd

maintenance of life activity and the growth of individual organisms as well as their multiplication.

On the world scale, it is not an overstatement to say that photosynthesis has played and still plays overwhelming roles:

1. It has provided the energy for the progressive evolution of all biological species. Nowadays, the energy captured by photosynthetic organisms is eventually used by non-photosynthetic ones. In that respect, it can be reminded that living organisms can be grossly divided in the autotrophs (thanks to photosynthesis, they are energetically autonomous) and the heterotrophs (their energy source comes from the oxidation of biomolecules synthesized by autotrophs). Plants are autotrophs and animals are heterotrophs. Bacteria are either autotrophs or heterotrophs.
2. Photosynthesis has permitted the building up of an atmosphere with a sufficient content in oxygen for animal respiration to operate.
3. Very active photosynthesis took place on earth at the Carboniferous period (about 300 million years ago), leading progressively to the accumulation of enormous amounts of organic materials which are now used as fossil fuels (coal, lignite, natural gas, petroleum).

As it takes place now in nature, photosynthesis can be grossly divided in two categories (see refs 1–4 for reviews on the subject):

1. Oxygenic photosynthesis, performed by plants and cyanobacteria, in which water is the source of electrons for the reduction of carbon dioxide. This type of photosynthesis is the most complex one. It necessitates the concerted function of two types of initial energy converters named reaction centers; these two types are named Photosystem-1 (PS1) and Photosystem-2 (PS2) (Figure 10.1).
2. Anoxygenic photosynthesis, performed by purple and green bacteria. These organisms have only one type of reaction center and they are not able to use water as a source of electrons: they use more reducing mineral compounds such as sulfur derivatives or organic molecules such as organic acids.

Quantitatively, oxygenic photosynthesis is undoubtedly the most important category since it takes place in forests, in cultivated land and in the main part of the oceanic biomass. It is characterized by its evolutionary stability: the process of photosynthesis for all organisms in these biotopes is nearly identical. On the other hand, the world of anoxygenic photosynthesis provides a much more modest contribution to the world-scale energy storage, but it is characterized by its diversity (which is very important for research, in order to understand the relations between molecular structures and functions) and by its simplicity: for instance, energy converters from purple bacteria provide good models for understanding the initial steps of photosynthesis in general.

PHOTOSYNTHESIS, A NATURAL MODEL FOR PHOTOCATALYSIS

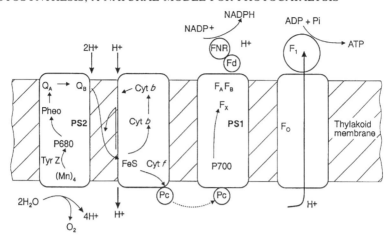

Figure 10.1. A schematic presentation of electron and proton transfers which take place in the photosynthetic membrane of oxygenic organisms. Endergonic electron transfer from water to $NADP^+$ is empowered by light-induced reactions occurring in reaction centers named PS2 and PS1 (see Figures 10.22 and 10.23 for further details on PS1 and PS2). FeS: iron–sulfur center; cytb, cytf: cytochromes b and f; Pc: plastocyanin; Fd: ferredoxin; FNR: ferredoxin-$NADP^+$-reductase; F_0, F_1: two protein complexes making up the ATP-synthase

Photosynthesis is usually defined as the complex process starting with the absorption of light and "finishing" with the synthesis of carbohydrates and nitrogen compounds. In its description, it is classical to distinguish the primary (or "light") reactions and the secondary (or "dark") reactions. This distinction has no vital significance since there is no clear barrier between both kinds of reactions. It may be more meaningful to distinguish between membrane reactions (all those "primary" reactions which take place in the photosynthetic membranes) and nonmembrane reactions (secondary reactions which are part of the cell metabolism and where the products of membrane reactions, i.e. ATP and a reduced molecule such as NADPH, are used for the photosynthetic carbon and nitrogen metabolisms). In this Chapter I shall deal only with membrane reactions.

It may be worth mentioning that photobiology, the science of interaction between light and living matter, includes two types of phenomena: those where light acts as a vector of information and triggers a sequence of reactions which do not store light energy (for instance, photomorphogenesis or vision), and those where the energy of light is stored in a chemical form. The latter situation is found, obviously, in the case of photosynthesis, but also in the case of a process occurring in archaic species of bacteria (archaebacteria) which contain a light-sensitive protein named bacteriorhodopsin: following light absorption, this molecule is able to pump protons against their gradient across a membrane. The light-induced proton gradient is then used to synthesize ATP. In photosynthetic systems, too, ATP is synthesized, by a similar mechanism which involves a complex protein named ATP-synthase.

2 MOLECULES AND CONCEPTS

2.1 PROTEINS, MEMBRANES

The function of biological systems involves, in addition to their genetic material and to small molecules, two large classes of objects: proteins and membranes. Proteins are linear polymers of amino acids, assembled according to a specified sequence. Each protein adopts a well-defined three-dimensional structure with (mostly on the outside) sites for interaction with other molecules and (mostly inside) sites for binding small chemical group named cofactors. Two classes of cofactors will be considered in detail below: pigments (for light absorption and energy transfer) and redox-active groups (for electron transfer).

Membranes are bilayers of lipids with a thickness of about 35 Å. Their local geometry is planar, but on a larger scale they make up closed volumes named vesicles, cells, thylakoids, etc. thus defining two distinct compartments (outside and inside). Membranes are indeed impermeable to most biomolecules, especially to the most polar ones which can cross them only through special sites named channels or carriers.

Proteins can be divided in two classes: soluble proteins, which stay in the cellular fluid, and membrane proteins, which spontaneously incorporate into membranes, where they adopt a well-defined orientation. This is due to the fact that their outside area includes hydrophobic parts (which fit in the central hydrophobic part of membranes) and hydrophilic patches which stay at the hydrophilic exteriors.

The primary reactions of photosynthesis involve mainly membrane proteins which are inserted in the so-called photosynthetic membranes. These membranes make up vesicles named thylakoids in oxygenic systems and chromatophores in non-oxygenic bacteria (see Figure 10.6) As shown in Figures 10.1, 10.6 and 10.14, the complete electron transfer involves the cooperation of soluble proteins and of membrane proteins.

2.2 ANTENNA, REACTION CENTERS

The primary reactions of photosynthesis involve two classes of objects which act in cooperation (Figure 10.2):

1. The antenna, an ensemble of membrane proteins carrying pigment molecules which fulfill the function of absorbing photons and of carrying electronic excitation energy stepwise toward the reaction centers [5]. Each antenna system comprises many pigments: between 25 and 1000 molecules per reaction center. It is assumed that energy transfer takes place mainly by inductive resonance as described in the Förster theory. One interesting aspect

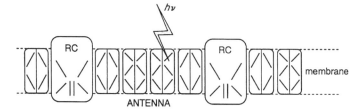

Figure 10.2. A model showing how reaction centers (RC) are associated in photosynthetic membranes with several types (two of them are schematically shown here) of pigment–protein complexes which serve an antenna function of capturing light. Excitation energy is thereafter transferred to the reaction centers where primary photochemistry takes place

of the antenna is its versatility (by contrast with the stability of reaction centers): its composition in proteins and in pigments, and its structure, vary greatly from one class of organisms to another. In plants, the major antenna is made of a small polypeptide which carries molecules of chlorophyll *a*, of chlorophyll *b* and of xanthophylls. In purple bacteria there are two categories of complexes, named LH1 and LH2, containing bacteriochlorophyll and carotenoids. Brown algae contain a highly specific complex of chlorophyll *a* and of peridinin, while green bacteria contain the so-called chlorosomes (with little protein and a dense packing of bacteriochlorophyll *c*) and cyanobacteria and red algae contain the so-called phycobilisomes (with a highly complex and variable organization of polypeptides and of bile pigments). The antenna will not be described in detail in this chapter.

2. Reaction centers which are large-membrane proteins where excitation energy arrives and induces a sequence of electron-transfer reactions which will be described below.

Reaction centers can work at an average rate of about $100\ s^{-1}$, which is about 100 times faster than the rate of their direct light absorption under usual natural conditions. This is why biological organisms have invented the antenna, an additional light-absorbing device which allows the reaction centers to function close to their optimal rate. The antenna size can vary greatly, often in inverse proportion to the light energy available in the natural habitat of the organism considered (it may be useful to recall that some photosynthetic organisms grow under full sunlight, while others grow in very dim light at the bottom of muddy ponds, with all kinds of intermediate situations).

2.3 PIGMENTS: CHLOROPHYLLS, CAROTENOIDS, LINEAR TETRAPYRROLES

Pigments are highly characteristic molecules of photosynthetic organisms. First, it should be clearly stated that pigments are not free in the photosynthetic apparatus:

they are bound to proteins, either in the antenna or in reaction centers, with a well-defined geometry suited for their function. Binding also insures a fine tuning of the functional properties of pigments, such as the energy level of the lowest singlet excited state in accordance with energy transfer, and the reduction potential for pigment molecules which are involved in electron transfer.

Chlorophylls [6] are cyclic tetrapyrrolic molecules with a central magnesium ion and a long saturated hydrocarbon chain, belonging to the class of chlorins (Figure 10.3). They are essential constituents of the photosynthetic apparatus and participate in energy and electron transfers. Bacteria possess various specific chlorophyll derivatives named bacteriochlorophylls, which are bacteriochlorins. Most reaction centers contain a small well-defined number of metal-less derivatives (pheophytins or bacteriopheophytins).

Carotenoids are linear polyenes which are widespread in all biological systems and have various detailed chemical structures (Figure 10.4). They

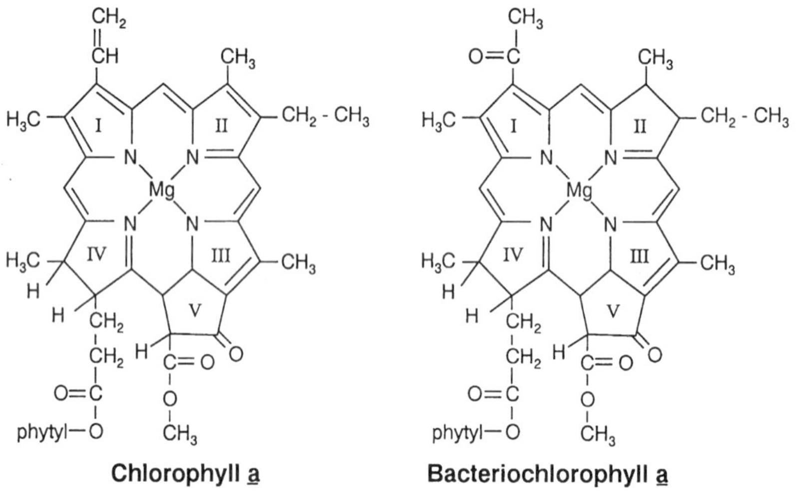

Figure 10.3. Chemical structure of chlorophyll a and of bacteriochlorophyll a (in the latter molecule the phytyl group, a saturated long-chain alcohol, is replaced by geranylgeranyl in many species of bacteria)

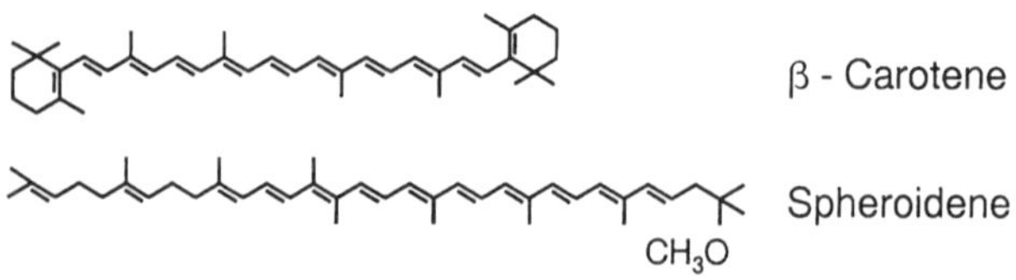

Figure 10.4. Chemical structure of two carotenoids found mainly in reaction centers of oxygenic organisms (β-carotene) and of purple bacteria (spheroidene)

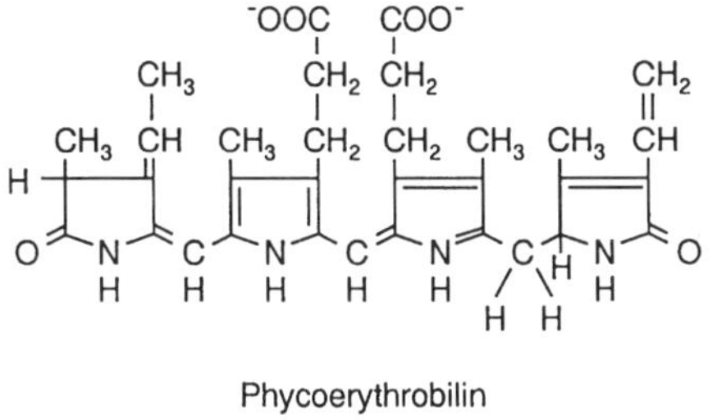

PHOTOSYNTHESIS, A NATURAL MODEL FOR PHOTOCATALYSIS 361

[Chemical structure diagram of Phycoerythrobilin]

Phycoerythrobilin

Figure 10.5. Chemical structure of phycoerythrobilin, a bile pigment which is one of the chromophores of phycobiliproteins. These proteins play an antenna function in several classes of bacteria and algae

participate in the antenna function by absorbing light and transferring excitation energy to (bacterio-)chlorophylls [7]. Their most important function, however, is undoubtedly to protect the photosynthetic apparatus against damaging effects of light, by quenching triplet excited states of other pigments and by quenching singlet oxygen ($^1O_2$), as discussed below.

Linear tetrapyrroles, belonging to the class of bilins or bile pigments, are the chromophores of phycobiliproteins which constitute a large part of the antenna system of cyanobacteria and of red algae (Figure 10.5). They make up large structures named phycobiliproteins comprising many bilin prosthetic groups and several classes of polypeptides.

All these pigments are characterized by an extended network of conjugated double bonds which insure a large cross-section for the absorption of light in the visible and near-IR spectrum. Two other properties are important for their function: they are stable (at least in their protein environment), and they are structurally organized in such a way that downhill electronic energy transfer can take place quickly and efficiently to the reaction centers.

2.4 TRANSFER OF ENERGY, ELECTRON OR PROTON

The function of photosynthetic systems is characterized by highly ordered flows of energy, of electrons and of protons. *Energy transfer* takes place in the antenna, immediately following light absorption. It is governed by the factors involved in Förster-type energy transfer: geometrical parameters of pigment organization (distance, orientation), fluorescence and absorption spectra, with the very essential factor of the O–O ground state to lowest singlet excited state energy transition.

Electron transfer is induced in reaction centers where it takes place from one redox center to another. A basic property of each redox center is its redox potential E_m (given versus the normal hydrogen electrode). Spontaneous (exergonic) electron transfer goes from a donor D with a low E_m (E_D) to an acceptor A with a higher E_m (E_A). Under standard conditions, the free energy change in the couple is: $\Delta G° = E_D - E_A$ (in volts). In the photosynthetic apparatus the redox potentials to be considered are comprised between approximatively +1.2 V and −1.2 V (Figure 10.8 p. 365). The most useful quantitative description of electron transfer in photosynthesis is provided by the Marcus theory [8,9]. In the reaction:

$$D\ (\text{red}) + A\ (\text{ox}) \Longleftrightarrow D\ (\text{ox}) + A\ (\text{red}), \tag{1}$$

which we assume to involve one electron, the final electron distribution is governed by the difference $E_D - E_A$, and the rate of electron transfer by $\Delta G°$, by the edge-to-edge intermolecular distance R, by the reorganization energy λ and by temperature T:

$$k = 4\pi^2/h \times V_R^2 \times FC \tag{2}$$

where h is the Planck constant, V_R^2 is an electronic term:

$$V_R^2 = V_o^2 \exp(-\beta R) \tag{3}$$

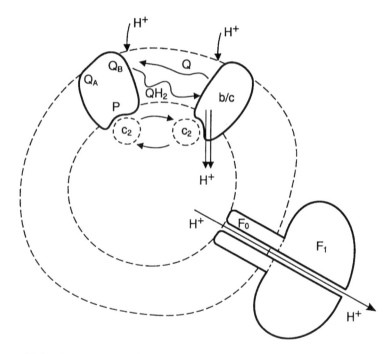

Figure 10.6. A scheme showing how the reaction centers of purple bacteria perform a cyclic electron transfer which drives protons from the outside to the inside of a volume limited by the photosynthetic membrane. This object, named chromatophore, is obtained by breaking intact membranes which afterwards reseal spontaneously. (P: primary electron donor; Q_A, Q_B: reaction center bound quinones; Q: mobile quinone; b/c: complex of cytochromes b and c, also containing an iron–sulfur center; c_2: cytochrome c_2. F_0 F_1: see Figure 10.1)

PHOTOSYNTHESIS, A NATURAL MODEL FOR PHOTOCATALYSIS

and FC is the Franck–Condon factor:

$$FC = (4\pi\lambda kT)^{1/2} \exp -[(\Delta G° + \lambda)^2/4\lambda kT] \qquad (4)$$

where k is the Boltzmann constant.

It is well established that distance is a key factor in electron transfer: in reaction centers, the redox centers are held very precisely by the protein, in a well-defined geometry. The protein plays another role in determining the redox potentials: these potentials are not those of the redox centers in solution since they are greatly influenced by the protein (electrostatic effects, hydrogen bonding, etc.). Another possible role of the protein is presently under debate: how much does the medium located between D and A influence the rate of electron transfer? Should the protein be considered as a nearly uniform medium, or are there specific features which allow fast electron transfer, following privileged paths (covalent bonds, hydrogen bonds, aromatic residues, etc.)?

Proton transfer is also very important in photosynthesis. In oxygenic photosynthesis, light-induced electron transfer fulfills two functions:

1. To provide a reductant needed for the reduction of CO_2 into carbohydrates; this reductant is NADPH, the product of reduction of $NADP^+$ by ferredoxin;

2. To provide ATP (Figure 10.7), the most important cellular energy currency, which is used to drive endergonic biochemical reactions.

The synthesis of ATP is realized by a membrane-spanning enzyme named ATP-synthase: $ADP + Pi \longrightarrow ATP$. The reverse reaction takes place when ATP is utilized, for many cellular functions. In the reaction of ATP synthesis, free energy is provided by a transmembrane gradient of proton concentration or $\Delta\mu H^+$. The mechanism by which $\Delta\mu H^+$ allows ATP synthesis is not yet understood, but it is well established that the proton gradient is the result of transmembrane proton transfer coupled to electron transfer, as illustrated in Figure 10.1. In purple bacteria, proton transfer is even more important, in functional terms, since it can be considered (in a first approximation) that electron transfer is cyclic, with

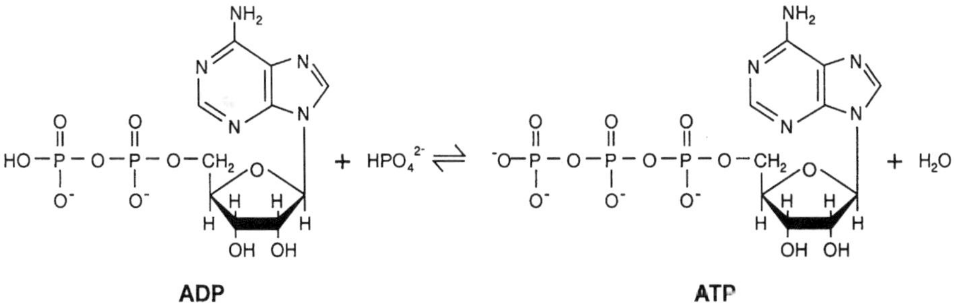

Figure 10.7. Chemical structures of adenosine diphosphate (ADP) and adenosine triphosphate (ATP)

no net accumulation of reduced molecules, and that the reaction centers are proton pumps which build up the membrane potential required for ATP synthesis (Figure 10.6).

This global description leaves out important molecular mechanistic aspects. Those related to reaction centers will be discussed below. One important protein of the photosynthetic membranes is specifically devoted to the coupling between electron transfer and transmembrane proton pumping: the cytochrome b/c complex. It receives electrons from a fully reduced quinone and donates electrons to a pure electron carrier: soluble cytochrome c or plastocyanin, a small copper protein (Figure 10.1). Since the overall reaction is

$$QH_2 + 2 \text{ cyt } c \text{ ox} \longrightarrow Q + 2 \text{ cyt } c \text{ red} + 2H^+ \tag{5}$$

one expects a proton to electron ratio of 1. The protein, however, functions in a more complex way and might transfer up to two protons per electron because it operates a complex recycling named Q cycle.

As mentioned above, transmembrane proton transfer is also effected by another light-sensitive protein, bacteriorhodopsin, found in archaic bacteria. This is a very interesting model system which explains how ions can be transferred through (overall hydrophobic) membrane proteins. It should be made clear, however, that proton transfer in bacteriorhodopsin is not coupled to electron transfer, by contrast with what happens in photosynthesis.

2.5 ENERGETICS AND YIELDS

In photosynthesis, one often considers the quantum yield, which is the number of product events divided by the number of photons absorbed. The same concept is frequently used in photochemistry. Considering simply the primary reactions, their quantum yield is the product of two yields:

1. The yield of trapping of excitation by reaction centers (Y_T). This yield can be written approximately:

$$Y_T = k_T/(k_T + k_F + k_{ISC} + k_Q) \tag{6}$$

   where the k's are rates of excitation trapping (k_T), of fluorescence (k_F), of intersystem crossing (k_{ISC}) and of various quenching sources (k_Q). This description is approximate because the antenna is not homogeneous. Experimental measurements, however, clearly show that Y_T is always very high (over 90%), because k_T, being of the order of 10^{10}–11^{11} s^{-1}, is much higher than the rates of other mechanisms of desexcitation.

2. The yield of charge separation in the reaction center (Y_{CS}). This yield is close to 100%, because all forward reactions (leading to the final state of charge separation) are much faster than back reactions and desexcitations.

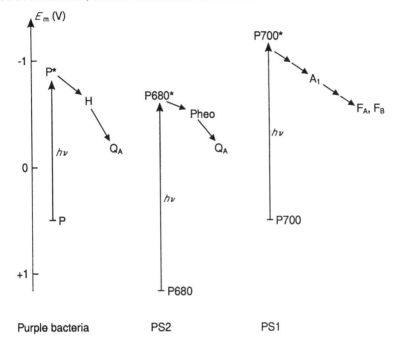

Figure 10.8. Energy diagram showing the redox potential of the primary electron donor P (P/P$^+$) and of several electron acceptors, in purple bacteria, in PS2 and in PS1. The diagram shows that light excitation makes P very reducing; it also shows a few differences in operational redox potentials in the three classes of reaction centers. (H: bacteriopheophytin; Pheo: pheophytin; Q_A: bound quinone; A_1: phylloquinone; F_A, F_B: iron–sulfur centers)

Finally the quantum yield is very close to 100% in all cases. The functional features which allow this high yield will be detailed below, inasmuch as they are understood. Another parameter to be considered is which fraction of the photon energy is saved in the reaction. A substantial fraction of that energy is lost in the pigments because of a very fast internal conversion to a low vibronic state of the lowest singlet excited state. In all reaction centers, the trap for energy is the species named P, a dimer of chlorophyll *a* or of bacteriochlorophyll, which also acts as primary electron donor:

$$P + \text{excitation} \longrightarrow P^*; \quad P^*A_1 \longrightarrow P^+A_1 \tag{7}$$

where A_1 is the primary electron acceptor. The electron is transferred from A_1 to a second electron acceptor, then to another one, etc. and P$^+$ is rereduced by a donor D. In the reaction center, the final state can be named D$^+$ A$_n{}^-$. If D and A_n have *in situ* redox potentials E_D and E_{A_n}, then the fraction of energy stored is

$$\eta = (E_D - E_{A_n}) \times e/h\nu(P) \tag{8}$$

where e is the charge of the electron, h the Planck constant, and $\nu(P)$ the frequency to populate the lowest excited state of P. The values of η are approximately the following: 0.40 in PS1, 0.30 in PS2, 0.21 in purple bacteria. The final yield of membrane reactions is lower, because several downhill reactions still have to occur after the state $D^+A_n^-$ has been reached. In oxygenic photosynthesis, considering that two photoreactions act in series, that electrons go from water to $NADP^+$, and that additional energy is stored as ATP synthesized thanks to the membrane electrochemical potential, the final energetic yield of the membrane reactions is approximately 0.20 (20%).

2.6 EXCITED STATES: SINGLET, TRIPLET

Pigment molecules like chlorophylls or carotenoids contain an extended system of conjugated double bonds. In their most stable electronic state, or ground state, their electrons are placed pairwise in molecular orbitals, two electrons with antiparallel spin occupying the lowest energy orbitals. The total spin S is zero and the spin multiplicity $(2S + 1)$ is unity: this state is thus named a singlet state. In the low-energy excited states of these molecules, one electron has left the lowest occupied orbital. Two classes of excited states are thus created, with the two electrons placed on different orbitals having opposite spins ($S = 0$, singlet state) or identical spins ($S = 1$, triplet state).

Direct light excitation of pigment molecules does not change the spin state, and thus populates a singlet excited state (Figure 10.9). For chlorophyll a molecules

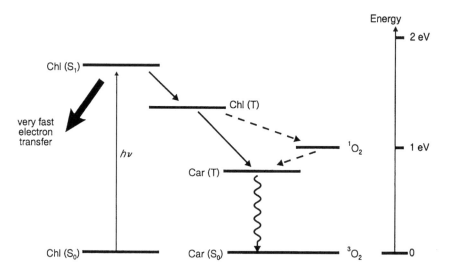

Figure 10.9. A scheme of energy levels of ground and excited states of chlorophyll, carotenoids and oxygen, showing a few important energy-transfer pathways. In photosynthetic reaction centers, formation of the chlorophyll triplet state is strongly quenched by the very fast electron-transfer reaction

in solution, that state decays by two major mechanisms: fluorescence with a rate of 5×10^7 s^{-1}, and intersystem crossing to the triplet state, of lower energy, with a rate of 1×10^8 s^{-1}. The triplet state returns slowly to the ground state, with a rate of 10^3 s^{-1}. It thus happens that the triplet state is populated with a high yield of about 66% and decays more slowly than the singlet excited state. For these reasons, it has long been thought that the reactive state of chlorophyll in photosynthesis was the triplet state. In fact, the primary electron transfer in reaction centers takes place in about 1 ps, a time at which the triplet state has had no time to become populated: the reactive state is the lowest singlet excited state of chlorophyll (or bacteriochlorophyll).

The triplet state is populated in two cases:

1. In the antenna, where the overall lifetime of the singlet excited state is around 100 ps, leading to a triplet yield of about 1%.
2. In reaction centers, in cases where electron transfer cannot proceed normally and where some charge recombination takes place, as discussed later.

In all living systems, the triplet state of tetrapyrrolic molecules (TP) is always the source of cellular damages. The reason is that, after its photochemical formation, that state reacts with the ground state of oxygen (a rare case of a molecule with a ground triplet state) to form singlet oxygen. This is a triplet energy transfer reaction:

$$^3TP + {}^3O_2 \longrightarrow TP + {}^1O_2 \qquad (9)$$

Singlet oxygen is a highly reactive species which can oxidize many essential biomolecules. The photosynthetic apparatus is full of tetrapyrrolic molecules and it is exposed to strong light. How can it resist the effects of $^1O_2$?

Three built-in features insure a very efficient protection (Figure 10.9):

1. Primary electron transfer is very fast and leaves little time for a high yield of triplet to be populated.
2. Pigment–protein complexes of the photosynthetic apparatus contain carotenoid molecules. These polyenes have a triplet state of very low energy which deactivates ^3TP by energy transfer:

$$^3TP + Car \longrightarrow TP + {}^3Car \qquad (10)$$

The state ^3Car decays quickly to the ground state, in a few microseconds, and its energy is so low that it is unable to form $^1O_2$.

3. These carotenoid molecules exert another function: if some $^1O_2$ has been formed, it is deactivated by triplet energy transfer:

$$^1O_2 + Car \longrightarrow {}^3O_2 + {}^3Car \qquad (11)$$

and ^3Car decays harmlessly to the ground state.

2.7 SUMMARY OF REACTION CENTERS

Simplified general structures of all reaction centers are shown in Figure 10.10. Two different situations are depicted. In both cases the structure (simplified, see below) includes two polypeptides which are tightly associated and located in the membrane. At their interface is located the species P, a dimer of chlorophyll *a* or of bacteriochlorophyll, which is the trap for energy, the excited state of which starts the electron-transfer reactions. Electron carriers are located dissymmetrically within the pair of polypeptides. In a first class of reaction centers (PS2 and purple bacteria), the very primary acceptors are chlorins or bacteriochlorins. They are followed by two quinones named Q_A and Q_B which function sequentially; the final product is a doubly reduced quinone (QH_2) which leaves the protein and is replaced by another quinone. In the second class of reaction centers (PS1, green sulfur bacteria, heliobacteria), the sequence of electron acceptors includes a (bacterio-)chlorin, then a quinone, then a very low potential iron–sulfur center, finally a small polypeptide including two iron–sulfur centers. The final step is the reduction of a soluble redox protein, usually a ferredoxin (with a 2Fe–2S center) or a flavodoxin (the redox center of which is a flavin) which serves to reduce $NADP^+$ or other cellular molecules.

The two classes of reaction centers show common features: a core of two hydrophobic polypeptides, the chemical nature of the primary donor P and of the

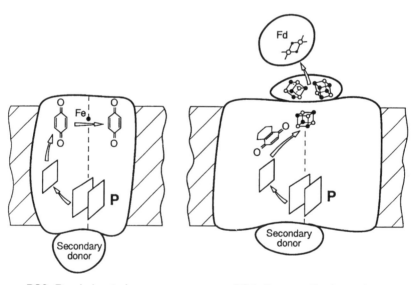

PS2, Purple bacteria   PS1, Green sulfur bacteria

Figure 10.10. A model of structure for the two broad classes of reaction centers: those with two quinones Q_A and Q_B (PS2 and purple bacteria) and those containing several iron–sulfur centers and transfering electrons to ferredoxin (Fd). P: primary electron donor

primary acceptors, their position in a membrane, etc. (Figure 10.10). Their main difference resides in the redox potentials of the final acceptors (and whence of intermediate ones): Q_B has an Em around 0 mV and ferredoxin an Em around −420 mV. Consequently the PS1-type reaction center reducing is more by 400–500 mV than its PS2 counterpart (Figure 10.8).

3 THE REACTION CENTER OF PURPLE BACTERIA

The reaction center of purple bacteria is by far the best known of all reaction centers in terms of function and of structure. Its overall function is to transfer electrons from cytochrome c, on one side of the membrane, to ubiquinone Q_B bound to the protein, on the other side (Figure 10.10) [10–12]. The redox potential of cytochrome c is around +340 mV and that of ubiquinone is around 0 mV. Electron transfer cannot occur spontaneously: this endergonic reaction is made possible by the energy brought about by light excitation.

3.1 FUNCTIONAL PROPERTIES

After photon absorption, electronic excitation energy is transferred to a bacteriochlorophyll dimer P which is brought to its lowest excited state (P*). This excited state is highly reactive and an electron is very rapidly transferred to the primary electron acceptor (Figure 10.11). It has long been thought that a quinone (Q_A) was the primary acceptor, until it was shown

1. That when Q_A^- is reduced there is still a light-induced charge separation which recombines in a few nanoseconds. Flash-induced difference spectra showed that the transient state is made of the oxidized donor (P$^+$) and of reduced bacteriopheophytin.

2. Under normal conditions, picosecond spectroscopy revealed that Q_A is reduced in about 200 ps by a bacteriopheophytin anion.

The following reactions have thus been shown to occur (H is bacteriopheophytin) where the first line describes the normal reactions and the second the reactions taking place when Q_A is reduced:

$$PHQ_A \xrightarrow{h\nu} P^*HQ_A \longrightarrow P^+H^-Q_A \xrightarrow{200\ ps} P^+HQ_A^- \quad (12)$$

$$PHQ_A^- \xrightarrow{h\nu} P^*HQ_A^- \longrightarrow P^+H^-Q_A^- \xrightarrow{10\ ns} PHQ_A^- \quad (13)$$

Further progress in time-resolved flash absorption spectroscopy has led to evidence that the bacteriopheophytin is reduced in about 2 ps, a time almost identical to the time of decay of P*. With further increase in time resolution, and in the sensitivity of femtosecond spectroscopy, serious indications appeared, however, that another species (a bacteriochlorophyll indicated as B in Figure 10.11) acts

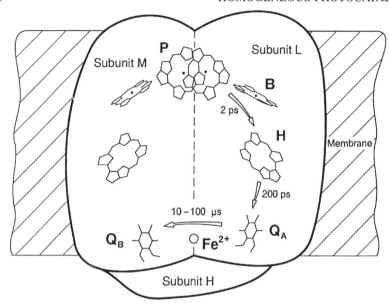

Figure 10.11. Arrangement of pigments and redox centers in the reaction center of the purple bacterium *Rhodobacter sphaeroides*. P: primary electron donor (dimer of bacteriochlorophyll *a*); B: bacteriochlorophyll; H: bacteriopheophytin; Q_A, Q_B: ubiquinone. The three protein subunits named L, M, H are sketched. Half-times of three electron-transfer steps are also indicated. The scale is given by the membrane thickness: 40 Å

as a very transitory intermediate in electron transfer between P* and H:

$$P^*BH \xrightarrow{2 \text{ ps}} P^+B^-H \xrightarrow{0.5 \text{ ps}} P^+BH^- \tag{14}$$

The question of the role of bacteriochlorophyll B raised considerable discussions which are not fully settled. The three-dimensional structure of the reaction center of *Rhodopseudomonas viridis* (and also of *Rhodobacter sphaeroides*) clearly showed that bacteriochlorophyll B is located between P and H. Many experiments, however, did not reveal that B became transiently reduced upon flash excitation, although it was felt that 2 ps was a very short time for the electron to go directly from P to H at a distance of 11 Å (edge-to-edge) [13]. Two explanations were proposed: (1) B is reduced more slowly than B^- is reoxidized; (2) electron transfer from P* to H is accelerated by super-exchange through B. This can happen if the energy of the state P^+B^- lies above that of P*. Recent experiments have provided an indication for a transient and partial reduction of B after a flash (see e.g. ref. 14). It is tentatively proposed that both mechanisms for the role of B contribute to variable extents, according to the energy level of P^+B^-, in reaction centers from various bacteria and in dependence of temperature (Figure 10.12).

PHOTOSYNTHESIS, A NATURAL MODEL FOR PHOTOCATALYSIS

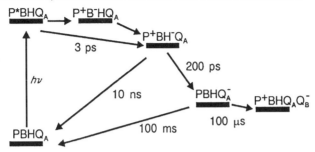

Figure 10.12. Energy levels and primary electron-transfer reactions in the reaction center of purple bacteria (adapted from ref. 34). In addition to forward reactions, the approximate $t_{1/2}$ of two back reactions are indicated (see Figure 10.11 for an explanation of the symbols)

Electron transfer from H^- to the first quinone Q_A raises no major problem, and the next step is electron transfer to the secondary quinone Q_B, which takes place in the order of 100 μs. In some bacteria Q_A and Q_B are chemically different: menaquinone (a substituted naphthoquinone) and ubiquinone (substituted benzoquinone), respectively, in *R. viridis*, for example (Figure 10.13). In many others, however, such as *R. sphaeroides*, Q_A and Q_B are both ubiquinone. The function of these two quinones raises a number of problems:

1. Functional model. A flash of light brings one electron on Q_A as a result of the primary charge separation and the effect of two successive flashes is as follows:

   1st flash: $Q_A Q_B \longrightarrow Q_A^- Q_B \longrightarrow Q_A Q_B^-$

   2nd flash: $Q_A Q_B^- \longrightarrow Q_A^- Q_B^- \longrightarrow Q_A Q_B^{2-}$ (with uptake of two H^+)

   (15)

   For the formation of doubly reduced Q_B, electron transfer is probably concomitant with proton transfer. The fully reduced quinone (hydroquinone) leaves its site on the protein and diffuses in the membrane, where it eventually

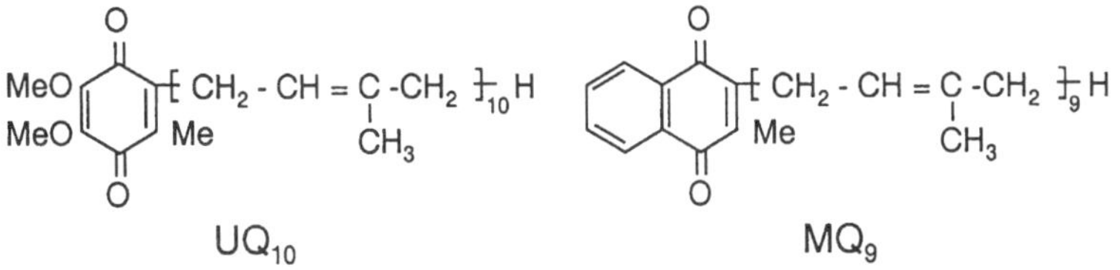

Figure 10.13. Chemical structures of two quinones found in reaction centers of purple bacteria: ubiquinone 10 (Q_B in all species; Q_A in most species) and menaquinone 9 (Q_A in some species, e.g. *R. viridis*)

reaches the site at which it becomes reoxidized at the cytochrome b–c complex. In the meantime another quinone has bound at the Q_B site on the reaction center and it is ready for receiving an electron from Q_A. In a sequence of flashes the properties of Q_B display a remarkable period-2 behavior with flash number:

$$Q_B \text{ - one flash} \longrightarrow Q_B^- \text{ - one flash} \longrightarrow (Q_B H_2) \longrightarrow Q_B \text{ - one flash} \longrightarrow Q_B^- \longrightarrow \ldots \qquad (16)$$

2. Binding of Q_A and Q_B quinones. The functional model described above implies that Q_A functions only as a one-electron carrier, while Q_B is a two-electron/two-proton carrier. Since Q_A and Q_B are chemically identical in many organisms, the difference in behavior must originate from a difference in the sites of binding, which are specialized "pockets" in the protein. At the Q_A site the quinone appears to be permanently bound, whereas at the secondary site it was shown that the binding strength is $Q_B^- \gg Q_B > Q_B H_2$.

3. Redox potential. Since electrons flow from Q_A^- to Q_B (or Q_B^-), the redox potential of the Q_A/Q_A^- couple has to be lower than that of the Q_B/Q_B^- couple. The measured difference is around 100 mV in *R. sphaeroides* (it is larger in those bacteria where Q_A is a menaquinone and Q_B an ubiquinone). The difference in the redox potential of Q_A and Q_B is a good illustration of a general feature which is the modulation of the thermodynamic properties of the redox cofactors by their surroundings in the protein.

4. Protonation. It is well established that protons are involved in electron transfer from Q_A to Q_B, but not in a simple manner. It seems that both singly reduced species Q_A^- and Q_B^- are not protonated: the semiquinone species remain anionic. Electron transfer from Q_A to Q_B, however, is accompanied by proton uptake, presumably by protonation of one or several amino acid residues in the vicinity of Q_B. Full reduction of Q_B takes place with the uptake of two protons:

$$Q_B^- + e^- + 2H^+ \longrightarrow Q_B H_2 \qquad (17)$$

The rate of the reaction is directly controlled by proton availability. An interesting problem is to know how protons reach Q_B: is there one specific route through the protein, or several routes or is there a channel with water molecules conducting protons?

Once it is fully reduced to the quinol form, the quinone at the Q_B site unbinds from the protein and diffuses away through the lipid bilayer for reacting with the cytochrome b–c complex (Figure 10.6). The Q_B site on the protein is available for binding an oxidized quinone, after which the reaction center is ready for another light-induced reaction. The turnover half-time is about 3 ms.

On the other side of the reaction center, the primary donor P, which has been oxidized in the primary reaction, has then been rereduced by a c-type

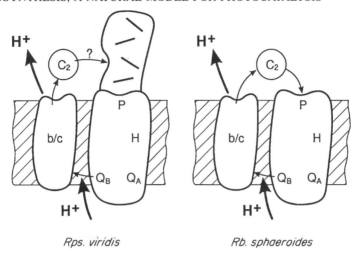

Figure 10.14. Comparison between the function of c-type cytochromes in two species of purple bacteria: both contain a soluble c_2 cytochrome, but *R. viridis* also possesses a reaction center-bound tetraheme c cytochrome

cytochrome (Figure 10.15). This reaction is fairly fast: its half-time is in the range 100 ns–10 µs. Two widely different situations are found among purple bacteria (Figure 10.14). In some species, P^+ is reduced by a small monoheme cytochrome (cyt. c_2) which docks transiently to the reaction center (when it is reduced), then transfers an electron, unbinds and migrates in the external medium to pick an electron at the cytochrome $b-c$ complex. In other cases (apparently the majority of species), P^+ is rereduced by a more complex cytochrome with four hemes (tetraheme cytochrome) which remains permanently bound to the reaction center. A soluble cytochrome c_2 shuttles electrons between the cytochrome $b-c$ complex and the tetraheme cytochrome.

3.2 ISOLATION — COMPOSITION

The existence of reaction centers was inferred from flash experiments before these entities could be effectively isolated. The membranes of photosynthetic bacteria contain many different proteins. The first step in their isolation is to dissociate the membranes by a detergent: the proteins become dissociated, each with a surrounding of detergent molecules. They are then separated by various methods, in which chromatography plays a key role.

In *R. sphaeroides*, isolated reaction centers have the following composition (Figure 10.11 p. 370): three polypeptides (named L, M, H with respective molecular weights of 32, 34 and 29 kDa), four molecules of bacteriochlorophyll a, two molecules of bacteriopheophytin a, two ubiquinone 10 (UQ10; this is a benzoquinone substituted with a methyl group in 2, with methoxy groups in

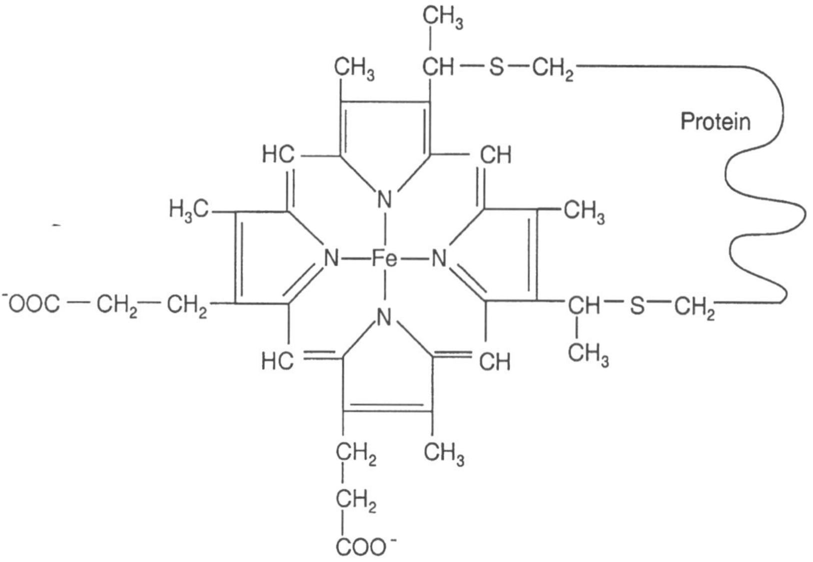

Figure 10.15. Chemical structure of the heme group of c-type cytochromes and its covalent binding to the protein

5 and 6, and in 3 with a long hydrocarbon chain comprising ten isoprenoid units) (Figure 10.13). The reaction center also contains one Fe^{2+} ion and one carotenoid, spheroidene Figure 10.4 (at least in the wild-type strain, although most of isolation work has been done on the R26 mutant which is devoid of carotenoid).

In *R. viridis*, isolated reaction centers differ in four respects. Firstly, bacteriochlorophyll *a* and bacteriopheophytin *a* are replaced by Bchl *b* and BPheo *b* which have a vinyl group instead of a methyl on carbon 3 [6]. Secondly, the reaction centers have one UQ10 and one substituted menaquinone, MQ9 (Figure 10.13 p. 371). Thirdly, there is one additional protein subunit, of 37 kDa, namely the tetraheme cytochrome *c*. Finally, the carotenoid molecule is a spheroidenone.

Reaction centers have been isolated from many species of purple photosynthetic bacteria. Their composition resembles more or less to that obtained in *R. sphaeroides* or *R. viridis*. It seems, however, that a tetraheme cytochrome *c* exists in a majority of species. In *R. viridis*, the reaction center becomes isolated with the firmly bound cytochrome because this polypeptide contains a post-synthesis modification (diglyceride extension of the *N*-terminal cysteine residue) which results in a strong anchoring of the cytochrome in the hydrophobic core of the reaction center. In other species the tetraheme cytochrome is often less firmly bound and it may either get lost or have a modified binding geometry upon isolation of the reaction centers.

3.3 STRUCTURE

The reaction centers of *R. viridis* and of *R. sphaeroides* have been crystallized and their atomic structure determined by X-ray crystallography. Useful crystals have also been obtained from *Chromatium tepidum*. Before going into a detailed examination of the structure, three remarks are of interest:

1. The reaction center structures of *R. viridis* and *R. sphaeroides* are as similar as possible (knowing that their compositions are not identical).

2. A large amount of valuable structural information was already available prior to the X-ray structure. Crystallography provides a lot more information on polypeptides and cofactors; it is also very reliable. Other methods, however, cover many interests, because they can be used for comparative studies of various bacterial species, for the study of mutated reaction centers, or for studies *in vivo* or in isolated membranes without the need to isolate pure reaction centers. A good example of such methods is linear dichroism, which precisely determines the orientation of the various cofactors: for example it permits quite quickly the study of pigment geometry in mutants.

3. Several levels of structural knowledge are of interest. The most macroscopic one describes the general shape of the protein, the way it is positioned in the membrane and the way different subunits are organized. The present stage of X-ray cristallography, with a resolution of 2.3–3.0 Å, provides a quite precise description of the polypeptide residues, of the cofactors, permitting reasonable predictions of hydrogen bonds, etc. This resolution, however, is still not sufficient for a good description of parameters which are determined by local properties: precise mode of cofactor binding, protein properties which might be involved in electron transfer rates, control of cofactor electrochemical properties, etc. A few spectroscopic methods are used more and more in order to obtain this local information: EPR (and its variations), vibrational spectroscopy (FT–IR and resonance Raman) and others. A few examples will be given below.

As results from crystallography, the reaction center structure has been precisely described by the groups involved in original researches. A few salient features of the *R. viridis* reaction center structure [15] can be summarized here.

1. Organization of the protein subunits. The reaction center core is made up of the polypeptides L and M; they are tightly packed together, and each of them includes five transmembrane helices with small interconnecting segments. The H subunit is located essentially outside, like a cap on the cytoplasmic side (see Figure 10.11), with only one α-helix extending through the reaction center core. The cytochrome subunit is also located externally, on the periplasmic side.

376   HOMOGENEOUS PHOTOCATALYSIS

The subunits L and M are nearly symmetric with respect to an axis running perpendicular to the membrane plane; this C_2 pseudo-symmetry axis also concerns the cofactors (Figure 10.11).

2. Organization of the cofactors. The primary donor P (a dimer of bacteriochlorophyll *b*) is located at the interface between L and M which both provide ligands for the bacteriochlorophylls (Figure 10.16). The C_2 axis runs through P, with the bacteriochlorophyll planes being separated by about 3.5 Å and being both perpendicular to the membrane plane. The two other bacteriochlorophyll *b* are located close to P, also in a symmetric positioning, and the two bacteriopheophytin are located somewhat further apart (see Figure 10.11 p. 370). The two quinones are also held in the L, M core, close to the H subunit. In the original structure, only Q_A was present (in *R. viridis*, it is recognizable from its structure since it is a naphthoquinone) and Q_B was absent (Q_B is weakly bound and easily becomes lost during the isolation of reaction centers). It has, however, been possible to reconstitute purified reaction centers with ubiquinone which binds at the Q_B site, and thus to have a complete geometry of the quinones. These two cofactors appear to be located on either side of the symmetry axis, which runs through the Fe^{2+} atom always found in reaction centers. It has been possible to replace Q_B by inhibitors (i.e. terbutryn) and to obtain by X-ray crystallography the proof that these inhibitors bind in the same pocket as the quinone. Details on the positions of the cofactors and on their interactions with the polypeptides are of considerable interest. They still raise many unanswered questions.

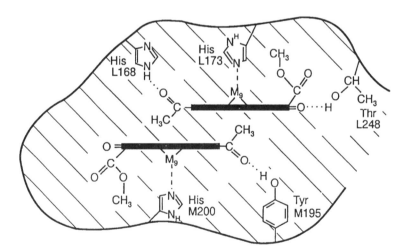

Figure 10.16. A scheme showing the primary donor P in the reaction center of *R. viridis* (side view) and a few of the interactions which permit its anchoring by polypeptides L and M. (Adapted from a model kindly provided by Dr B. Robert.)

The four hemes of the cytochrome (Figure 10.14) are located in a more or less linear geometry which does not follow the overall C_2 symmetry. They may have, however, an approximate local C_2 symmetry axis which runs parallel to the membrane plane. Also non-symmetric is a carotenoid molecule (spheroidenone) which is close to one of the monomeric bacteriochlorophylls.

3. Comparison with the reaction center of *R. sphaeroides*. Four laboratories have obtained three-dimensional structures from crystals of *R. sphaeroides* reaction centers. Early structures were not highly resolved and they differed significantly between each other, and also with *R. viridis*. It seems, however, that the results of progressive resolution and refinement converge to nearly identical structures, except that there are some differences due to minor differences in chemical structures and to the absence of a bound cytochrome in *R. sphaeroides*.

3.4 RELATION BETWEEN STRUCTURE AND FUNCTION: OVERALL FEATURES

Following excitation by light, an electron goes from P* to bacteriopheophytin, there to Q_A and then to Q_B. The reaction center structure is obviously consistent with that electron path. The major surprise in the structure is the near symmetric arrangement of the cofactors (Figure 10.11 p. 370): the question was raised of the functional significance of the two "branches", i.e. the two sequences of redox cofactors which go from P to a quinone through a bacteriochlorophyll and a bacteriopheophytin, and it was concluded that only one of these branches, which goes to Q_A (which can be identified unambiguously in *R. viridis*), is effectively functional.

The structure is also consistent with the earlier proposal of "vectorial" electron transfer, i.e. electrons go from one side of the membrane to the other, creating a dipolar state which gives rise to a membrane potential. In earlier experiments the membrane electric potential was measured by its effect on the absorption spectrum of carotenoids, the so-called "electrochromic effect", which is a Stark effect due to the membrane electric potential. The time-dependent development of the membrane potential with transmembrane electron transfer can be measured nowadays by the powerful method of flash-induced photovoltage [16].

3.5 RELATIONS BETWEEN STRUCTURE AND FUNCTION: DETAILED CONSIDERATIONS

It is of great interest to understand how many aspects of reaction center function are related to its structure. Many tools are now available to facilitate this understanding: advanced spectroscopic methods, site-directed mutagenesis, chemical modifications, synthesis of artifical models, theory, etc.:

(a) P, the primary donor. At the heart of the process, why is there a dimer of bacteriochlorophyll and how does it work? These questions have no definitive, simple answers. A few ideas have been put forward. Firstly, the two partners

in the dimer experience excitonic interaction, which leads to an absorption band located at a lower energy than in monomers, a property which facilitates energy trapping by P. Secondly, the dimer is not fully symmetric as shown by analysis of spin density distribution in P^+ by Electronic and Nuclear Double Resonance (ENDOR) [17]: this property may facilitate forward electron transfer along the functional branch and decrease the probability of back reaction. Thirdly, the bacteriochlorophyll pair, at the interface between L and M subunits, may be involved in the reaction center biosynthesis and in the stability of the edifice. None of these proposals is fully satisfying, but no site-directed mutant has been made, where P is not a pair of bacteriochlorins.

The redox potential of the P/P^+ couple is of obvious importance since it is a key factor in the energetics of primary reactions. It is largely determined by the protein surrounding P, and particularly by the hydrogen bonds. It has recently been possible to modify it substantially (from $+410$ to $+765$ mV versus NHE) by changing the number of hydrogen bonds [18].

(b) Can reaction rates be accounted for by electron-transfer theories? Reaction rates are key factors to determine the high quantum efficiency of photoinduced charge separation in photosynthesis. This efficiency, which is close to 100%, indeed requires that all forward steps are much faster than back reactions (charge recombinations). According to the Marcus theory [8], the rate of electron transfer is determined by three major factors (at physiological temperature): the overlap of electronic wave functions of donor and acceptor, the free energy change $\Delta G°$, and the reorganization energy λ. In all cases it is clear that a forward reaction $(A^- + B \longrightarrow A + B^-)$ which has a negative $\Delta G°$, is faster than the corresponding reverse reaction $(A + B^- \longrightarrow A^- + B)$ which is always endergonic. The solution is not obvious for the electron return to P^+ as exemplified in the following case: once the P^+H^- state has been created, what will be the relative rates of the two reactions:

$$\text{forward reaction (1)}: P^+H^-Q_A \longrightarrow P^+HQ_A^-$$
$$\text{back reaction \quad (2)}: P^+H^-Q \longrightarrow PHQ_A \qquad (18)$$

In reaction 2, the return is not to the excited state of P (endergonic), but to the ground state (exergonic reaction). In a first approximation, the distances between H and P or Q_A are nearly the same, but the forward reaction is much faster than the back reaction (200 ps instead of 10 ns). The difference can be understood on the basis of the Marcus theory and of the respective values of $\Delta G°$ and of λ for the two reactions (Figure 10.17). The redox potentials are approximately: -600 mV for the H/H^- couple, $+500$ mV for the P/P^+ couple and -200 mV for the Q_A/Q^- couple.

$$\text{Reaction 1}: -\Delta G° = 400 \text{ meV}, \lambda = 600 \text{ meV}$$
$$\text{Reaction 2}: -\Delta G° = 1100 \text{ meV}, \lambda = 200 \text{ meV} \qquad (19)$$

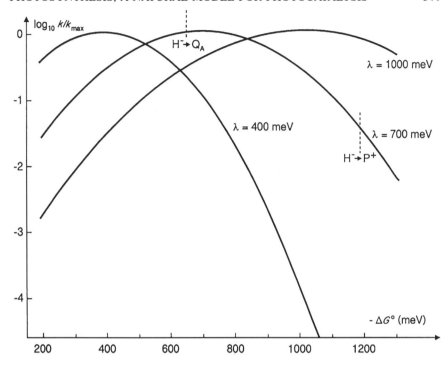

Figure 10.17. Effect of free energy variation $\Delta G°$ on the relative rate of electron transfer for three values of the reorganization energy λ. The situations corresponding approximately to electron transfer from H^- to Q_A and from H^- to P^+ are indicated

If we assume that both reactions have equal rates under optimum conditions ($-\Delta G° = \lambda$), the Marcus plot of $\ln k$ versus $\Delta G°$ (Figure 10.17) clearly shows that the rate is close to the maximum for reaction 1: $-\Delta G°$ is not very different from λ and we are close to the top of the curve where k varies little with $-\Delta G°$. For reaction 2, however, $-\Delta G° > \lambda$: we are in the so-called "inverted region", far from the optimum, in a region where k decreases when $-\Delta G°$ increases. To be fully conclusive, this discussion should consider precise values of λ (which are not known) and of the electronic parameters for electron transfer: with bacteriochlorophyll B located between H and P (Figure 10.11), the effect of distance may be more complex than the simple exponential law.

Many groups have used the structure–function properties of reaction centers as a model to check the application of electron-transfer theories in proteins. Only a few steps, however, are quite simple and they can be understood relatively correctly: electron transfer from H^- to Q_A, electron return from Q_A^- to P^+ and (perhaps) electron transfer from the cytochrome to P^+. The initial step offers several complicating features such as its rapidity: thermal relaxation cannot be considered as granted in the sub-picosecond time domain [19]. Electron transfer from Q_A^- to Q_B is fairly complex, as discussed below.

(c) Why is electron transfer much more efficient on one branch? The approximate symmetry of the reaction center, in terms of polypeptides and of cofactors, leads to the expectation that electrons leaving P* can have an equal probability of going on either side. In fact the probability of going through the "wrong" branch is less than 5%. There is no known overwhelming factor to explain this difference. Various theoretical treatments have shown that the difference can only be explained by a conjunction of several factors such as electrostatic effects of the protein on the redox potentials of the two monomeric bacteriochlorophylls (see e.g. ref. 20).

Another related question: why is the inactive branch present if it plays no role? It may be argued that it is a left-over of evolution — a monomeric reaction center, with one core polypeptide and one chain of redox cofactors, may have dimerized during the course of evolution. Dimerization may have several advantages such as a greater stability of the edifice or an interest for the efficiency of having two different quinones, one permanently bound Q_A and one able to leave its site (Q_B). With specialized quinones, one of the branches becomes useless, however, and a series of mutations may have rendered it inactive. It remains surprising, then, that reaction centers have kept the inactive bacteriochlorophyll and bacteriopheophytin.

(d) Functional properties of Q_B. The primary quinone Q_A serves as a one-electron relay between bacteriopheophytin and Q_B. Its functional properties are not exceptional: what is more exceptional is that it always remains bound to the protein. The secondary quinone Q_B presents the normal three redox states of a quinone, but it displays two interesting properties: proton uptake and a cycle of binding and release.

Proton uptake would be a fairly trivial process in a protic solvent; it is not so, however, in a hydrophobic protein. The phenomenon can be measured experimentally by several means: for the semiquinone, by flash absorption (since the UV spectra of Q^- and $QH^{\bullet}$ are different); for the doubly reduced quinone, by following the kinetics of electron transfer, which is controlled by protonation; also by measuring the pH change in the medium where reaction centers are placed. From several studies, it was concluded that protons are provided by donating groups in the protein. This proposal is reinforced by site-directed mutagenesis which showed the important role played by a few amino acids such as L212 (glutamate) and L213 (aspartate) (Figure 10.18) [21,22]. Two possibilities have been envisioned: a unique path for protons in the protein, involving a well-defined series of amino acids, or multiple paths. There is evidence, also, for a "water channel": in a recent analysis of the *R. sphaeroides* reaction center structure, it was shown that 14 molecules of water are organized between the Q_B site and the exterior. These molecules are precisely structured since they are seen by X-ray crystallography. It was hypothesized that they may serve for proton conductance

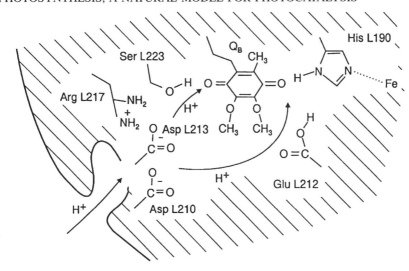

Figure 10.18. A model of the site where quinone Q_B is bound in the reaction center of purple bacteria, showing some of the interactions which hold the quinone in a pocket of the L subunit, and also showing possible paths for protons which are taken upon quinone reduction (see refs 21, 22). Fe is the Fe^{2+} atom located between Q_A and Q_B

from the medium to Q_B [23]. In fact, the roles of protein residues and of bound water molecules are not exclusive: they might cooperate for proton migration in a process similar to percolation.

An analogous problem is presented by the quinone itself: after its full reduction, the hydroquinone has a weak affinity for the site and it diffuses out of the reaction center, in the lipidic bilayer. How does that diffusion take place? How does another (oxidized) quinone migrate from the lipidic bilayer to the Q_B site? These questions are not trivial since the quinone, with its long isoprenoid tail, is a fairly large molecule (Figure 10.13 p. 371).

(e) Electron transfer from cytochrome to P^+. The rereduction of P^+ is insured by a cytochrome. As mentioned above, two different situations are found in purple bacteria: a tetraheme cytochrome *c* or a monoheme cytochrome c_2 (Figure 10.19). The latter situation is relatively simple to describe. Soluble c_2 is reduced by the cytochrome *b-c* complex; it then diffuses in the periplasmic space and docks to the reaction center. Electron transfer from docked cytochrome c_2 to P^+ has a $t_{1/2}$ of 600 ns in *R. sphaeroides*. A major problem here is to understand the geometry of the complex. Two models have been made on the basis of crystal structures of reaction center and cytochrome c_2 separately [24]. Progress is underway to elucidate the structure of the complex by X-ray analysis of co-crystals. The former situation has been studied more thoroughly, specially from the structural point of view, since it is encountered in *R. viridis* [15]. In that species, the four hemes have

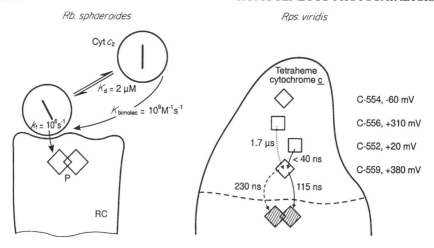

Figure 10.19. A detail of electron-transfer reactions on the donor side of the reaction center of purple bacteria. Left: situation in *R. sphaeroides* where a soluble cytochrome c_2 docks transiently to the reaction center. Right: situation in *R. viridis* where a tetraheme cytochrome is permanently bound to the reaction center (on the right are given the redox potentials and the wavelengths of the α-absorption bands of the four hemes)

different spectroscopic and redox properties. Detailed studies by linear dichroism in the α-absorption bands and by EPR spectroscopy have permitted to attribute the properties of each heme. From the proximal heme (closest to P) to the distal one, Em and $\lambda_{\max}$ of the α-band are as follows: c-559, +380 mV, 559 nm; c-552, +20 mV, 552 nm; c-556, +310 mV, 556 nm; c-554, −60 mV, 554 nm. Electron transfer to P$^+$ takes place from c-559; it is quite fast, with a $t_{1/2}$ of 230 ns when only c-559 is reduced and 115 ns when all three c-559, c-556 and c-552 are reduced. The acceleration has been attributed to an electrostatic effect [25]. A soluble cytochrome c_2 insures electron transfer from the cytochrome b–c complex to the tetraheme cytochrome (Figure 10.14 p. 374). The accepting heme is presumably c-556.

Electron transfer from cytochrome to P$^+$ deserves two major comments:

1. Since a monoheme cytochrome works perfectly in some bacteria, why is there a much more complex system in others, largely more numerous? A possible rationale is that the bound cytochrome may act as a buffer and insure permanently a rapid rereduction of P$^+$, but the need for that rapid reduction is not understood. In any case, the need for multiple hemes is not explained. There may be electron donors other that cytochrome c_2, bringing electrons from low potential compounds, such as sulfur compounds. A tetraheme cytochrome would thus facilitate the entry of electrons coming from various sources.

2. The complex between cytochrome (bound tetraheme or transiently docked c_2) and the reaction center is a good model for studying electron transfer between a small metalloprotein and a large membrane protein complex. This type of electron transfer is very widespread in biology, with examples such as cytochrome c/cytochrome oxydase, cytochrome c/cytochrome peroxydase, plastocyanin/ PS1, etc. The complex under consideration is a good model for the following reasons:

- The three-dimensional structure is well known in *R. viridis*.
- The structure can be modified by site-directed mutagenesis.
- Electron transfer kinetics can be measured very precisely since the reaction can be triggered by a short flash of light.
- The reaction center works down to a few K, and so the effect of temperature can be studied in detail. Temperature is an important parameter in the study of electron transfer.
- Other aspects of the system can be modified: water content, solvent properties, free energy change. For instance, this last parameter ($\Delta G°$) has been modified in *R. sphaeroides* by site-directed mutagenesis of amino acids around P: changing the number of histidine residues which establish hydrogen bonds with the bacteriochlorophylls changes the Em of P/P$^+$ between +410 and +765 mV. Since the Em of cytochrome c_2 is not modified, the $\Delta G°$ of the reaction changes accordingly, and it has been shown that the rate of electron transfer varies in agreement with the Marcus theory [26].

(f) Effect of temperature on electron transfer in reaction centers. The rate of many biological reactions is strongly dependent on temperature. By contrast, in reaction centers, the rates of several electron-transfer steps are practically temperature independent; in several cases the rate is even faster at low temperature. To understand this behaviour, we should remember that the redox partners in reaction centers have well-defined positions and they do not have to move for reaction to take place. Secondly, under these conditions, the Marcus theory predicts a temperature-dependent rate, unless $\Delta G° + \lambda = 0$ (see eq. 1). These conditions are also those which give rise to a maximum rate for a given $\Delta G°$ (Figure 10.17 p. 379). This would mean that reaction centers are built so as to tune free energy change and reorganization energy perfectly. Is it so? In an important series of experiments, Gunner and Dutton have changed the $\Delta G°$ for two reactions (electron transfer from H$^-$ to Q$_A$ and from Q$_A^-$ to P$^+$). They found that temperature independence is maintained when $\Delta G°$ is varied in a broad range [27]. They proposed that one aspect of the Marcus treatment does not apply in these reactions: coupling to low-energy vibrations, such that their energy levels are effectively densely populated at all temperature in the reactive state. They suggested instead that electron transfer involves (at least in a large

part) high energy vibrations (over 300 cm^{-1}) which are not populated at the temperatures considered and which thus cannot lead to a temperature-dependent rate. In recent experiments of Ortega et al, the $\Delta G°$ of $P^+Q_A^-$ back reaction was changed as said in paragraph (e) above, and it was found that λ decreases when temperature decreases [28]. This may explain why some reactions are accelerated when temperature decreases. It has also been proposed that acceleration at low temperature is due to the contraction of the protein which decreases the distances between redox centers.

Two reactions have a very different behavior. Firstly, electron transfer from Q_A^- to Q_B (or Q_B^-) has a high activation energy which may be related to the involvement of protons. Secondly, electron transfer from cytochrome to P^+ is considered as a classic in biological electron transfer, since the pioneer work of De Vault and Chance [29], who showed that the reaction slows down slowly when temperature is decreased and then becomes temperature independent below 100 K. Recent experiments with *R. viridis* reaction centers led to the conclusion that there is indeed a small activation energy between 300 K and a temperature included between 80 and 240 K (according to the redox state of the tetraheme cytochrome) where electron transfer from the proximal heme to P^+ becomes blocked [25]. Inhibition of electron transfer is attributed to the freezing of a structural change which normally accompanies cytochrome oxidation.

(g) Structural changes coupled to electron transfer. The reaction center structure is well defined and eventual structural changes coupled to its function are certainly of small amplitude. They may nevertheless be highly significant and they are actively searched. Three of them have been briefly mentioned above:

1. Cytochrome heme oxidation, for which a good model exists with isolated cytochrome *c*, the structure of which has been accurately measured in the reduced and oxidized states, showing significant differences [30];
2. Proton movement associated with Q_B reduction (and perhaps also with Q_A);
3. Movement of Q_BH_2 out of its site.

Differential X-ray crystallography has not yet been possible for reaction centers, but vibrational spectroscopies are good tools to investigate local changes in the structure. These approaches are still at their beginning, but two observations may be mentioned. By Fourier transform differential infrared spectroscopy, it was shown that Q_A reduction causes large changes in the bonding interactions between the quinone carbonyls and the protein. In the oxidized state of Q_A, the two bonds are nonsymmetrical, but they become symmetric for Q_A^-. These changes were attributed to micro-conformational changes of amino acids in the vicinity of Q_A [31]. There are now methods being developed to study these changes in a time-dependent manner, even in a sub-picosecond time scale. A second example is a change in the interactions with the protein, experienced by the bacteriochlorophyll *B* upon its reduction, which was detected by resonance Raman spectroscopy [32]. Structural changes have also been probed, although less directly, by kinetic measurements.

(h) Radical-pair; formation and fate of ^3P. When reaction centers are excited by a flash of light, the state (P^+H^-) is formed in a few picoseconds. It normally decays in about 200 ps by electron transfer from H^- to Q_A. There are cases, however, where this transfer cannot occur, either because Q_A is already reduced (this can occur under physiological conditions) or because Q_A is missing (a pathological case which can be induced with isolated reaction centers). Under these conditions, the state (P^+H^-) lasts for a time of around 10 ns, and it eventually decays by charge recombination (Figure 10.20). These abnormal events have been studied in great detail because they are very informative on the properties of reaction centers [33].

The state (P^+H^-) is named a radical pair. Since it is formed from a singlet state P^*, the radical pair initially possesses a singlet character, i.e. the unpaired electrons localized on P and H have opposite spins. With time, however, the system oscillates between this singlet state and a triplet state where the unpaired electrons have identical spins (Figure 10.20). Charge recombination can take place when the radical pair is in a singlet state (creating a neutral state, either P^* or a vibrationally excited state of the ground state P) or in a triplet state, creating the triplet state of P:

$$^1(P^+H^-) \longrightarrow {}^3(P^+H^-) \longrightarrow {}^1(P^+H^-) \longrightarrow \ldots$$
$$^1(P^+H^-) \longrightarrow P^*H \text{ or } PH \tag{20}$$
$$^3(P^+H^-) \longrightarrow {}^3PH$$

These events present a number of interesting features, some of which will be briefly mentioned:

1. The study of the relative importance of the routes of P^+H^- evolution, for instance versus temperature, permits the determination of the relative energy levels of P^*, of (P^+H^-) and of ^3P, which are important parameters of the primary photochemical events (see e.g. ref. 34)

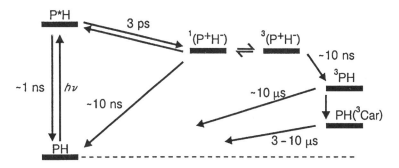

Figure 10.20. Reactions taking place in reaction centers of purple bacteria when electron transfer from H^- to Q_A is not possible (adapted from ref. 34)

2. Formation of ^{3}P can be measured by flash-induced absorption and, more specifically, by EPR at low temperature. In the magnetic field of an EPR spectrometer, the three sublevels of a triplet state have different energies and EPR records the transitions between these sublevels. When the triplet state is formed by recombination of a radical pair, according to the so-called radical-pair mechanism, as happens in reaction centers, its middle sublevel is highly populated and it is said to be "polarized". This property allows a clear identification of the triplet state ^{3}P.

3. The properties of ^{3}P are a good tool for understanding the properties of the bacteriochlorophyll dimer P. At a local level, it is possible for instance to determine the dissymmetry of the dimer, in terms of spin density. At a macroscopic level, the EPR spectrum of P can be used as a structural probe, for instance to determine the orientation of the dimer in membranes.

4. In reaction centers of wild-type *R. sphaeroides*, which contain one molecule of the carotenoid sphaeroidene (p. 360), energy transfer takes place rapidly from ^{3}P to sphaeroidene, in about 20 ns, to form the sphaeroidene triplet state, which decays in a few microseconds (Figure 10.20). This process is considered as a protective mechanism, as discussed above, to avoid the formation of singlet oxygen (Figure 10.9 p. 366). Remarkably, this triplet energy transfer stops at low temperature, below 77 K, so that ^{3}P becomes rather long-lived (around 100 μs). This property allows the study of ^{3}P by EPR at low temperature.

4 OTHER TYPES OF REACTION CENTER

4.1 PS2 REACTION CENTER OF OXYGENIC PHOTOSYNTHESIS

Plants, algae and some bacteria named cyanobacteria perform a more complex photosynthesis where water serves as a source of electrons to reduce carbon dioxide into organic compounds. These organisms are named "oxygenic" because they evolve dioxygen from water. Their photosynthetic membranes include two types of reaction centers named PS2 and PS1 (Figure 10.1) [1–4].

The PS2 reaction center can be called a light-activated water oxidase/plastoquinone reductase. It brings about the oxidation of water into molecular oxygen and transfers electrons to a quinone named plastoquinone, analogous to the ubiquinone of purple bacteria. The photoinduced reaction abstracts electrons and pulls the redox couple to the right:

$$2H_2O \Longleftrightarrow O_2 + 4e^- + 4H^+ \qquad Em, 7 = +0.81 \text{ V} \qquad (21)$$

It is well known that the inverse reaction takes place in respiration. This basic function immediately imposes two important characteristics of PS2, if we consider it as a microphotocell:

1. It needs to have a positive pole at a very oxidizing potential, much higher than +0.81 V in order to oxidize water practically irreversibly.

2. It needs to include some storage device in order to accumulate the successive loss of four electrons, due to four successive light-induced reactions (Figure 10.21). It has indeed been found experimentally, in agreement with thermodynamic predictions, that water is not oxidized in four one-electron reactions but rather in one unique four-electron reaction.

Many experimental features led to the recognition in the 1980s that the PS2 reaction center has many aspects in common with its counterpart of purple bacteria [35]. It has an inner core made of two hydrophobic polypeptides named D_1 and D_2 (equivalent and somewhat homologous to L and M). Its sequence of electron acceptors includes, as in purple bacteria, a pheophytin, a quinone Q_A (one-electron carrier) and a quinone Q_B (two-electron carrier) displaceable by competitive inhibitors, and with a Fe^{2+} atom presumably located between Q_A and Q_B. These analogies lead to the conclusion that the PS2 structure, for which there are no crystallographic data available, is similar to that of purple bacteria (Figure 10.22).

It should be realized, however, that PS2 displays very important differences form its bacterial counterpart, differences which can probably be traced as essential implications of its ability to oxidize water [36–38]. The aspects similar to those found in purple bacteria (including rates of electron transfer on the acceptor

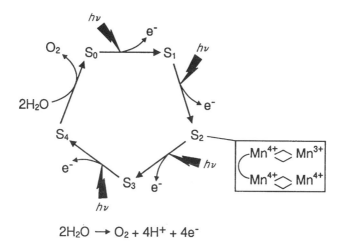

Figure 10.21. Phenomenological scheme of the five states of the water-oxidizing system in PS2 ("Kok–Joliot clock"). The reaction center reaches different states: from S_0 to S_4 one goes from one state to another by one electron abstraction due to a single photochemical event; the transition from S_4 to S_0 is spontaneous and is accompanied by evolution of one molecule of dioxygen. H^+ displacements are not shown (they are still poorly understood). A tentative structure of the manganese cluster is shown for state S_2

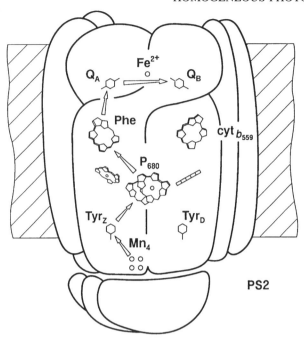

Figure 10.22. A tentative model of the PS2 reaction center, drawn by analogy with the reaction center of purple bacteria. The round shapes represent different polypeptides. Mn4: cluster of four manganese atoms. Tyr_Z, Tyr_D: redox active tyrosine residues. (Adapted from a model kindly provided by Dr A. W. Rutherford.)

side) need not to be described again and I shall, rather, insist on some specific properties.

1. Polypeptides. PS2 has no subunit directly equivalent to the H subunit of purple bacteria. It contains, however, a large number of polypeptides which are encoded in part by the nuclear genome and in part by the chloroplastic one (in eukaryots). In addition to D_1 and D_2, at least 11 polypeptides are present in the membrane part of the reaction center (without considering the PS2 antenna) and at least three polypeptides (of 16, 23 and 33 kDa) are peripheral and involved in maintaining the water oxidizing catalytic site. It is possible to isolate an inner core of PS2, which might be called "reaction center". It includes five polypeptides: D_1 and D_2, the two subunits of cytochrome b_{559}, and a very small hydrophobic polypeptide (4 kDa) named subunit I. This inner core, however, has a limited capacity in terms of electron transfer since Q_A is lost.

2. Non-hemic iron, Fe^{2+}. By contrast with what happens in purple bacteria, this atom can be oxidized to Fe^{3+}. The redox couple has an Em around +400 mV at pH7. Fe^{3+} is a very good electron acceptor for Q_A^-. There are also differences in the liganding, which may be of physiological relevance: instead of four histidines and one glutamate (bidentate), there are only four histidines available in PS2. The other ligands are perhaps provided by bicarbonate (HCO_3^-). This bicarbonate can be displaced, resulting in a block of electron transfer from Q_A to Q_B. There is no evidence, however, for a direct involvement of the iron atom as a redox center functioning between Q_A and Q_B.

3. Properties of P-680. Dimer or monomer? P-680 is very difficult to study and it should be stressed that most spectroscopic data concerning that species are very ambiguous. The redox potential of the P-680/P-680$^+$ couple can only be roughly estimated since no redox titration can be made in reaction centers above $+1$ V. The oxidized form can only be detected as a transient species, in flash experiments. In all attempts to oxidize P-680 in steady state, under continuous light, it is not certain whether P-680$^+$ does accumulate or oxidizes neighbour pigment molecules. In flash experiments, oxidation of P-680 can be detected by absorption changes around 680 nm (disappearance of chlorophyll *a* absorption) or 820 nm (appearance of a chlorophyll radical cation).

Another way to study P-680 is to populate its triplet state by the same mechanism as presented above for P in purple bacteria: when electron transfer from pheophytin to Q_A is not possible, recombination within the radical pair populates a triplet state which can be studied by EPR at low temperature. Two aspects of this reaction are fairly surprising:

1. The yield of ^{3}P-680 is very low when Q_A is singly reduced to the radical-anion state and it rises greatly when Q_A is either fully reduced (two electrons) or displaced from its site. These two situations are pathological states which are not encountered in the normal function of PS2.
2. The orientation of P-680 has been studied by measuring the EPR spectrum of ^{3}P-680 in oriented membranes. It appears that the g_z axis, which is perpendicular to the macrocycle plane, is oriented rather close (at 30°) to the normal to the membrane, which means that the tetrapyrrolic plane is close (30°) to the membrane plane. In purple bacteria this plane is perpendicular to the membrane.

Many properties of P-680 are not yet understood. For instance, we do not understand why P-680 has such a high redox potential, and why its absorption maximum is close to that of chlorophyll *a* in the antenna. Several hypotheses can be envisioned, ranging from P-680 having a structure similar to that of P in purple bacteria (i.e. a dimer of two chlorophyll *a*), or being a multimer of chlorophyll *a* with a weak intermolecular interaction, or even a monomeric chlorophyll *a* with the same position as the "accessory" bacteriochlorophyll located on the L branch in bacteria.

4. Electron donor to P-680$^+$. Tyrosine Z. The immediate donor to P-680$^+$ has been identified as a tyrosine residue (named TyrZ) of the polypeptidic backbone of D_1. It donates an electron to P-680$^+$ in about 50 ns. It is assumed that TyrZ experiences a large decrease in pK upon oxidation and that a proton is transferred to a neighbour amino acid. The oxidized species, the radical TyrZ•, is mainly known from its EPR properties which have been studied in great detail. TyrZ has a counterpart on D_2, named TyrD, which is not an efficient electron donor to P-680$^+$ in the main flow of electron. It is usually kept oxidized (a fairly unusual property for a high-potential species): it is assumed to be oxidized by P-680$^+$ and to play a role in the biosynthetic pathway which builds the manganese cluster, presumably by oxidizing Mn^{2+}.

5. Catalytic site for water oxidation. Water oxidation is the major functional property of the PS2 reaction center: it is a process of fundamental importance but which is still very poorly understood. The water-oxidizing machinery includes the following elements:

1. Four manganese atoms, which are the major site for the storage of holes (electron deficits) produced by successive electron transfers to TyrZ•.

2. Parts of the D_1 and D_2 subunits which serve three functions: scaffolding the cluster of four Mn, electron donation by means of at least one residue (presumably histidine), and proton movements which compensate the positive charges on redox active centers.

3. Ca^{2+} and Cl^- ions, which are essential for the function of the catalytic site.

All these elements are important and their structure/function properties are a field of very active research. Of special interest is the structure of the manganese cluster and the state of oxidation of the different atoms (a tentative proposal of structure is shown in Figure 10.21). These questions are mostly studied by EPR and by spectroscopy of X-ray absorption, together with the synthesis of artifical clusters which attempt to mimic the biologically active system. The mechanism of water oxidation itself is totally unknown.

6. Cytochrome b_{559}. In all organisms, the PS2 reaction center includes a tightly bound cytochrome, named cytochrome b_{559}. This cytochrome remains present in the smallest PS2 reaction centers which have been obtained, so it is certainly a *bona fide* component, closely associated with the D_1–D_2 core. Flash experiments did not show any evidence for any role of cytochrome b_{559} in the normal flow of electron transfer from water to plastoquinone. There is no counterpart of that species in other types of reaction centers.

What is the function of cytochrome b_{559}? A role for this cytochrome can be investigated along several lines:

1. A purely structural function: it would be required for the assembly of the reaction center;

2. A redox function associated with the process of building up the manganese cluster from Mn^{2+} ions present in the cellular fluid;

3. A protective role, associated with the reduction of an excess of positive charges, specially when the water-oxidizing machinery is not working perfectly.

7. Photodamage to PS2. It is well established that light has damaging effects on PS2 [39]. These effects take place already under low light; they are increased under high light and UV light has additional effects. Polypeptide D_1 is a major site of damage, resulting in its proteolytic cleavage either on the stromal or on the lumenal side of the membrane. The molecular mechanisms of damage are the subject of considerable controversy, and there is also much debate concerning the biological significance of these processes. It would seem natural that damages originate in the donor side of PS2, because of the very oxidizing character of P-680$^+$ and TyrZ$^\bullet$, of the production of oxygen, and of possible mistakes in the function of the water-oxidation process which could induce the formation of peroxides or radicals. This view, however, is contradicted by many data showing that damage can also result from acceptor side effects.

4.2 PS1 AND BACTERIAL ANALOGS

The PS1 reaction center of plants follows the same principles of construction and operation as previously discussed for other types of reaction centers, with a pair of hydrophobic polypeptides (named PS1-A and PS1-B) carrying the redox centers involved in primary electron transfer (Figure 10.23) [40,41]. The PS1 reaction center of a thermophilic cyanobacterium, *Synechococcus elongatus*, has been crystallized and its structure has been elucidated at 6 Å resolution, permitting the location of a few redox centers in the protein [42]. The major characteristic of PS1 is that it operates at low redox potentials in order to transfer electrons from small proteins at an Em of about +360 mV to other small proteins, ferredoxin or flavodoxin, which are very reducing, with an Em around −350 mV. Similar properties are found in several classes of bacteria [43], the green sulfur bacteria and the heliobacteria, which will not be described in detail.

4.2.1 Sequence of Electron Carriers

The primary donor, named P-700, is a dimer of chlorophyll *a*. The P-700/P-700$^+$ redox couple has a relatively low Em, +500 mV. The dimer is held at the interface between the PS1-A and B subunits. Close to P-700, there are several chlorophyll *a* molecules which are presumably organized similarly to monomeric bacteriochlorophylls and bacteriopheophytins in purple bacteria. They probably play a similar role, two of them being the primary electron acceptors and two being on an inactive branch. These assignments are very tentative, however. The reaction center also contains two molecules of phylloquinone which have not been located in the X-ray electron density map. One of the phylloquinones is the next electron acceptor while the other has no known function. From its function, it can be estimated that the redox potential of the active phylloquinone is around −800 mV, which is very low for this class of molecules. The low potential could be due to destabilisation of the radical anion in its highly hydrophobic environment.

The following electron acceptors are three 4Fe−4S centers, named F_X, F_B, F_A. The most primary of them, F_X, has a very low potential (about −730 mV). It has the very unusual property that its cysteine ligands are brought by two different

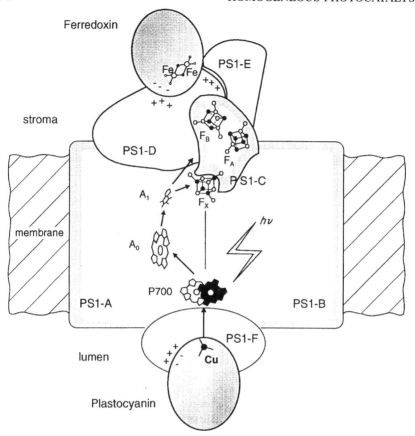

Figure 10.23. A tentative model of the PS1 reaction center. PS1-A,B,C,D,E,F: subunits involved in electron transfer. The active redox cofactors A_0 and A_1 might be duplicated by symmetrically located cofactors. (Adapted from a model kindly provided by Dr P. Sétif.)

protein subunits (the large subunits PS1-A and B). F_B and F_A (with Em around -590 and -540 mV respectively) are held by another subunit, named PS1-C. The path and kinetics of electron transfer between the phylloquinone, F_X, F_B and F_A have not been elucidated with certainty, in spite of much effort, probably because there is not a unique well-defined sequence for electron transfer. It is nevertheless clear that the terminal acceptors in the reaction center are the Fe−S centers F_A and F_B, where an electron presumably arrives in 1 μs or less.

4.2.2 Large Core Antenna

The couple of polypeptides PS1-A and B which make up the reaction center core have a large size (molecular weight around 83 kDa for each). In addition to the redox centers, they include a large number of pigment molecules with an antenna function (about 100 chlorophyll a and 10 β-carotene). It has been suggested that the large size of PS1-A and B is due to their genes being the result, during evolution, of a fusion of genes of the core polypeptides and of the core antenna in a PS2-like precursor [43].

4.2.3 Electron Donor and Electron Acceptor to PS-1

The PS-1 reaction center has the peculiarity of interacting with soluble proteins on both sides of the membrane (Figure 10.23). The electron donor to P-700$^+$ is usually a small copper protein named plastocyanin (it can be replaced in some cases by a cytochrome c). The docking of plastocyanin, which permits a fast electron transfer ($t_{1/2}$ about 12 μs), may be facilitated by one subunit of the reaction center, PS1-F, which extends into the lumenal volume.

On the stromal side, electrons are picked from the F_A, F_B couple by a small soluble ferredoxin (a protein with a 2Fe–2S center) (this protein also has a substitute, flavodoxin). After its reduction, ferredoxin fulfills several cellular functions, the principal of which is to reduce NADP$^+$ with the help of an enzyme named ferredoxin–NADP$^+$–reductase. The docking of ferredoxin involves two subunits of the reaction center, PS1-D and PS1-E.

5 CONCLUSIONS

Photosynthetic reaction centers are already relatively well known. A great deal remains to be done, however, to achieve a satisfying understanding of their structure and of their functional properties. This will require the concerted use of many methods:

1. Biochemical methods, to isolate reaction centers, to know their composition in terms of proteins and cofactors, to modify them in a desired manner, to identify protein-protein interactions, etc. When possible, X-ray crystallography provides essential structural information.
2. Spectroscopic methods of all kinds, with structural or functional objectives, find with reaction centers a field to demonstrate and check their usefulness. Of special interest are: flash absorption spectroscopy, vibrational spectroscopies, magnetic resonance (EPR, NMR and their cross-products), X-ray absorption spectroscopy, etc.
3. Molecular biology is of special interest in permitting to obtain the sequence of polypeptides and to make selective modifications by site-directed mutagenesis.
4. Chemists bring essential contributions in synthesizing models which permit the understanding of the properties of redox centers and of light-induced electron transfer. Theoretical aspects are also important for electron transfer, energy transfer, molecular dynamics, etc.

A survey of all reaction centers clearly shows that they can be classified in a few classes, which obey common principles for structure and function: a core of two hydrophobic polypeptides; transmembrane positioning; successive electron transfer steps, etc. They also make use of a small variety of pigments and redox-active components. Each class of reaction center displays little variation, an indication that the evolutionary constraints are very strong at their level. This evolutionary stability is particularly striking when one compares the PS1 and PS2 reaction centers of plants and of cyanobacteria: these organisms have a wide evolutionary difference but very similar reaction centers. An important recent finding has been the observation that the PS-1 type reaction center of green sulfur bacteria and of heliobacteria has a homodimeric protein structure: it is made of two identical subunits, instead of two homologous ones [44]. It will be exciting to learn whether this homodimeric structure goes with complete symmetric properties for the cofactors, or if there is a dissymmetry, as has been well indicated in purple bacteria.

Photosynthetic reaction centers are focal points where an enormous amount of excitation energy arrives permanently during daytime. How can they resist such a high flow of energy? Two reasons may be proposed for their stability:

1. In most cases where photodamages have been studied *in vitro*, the triplet state of a pigment was found to be the source of damage, probably by reacting with O_2 to form $^1O_2$, according to the series of reactions (D is the photosensitizing pigment):

$$D + h\nu \longrightarrow {}^1D^* \; ; \; {}^1D^* \longrightarrow {}^3D^*$$
$$^3D^* + {}^3O_2 \longrightarrow D + {}^1O_2 \; ; \; {}^1O_2 : \text{ source of damage} \tag{22}$$

   The high yield of charge separation in reaction centers results in a very short life for the singlet excited state of chlorophyll-like pigments and a low yield of triplet state formation. In the antenna, any triplet state formed is rapidly quenched by carotenoids for harmless internal conversion into heat.

2. Radicals are also a frequent source of degradation in biological systems. They are necessarily formed during the normal function of reaction centers. They are, however, restricted to specific volumes which are screened from unwanted access by the protein thickness.

For both of these potential mechanisms of photodamage, PS2 displays unique properties because the lifetime of chlorophyll *a* singlet excited state is relatively long, and because radical species at the water-oxidizing site are not completely screened from the lumenal medium. These are probably the reasons why PS2 degradation is prevented by additional mechanisms of excited states quenching and why it is efficiently repaired in case of damage.

It may be of interest to raise the question: Are reaction centers ideal energy converters? They are certainly well adapted for their specific functions such as cyclic electron flow for transmembrane proton pumping in the case of purple bacteria, oxidation of water in the case of PS2, and formation of reduced low-potential molecules in the case of PS1. But what about energy conversion? A thesis can be defended that reaction centers are optimized for a high quantum efficiency, and not for a maximum energy storage efficiency. If we do not consider energy losses in the antenna, the quantum efficiency of all reaction centers is close to 100%: all excitations of P, under physiological conditions, result in a charge separation leading to longlived products. Biological systems are usually well optimized, and it can be argued that there are two possible reasons for this high quantum efficiency:

1. It might permit a high yield of energy storage. In fact that reason is not correct: a high quantum efficiency is made possible by forward reactions having a rate much higher than back reactions, and in turn this requires a large $\Delta G°$, i.e. a large free energy loss. It seems probable (at least in PS2 and in purple bacteria) that a higher energy storage could be achieved by decreasing the $\Delta G°$ and by allowing a probability of back reaction of about 10%.

2. It seems more probable that the high quantum efficiency is required for the stability of reaction centers, as a built-in mechanism for their protection against photodamages. Charge recombination in the early stages following the primary charge separation would lead to a significant formation of triplet state of P, leading to the possible damage which has been discussed earlier. It can thus be proposed that reaction centers are optimized for stability against photodamage, while having a reasonable energetic yield.

It is tempting to think of artificial molecular assemblies which would efficiently convert the energy of light into chemical energy. What lessons can we learn from the way biological reaction centers are organized and function? Many chemists are working on the synthesis and study of supramolecular systems which would achieve artificial photosynthesis, and the number of achievements is already quite impressive [45].

REFERENCES

[1] M. D. Hatch and N. K. Boardman, (editors) *Biochemistry of Plants*. Vol 8: *Photosynthesis*. (Academic Press, New York, 1981).
[2] Govindjee (editor) *Photosynthesis: Energy Conversion by Plants and Bacteria* (2 vols.) (Academic Press, New York, 1982).
[3] J. Amesz, (ed.) *Photosynthesis* (Elsevier, Amsterdam, 1987).
[4] J. Barber, (editor) *The Photosystems: Structure, Function and Molecular Biology* (Elsevier, Amsterdam, 1992).
[5] R. Van Grondelle, Biochim. Biophys. Acta, **811**, 147-195 (1985).
[6] H. Scheer, (editor) *Chlorophylls* (CRC Press, Boca Raton, Florida 1991).
[7] H. A. Frank and R. J. Cogdell, Photochem. Photobiol., 63, 257-264 (1996).
[8] R. A. Marcus and N. Sutin, Biochim. Biophys. Acta, **811**, 265-322 (1985).
[9] C. C. Moser, J. M. Keske, K. Warncke, R. S.Farid and P. L. Dutton, Nature, **335**, 796-802 (1992).
[10] G. Feher, J. P. Allen, M. Y. Okamura and D. C. Rees, Nature, **339**, 111-116 (1989).

[11] J. Breton, *ISI Atlas of Science, Biochemistry* Vol. 1, pp. 323-328 (1988). CRC Press, Boca Raton.
[12] J. Deisenhofer and J. R. Norris, (editors) *The Photosynthetic Reaction Center*, Vols. 1 and 2, (Academic Press, 1993).
[13] J. L. Martin, J. Breton, A. J. Hoff, A. Migus and A. Antonetti, Proc. Natl. Acad. Sci. USA, **83**, 957-961 (1986).
[14] T. Arlt,, S. Schmidt, W. Kaiser, C. Lauterwasser, M. Meyer, H. Scheer and W. Zinth, Proc. Natl. Acad. Sci., USA, **90**, 11757-11761 (1993).
[15] J. Deisenhofer and H. Michel, EMBO J., **8**, 2149-2169 (1989).
[16] W. Leibl and J. Breton, Biochemistry, **30**, 9634-9642 (1991).
[17] F. Lendzian, M. Huber, R. A. Isaacson, B. Endeward, M. Plato, B. Bönigk, K. Möbius, W. Lubitz and G. Feher, Biochim. Biophys. Acta, 1183, 139-160 (1993).
[18] X. Lin, H. A. Murchison, V. Nagarajan, W. W. Parson, J. P. Allen and J. C. Williams, Proc. Natl. Acad. Sci., USA, **91**, 10265-10269 (1994).
[19] M. H. Vos, M. R. Jones, C. N. Hunter, J. Breton and J. -L. Martin, Proc. Natl. Acad. Sci. USA, **91**, 12701-12705 (1994).
[20] W. W. Parson, Z. T. Chu and A. Warshel, Biochim. Biophys. Acta, **1017**, 251-272 (1990).
[21] M. L. Paddock, S. H. Rongey, P. H. McPherson, A. Juth, G. Feher and M. Y. Okamura, Biochemistry, **33**, 734-745 (1994).
[22] V. P. Shinkarev and C. A. Wraight, in *The Photosynthetic Reaction Center*, Vol. 1, edited by J. Deisenhofer and J. R. Norris (Academic Press, New York, 1993) pp. 193-255.
[23] U. Ermler, G. Fritzsch, S. K. Buchanan and H. Michel, Structure, **2**, 925-936 (1994).
[24] D. M. Tiede, A. C. Vashishta and M. R. Gunner, Biochemistry, **32**, 4515-4531 (1993).
[25] J. M. Ortega and P. Mathis, Biochemistry, **32**, 1141-1151 (1993).
[26] X. Lin, J. C. Williams, J. P. Allen and P. Mathis, Biochemistry, **33**, 13517-13523 (1994).
[27] M. R. Gunner and P. L. Dutton, J. Am. Soc., **111**, 3400-3412 (1989).
[28] J. M. Ortega, P. Mathis, J. C. Williams and J. P. Allen, Biochemistry, **35**, 3354-3361 (1996).
[29] D. De Vault and B. Chance, Biophys. J., **6**, 825-847 (1966).
[30] A. M. Berghuis and G. D. Brayer, J. Mol. Biol., **223**, 959-976 (1992).
[31] J. Breton, J. R. Burie, C. Berthomieu, G. Berger and E. Nabedryk, Biochemistry, **33**, 4953-4965 (1994).
[32] B. Robert, Biochim. Biophys. Acta, **1017**, 99-111 (1990).
[33] D. E. Budil and M. C. Thurnauer, Biochim. Biophys. Acta, **1057**, 1-41 (1991).
[34] J. M. Peloquin, J. C. Williams, X. Lin, R. G. Alden, K. W. Taguchi, J. P. Allen and N. W. Woodbury, Biochemistry, **33**, 8089-8100 (1994).
[35] P. Mathis and A. W. Rutherford, In *Photosynthesis*, edited by J. Amesz, (Elsevier, Amsterdam, 1987) pp. 63-96.
[36] R. Debus, Biochim. Biophys. Acta, **1102**, 269-352 (1992).
[37] A. W. Rutherford, J. L. Zimmermann and A. Boussac, *The Photosystems: Structure, Function and Molecular Biology*, edited by J. Barber, (Elsevier, Amsterdam, 1992).
[38] A. W. Rutherford and W. Nitschke, In *Origin and Evolution of Biological Energy Conservation*, edited by H. Baltscheffsky (VCH, New York, Oxford) in press.
[39] N. R. Baker and J. R. Bowyer, *Photoinhibition of Photosynthesis*. (Bios Scientific Publishers, 1994).

[40] B. Lagoutte and P. Mathis, Photochem. Photobiol., **49**, 833-844 (1989).
[41] J. H. Golbeck and D. Bryant, Curr. Top. Bioenerg., **16**, 13-57 (1987).
[42] N. Krauss, W. Hinrichs, I. Witt, W. Pritzkow, Z. Dauter, C. Bretzel, K. S. Wilson, H. T. Witt and W. Saenger, Nature, **361**, 326-331 (1993).
[43] W. Nitschke and A. W. Rutherford, Trends in Biochemical Sciences, **16**, 241-245 (1991).
[44] U. Feiler and G. Hauska, In *Anoxygenic Photosynthetic Bacteria*, series *Advances in Photosynthesis*, edited by R. E. Blankenship, M. T. Madigan and C. E. Bauer (Kluwer Academic Publishers, Dordrecht) (1995) pp. 665-685.
[45] M. R. Wasielewski, Chem. Rev., **92**, 435-461 (1992).

Index

Absorbance, 18
Absorption
  band, 29
  intensity, 34
  of light, 17
  of photons, 229, 245, 246, 361
Absorption spectrum, 18, 251
  AgBr crystal, 179
  human retina, 178
  sensitizers, 180
Acceptor
  electron, 288
Acids
  carbon centered, 64, 91, 92
  oxygen centered, 64, 70
Acid-base
  strength and catalysis, 56
  equilibrium in an excited state, 62
Acid catalysis
  criteria for, 56
  general, 56, 65
  specific, 56, 61
Acidity constant
  change on excitation, 62, 64, 91, 89, 86, 73
Activated complex, 22
Activated state, 22
Activation
  associative, 6
  dissociative, 6
  of C-H bond, 162-165
  energy, 254
Active sites, 10, 241
Acridine, 86
Acridine dyes
  as photosensitizers, 273
Adiabatic approximation, 19
  limits, 26
Adiabaticity
  for highly exothermic reaction, 89, 91
  requirements for, 62

Adsorption
  of sensitizers, 180, 222, 224, 226, 245, 246, 247
Ag^+
  in zeolites, 287
Ag, photodeposition, 176
Aggregates of cyanines, 295
Aggregation
  in colloids, 253, 256, 257, 311
  of silica, 225
Alkenes
  amidation, 335
  cycloaddition, 100
  dimerization, 100
  photohydroformylation
  photohydrogenation, 153-156
  photohydrosilation, 153, 156-157
  photoisomerization, 100, 141, 158-161, 237, 248, 249, 253
  photooxidation, 145-146
Allylthiourea, 277
Allyltrimethylsilane, 111
Allowed transition, 32, 43
Alumina
  amberlites, 247
  as a support, 224, 225, 247
Amidation of olefins, 335
Alkaloids, 126
Amines
  as electron donors, 277
  photooxidation, 108
Aminoacids
  as electron donors, 277
Amplification of the image, 187
Amphiphiles, 321
Antenna, 358
Anharmonicity, 25
Anilines, 210
Anionic polymerization, 142-143
Anthracene, 35
Antiaromatic, 74
Antijuvenile hormone, 107

Arrhenius law, 22
Aromatic photosubstitutions, 106
Aromaticity, 74
Ascorbate, 277
Atom transfer
  reactions, 82
Autocatalytic electron transfer, 197, 201
ATP, 363
Aziridines
  bond cleavage, 116
Azomethinic dye, 211

Bach reaction, 298
Bacteria
  green, 368, 391
  photosynthetic, 356, 369, 391
  purple, 356, 369
Bacteriochlorophyll, 359, 360, 369
Bacteriopheophytin, 369
Band gap
  particle size dependence
    silver bromide, 182
  of emission, 37
Bases
  carbon centered, 64, 88, 89
Base catalysis
  criteria for, 56
  general, 56, 65
  specific, 56, 61
BEBO model, 77
Beer–Lambert Law, 17
  departure from, 18
Benzophenone
  sensitizer, 245, 248, 249, 250, 257
Benzopyrrolidines
  one-pot synthesis, 107
Bibenzyl
  photocleavage, 112
Bilayers, 314, 345
Biprotonic transfer, 93
Bipyridinium ions, 274
Bitumer, 173
Blue shift
  of absorption edge, 252, 257
Bohr's frequency, 19
  Bond activation, 162–165
Bond breaking
  C–C bond cleavage, 112
  energy for, 19
Bond dissociation energies
  tables, 83
  formation

Bond lengths
  tables, 83
Bond orders, 82
Born model of solvated clusters, 199
Born–Oppenheimer approximation, 19
Bosons, 16
Brönsted
  base catalysis, 66
  coefficient, 66, 67, 86
Brönsted plots
  curved, 89

Cadmium colloidal, 278
Cage complexes as electron relays, 294
Cage escape efficiency, 283, 284
Cascade
  electron transfer, 290, 297
Catalyst
  efficiency of 278
  homogeneous, 277, 286
  microheterogeneous, 274, 277, 278, 285
Camera obscura, 170
Carbon–Carbon bond, photoformation, 100
Carbon dioxide
  reduction, 356
Carbon–hydrogen bond, photoactivation, 162–165
Carbon monoxide, 154, 163
Carotenoid, 360, 367
Catalysed photochemistry, 7, 197
  photoreaction, 6, 253
Catalyzed photolysis, 143
Catalysis, 310
  acid–base, 55
  Bell's definition, 2
  Berzelius definition, 3
  definition, 56
  generalized definition, 4
  heterogeneous, 10
  photogenerated, 140
  photoinduced, 140
  static and dynamic, 93
  thermal, 137–139
  variety of contexts, 3
Catalytic constant, 56
Catalytic germs, 206
Catalytic nuclei, 206
Catalysed photoreaction, 197
Catalyst, 206
  photochemically generated, 3, 6, 140–143

INDEX

Catalytic sites, 10, 242, 246, 251, 257
Celite
  as a support, 224
Cellulose powder
  as a support, 224
Cellulose acetate
  membranes, 231, 232, 251, 252
Cellulose nitrate (collodion), 182
C-H
  benzylic heterolysis, 91
  C-N, bond breaking, 19
Chain reactions, 4, 59, 174, 196, 203
  and catalysis, 4
  photoinitiated, 5, 141-143
Charge
  charge-transfer species, 98
  recombination, 46, 378
  separation, 46, 369, 181
  translocation, 46
Charge recombination, 267, 268, 288, 291
Charge separated state, 289, 290, 292, 293
Charge separation, 266, 268, 288, 295, 296, 301
  quantum yield of, 290, 292
Charge transfer states, 150-151
Chromatic sensitivity, 177
Chromium, doping of TiO_2
CIDNP, 98
Cinematography, 214
Clays, 298, 247
Cleavage, *see also* Photo-degradation of H_2S
  oxidative, of arenes, 253
  of water, 265, 297, 301, 386
  requirements for, 266
Chlorophillin, 234
Chlorophyll, 234, 340, 359, 360, 366, 389, 391
Chloranil
  photosensitizer, 102
Clusters, 184, 186, 189, 202
  copper clusters, 190
  gas phase clusters, 190
  ionization potential, 190
  redox properties, 198
  silver clusters, 190
  soft landing of clusters, 191
  solvated clusters, 191
  thermodynamics, 189
Coal, 266
CO, *see* Carbon monoxide
Co(bpy)$_3$, 279

Coenzymes
  as electron donors, 277
Colloidal
  metals, 274, 277, 278
  particles, 285, 298
  platinum, 272, 279, 299
  systems, 298, 299
Color photography, 207, 209
  instant colors photography, 213
Complexes
  cage, 294
  transition metal, 271, 273
Confined systems, 287
Constrained systems, 295
Coordination complexes, 271, 273
Coordination compounds, energy levels, 149
Copper colloidal, 278
Coumarin, 450, 35, 228
  photoaquation
Critical nuclearity, 196, 201, 204
  coupler, 209, 210
Cross section, 18, 361
Crystal
  tabular, 177
  silver bromide, 177
  growth, 177
Crystal violet, 237
  reduction, 237
Cyanine dyes, 180
  as photosensitizers, 273
Cyanoacrylate polymerization, 142-143
Cyanobacteria, 356
1-Cyanonaphthalene-photosensitizer, 106
Cycloadditions, photo-, 100, 143, 157-162
Cycloheptatriene
  deprotonation, 92
  hydrogenation, 156
Cyclohexene
  cycloaddition, 160-162, 249, 251
Cyclopentadiene
  deprotonation, 92
Cysteine, 277
  c, 274, 374, 382
  b 559, 390

Daguerreotype, 175
Deactivating process, 5
Defects in crystals, 180
Definition of photocatalysis, 1

Deshydrogenation, photo-, 153
Development (photographic), 202
  threshold, 202
  of color photography, 210
  physical development, 189
  dry-silver development, 188
  model, 201
  offset process, 214
Dexter theory
  of electron-exchange, 45
  energy transfer, 45
Diabatic reactions, 62
  and preexponential factor, 62
Dibenzosuberenol, 91
Dicyano anthracene
  photosensitizer, 101, 107, 111, 117, 42
Dicyanobenzene
  photosensitizer, 101
Dicyanonaphthalene
  photosensitizer, 101
Dielectric saturation effect
  on e.t. rates, 50
Diels–Alder
  Lewis acid catalysis, 4
  radical cation catalysis, 101
  sensitized, 225, 229
  triplex, 103
Diffusion control, 41, 45, 98
  of protonation, 68
  of proton transfer, 249
Dihydrogen, *see* hydrogen
Dimer, 378, 389, 391
Dimerization, photo-
  4 + 2, 101, 103
  2 + 2, 101, 160, 249, 251
Dimethyl viologen, 345
  reduction, 233, 252, 256, 342
Dimethyl viologen, *see* Methyl viologen
DNA coil, 51
Dioxygen, 367
  singlet, 117, 225, 235, 236, 245, 246, 253, 394
Dipole moment, 27
  dipole interaction energy, 44
  induced, 28
  operator, 29
  oscillating, 28, 44
  transition, 28
Diquat, *see* Viologen
Discrimination
  between clusters, 202
Dissociation energy, 19

Dithiolene complexes, 300
Donor–acceptor complexes, 3
Donor emission, 44
Dowex, 225, 245
  supported sensitizers, 224, 225, 246
Dry-silver process, 188
Driving force
  for electron transfer, 47, 379
Dualistic theory, 16
Dyad (or diad)
  definition, 265
  systems, 288, 295
Dye photosensitization, 182, 183

Eburnamonine, 127
EDTA
  as electron donor, 271, 272, 274
  degradation of, 281
Efficiency
  cage escape, 283, 284
  of catalysts, 278, 279
  charge separation, 290, 292
  in photography, 181
  in solar energy conversion, 364, 395
Ehrenfest principle, 20
Einstein, 19
Einstein transition probability
  of absorption, 29
  for spontaneous emission, 34
  of induced emission, 34
  Electric dipole, 27
  transition, 28
Electric field
  oscillating, 28
Electric field vector, 17, 27
Electrochemical potential
  redox potential, 198
Electromagnetic radiation, 16
  infrared, 16
  near-ultraviolet, 16
  range of frequencies, 16
  visible, 16
Electrons
  acceptors of, 239, 241, 252, 253, 254
  acceptors sensiters, 119
  diffusion length, 182
  donor, 196, 197
  electrochemical potential, (of solvated electrons), 194
  motion, 19
  trapping, 182
Electron acceptors, 288

INDEX

Electron affinity
 oxidative, 152
 photoinduced, 152
 rate at junctions
 reductive, 152
Electron donors, 268, 274, 277
Electron-donor complexes, 252
Electron/hole pairs, 181
 quantum yield, 181
 recombinaison, 181
 solvated, 194
Electron relays, 268, 273, 274
Electron transfer, 339, 46
 adiabatic, 47
 bach, 296, 298, 301
 book, 128
 Carter-Hyne, 50
 cascade, 290, 297
 catalytic, 186, 196, 201
 classification based, 46
 distance dependence, 40, 363
 Hush theory, 47
 induced chain reactions, 107, 151
 inner sphere, 47
 intramolecular, 35, 150, 290
 Jortner-Bixon, 39
 Kakitani-Mataga, 39
 Levich-Dogonadé, 38
 Marcus equation, 47, 363, 378
 Marcus inverted region, 49, 379
 non adiabatic, 38
 outer sphere, 47, 229, 233,, 237, 239, 241, 254
 photoinduced, 99
 proteins, 51, 274
 quantum effects, 50
 quenching dynamics, 98
 reverse or back, 47, 99, 118, 364
 and solvent polarity, 48, 99, 118
 Taube's classification, 35
 Tachiya, 50
 thermodynamic stability, 98
 turn-over, 197
 unidirectional, 297
Electronic configurations,
 transition metal complexes, 149-152
Electronic coordinates, 19
Electronic devices, 217
Electronic factor, 57
Electronic motions, 20
Electronic states, 21, 25, 149-152
Electron transfer sensitized

anti-Markovnikov additions, 105
book, 128
by supported sensitizers, 229, 232, 233, 237, 252, 254
cleavage of bonds, 112, 114
cycloadditions, 100, 160-162
debenzylation, 113
deselenylation, 113
desilylation, 112, 113
deoxygenations, 115
deprotection of protected groups, 115
Diels-Alder, 101, 229
dimerizations, 100, 102
fragmentations, 112
heterocyclization, 111
isomerizations, 100, 145, 233
mixed cycloadditions, 103
NO insertion, 116
oxidations, 151
oxygenations, 122, 145
rearrangements, 100
reductions, 237, 252
reviews, 98
supported sensitizers, 229, 232, 233, 237, 252, 254
strained ring cleavage, 115
triplex Diels-Alder, 103
Electron transport, chain, 357
El-Sayed's rules, 40
Emission probability
 spontaneous, 34
Emission spectrum, 18, 256, 257
Enantioselectivity, 103, 120
Ene reaction
 of $^1O_2$, 225
Energy
 conversion, 395, 226
 gap law, 40
 kinetic, 21, 23
 migration
  on fractal, 46
  in polymeric sensitizers, 248, 249, 250, 251
 potential, 21, 23
 of reorganization, 48, 378
 or reorganization of solvents, 48
 storage capacity, 266
 states, 22, 27, 32, 38
 transfer
  coulombic mechanism, 44
  exchange mechanism, 45
  in fractals, 46

Energy gap law, 40
Energy transfer, 43, 141
  and high energy vibrations, 40
  between chromophores, 43, 226, 228, 229, 234, 236, 248, 249, 250, 252, 253, 254, 255, 361
  collisional, 43, 45
  collisional mechanism, 43, 45
  condition for, 59
  coulombic, 43, 44
  electron-exchange, 43, 45, 229
    orbital interactions, 45
  in disordered materials, 46
  intramolecular, 249
  rate constant, 44, 45, 228
  Rehm–Weller treatment, 47
  resonant, 44
  selection rules, 43
  sensitization, 98, 144, 225, 226, 228, 229, 234, 235, 236, 248, 249, 250, 251, 252, 253, 254, 257
Enzymatic process, 339
Enzymes
  or catalysts, 277
Eosin
  supported, 225, 234
Europium, 277
Evolution
  of oxygen, 356
Excimers, 97
Exciplexes, 42, 97, 329
  book, 128
  diastereomeric, 105
Excitation spectrum, 36
Excited states, 366
  acid–base properties, 7, 73
  bimolecular processes, 41, 226, 229, 247, 249, 250
  charge transfer, 150–151
  chemical reaction, 42, 368, 226, 235, 248, 249, 250
  kinetics, 61, 251, 253
  lifetime, 33, 232, 248, 252, 253
  ligand field, 149–150
  of metal complexes, 148–153
  metal–metal bonding, 151–152
  oxidation, 43
  processes
    efficiency, 252, 253, 256
    kinetic aspects, 33, 250, 252, 253
    mechanisms, 33, 247, 252, 256
    quantum yield, 33
  thermodynamic aspects, 226
  in sensitization, 43, 226
  unimolecular processes, 33
Exposure, 18
Extinction coefficient, 18
Eye, 178
  photoreceptors, 178
Eyring theory, 22

Fe_2O_3, 278
Fermi
  golden rule, 40
  level
    of normal hydrogen electrode, 198
Fermi's golden rule, 40, 44
Ferredoxin, 368, 393
Flash photolysis, 98, 118, 281, 393
Fluorene, 65, 74, 91
Fluorescein, 234
Fluorescence, 33, 98
  femtosecond, 37
  quantum yield, 34, 41, 255
  quenching, 41, 47, 364
  radiative lifetime, 34
  spectral distribution, 32, 257
  spectral narrowing, 37
  structural condition for, 40
  time dependent decay, 33
  time dependent spectrum, 37
Fogging, 188
Forbidden transition, 30, 32
Force constants
  tables, 83
Force field
  time dependent, 28
Formamide, 333–336
Förster cycle
  and pK_a^*, 71, 72
Forster theory
  of coulombic energy transfer, 44, 358
Fractal medium and lifetime, 46
Franck–Condon
  average density, 50
  factor, 31, 40, 363
  overlap integral, 31, 39
  principle, 30
Free rotor picture, 41
Frequency factor, 22
Fuels, 265
  energy storage capacity of, 266

INDEX

Gears
  photocatalytic, 8
Gelatine, 176, 182, 184
Germs, 206, 214
Glass
  as support, 226, 231, 247
Ground state complexes, 143–144, 157–162, 233, 254

Hadrons, 16
Hamiltonian, 19
  perturbation, 28
Harcourt
  increased-valence structures, 82
Harmonic oscillator, 23
  limits, 26
Hausdorff
  fractal dimension, 46
Heavy atom effect, 3, 31
Heisenberg uncertainty, 23
Heliography, 170, 173
Hematoporphyrin, 234, 235, 238
Hermite polynomials, 24
Heterocyclisation, 111
Heteropolycompounds, of W and Mo, 164–165
Hydrogen
  as a fuel, 265
  energy storage capacity, 265, 266
  explosive limit in air, 266
Hydrogen production, 271–284
  catalysts for, 277
  mechanism of, 281
  model systems for, 271, 297
  optimization, 282
  quantum yields, 273, 278, 284
Hydrogenase
  as catalyst, 277, 294
Hydrogenation
  of methyl viologen, 277, 281, 282
  poisons, 284
Hydrophilic sensitizer, 229, 231, 235, 236, 253, 256
Holes, 181
  diffusion length, 182
Holography, 215
Homolytic cleavage
  photoinduced, 6
Hydrocarbon conversion, 162–165
Hydroformylation, photo-thermal, 138–139

Hydrogen atom
  abstraction, 226, 227, 228, 236, 250, 251
  interaction with hν, 28
Hydrogen molecule
  interaction with hν, 28
Hydrogenase, 341
Hydrogen abstraction, 226, 227, 228, 236, 250, 251
Hydrogen evolution, 193
Hydrogen, 266
Hydrogenase, 277
Hydroquinone, 187
Hydrogenation, photo-, 153–156
Hydrophobic interactions, 311
Hydrosilylation, photo-, 153, 156–157

Indazolone dye, 211
Indophenol dye, 212
Initiators, 141–143
Interferential process in photography, 207
Intermolecular processes, 41
Internal conversion, 33, 40
Intersecting state model, 21, 57, 75
Intersystem crossing, 33, 40, 42, 367
  oxygen induced, 42
Intrinsic lifetime
  aromatics, 5, 58
Inversion-transfer processes, 205
Inverted region, electron transfer, 49, 50
Ion exchange resins
  complexed sensitizers, 224, 225, 245, 246
Ionization potential
  silver clusters, 190
  copper clusters, 190
Ion pairs, 150–151
Iridium colloidal, 278
  Ion pair charge transfer, 150–151
Iron-sulfur-center, 368, 392
Isomerization, photo-, 330, 331
  cis-trans, 115, 141, 158–161, 228, 231, 233, 236, 249, 250, 253
  valence-, 145–146, 223, 245, 250, 251, 257
Isotope exchange
  and acid-base photocatalysis, 65
IUPAC
  definition photocatalysis, 4
  quantum yield, 10

J-aggregates, 295

Kaysers, 17
Kinetic
　isotope effect, 91
Kinetics
　electron transfer, 96
　of excited states, 33
Kuhn model, 29

Laser
　Nd-Yag, 17
　properties, 17
　pulse laser, 193
Laser dyes, 29, 37
Latent image, 184
　latent image regression, 204
Lead colloidal, 278
Lifetime
　of excited states, 33, 364, 226, 232, 247, 251, 252, 253, 256
　fluorescence, 33, 34, 232, 253
　radiative, 33, 251
Ligands exchange, 348
Light
　absorption of, 358, 17, 229, 246, 251, 253, 255, 256, 257
　intensity, 17
　nature of, 16
　quantum energy, 16, 225, 228, 232, 235, 236, 245, 246, 251, 253, 254
　scattering, 256
Light intensity, 17
　transmitted, 17
　absorbed, 17, 225, 228, 232, 235, 236, 245, 246, 251, 253, 254
Lipids, 356
Liposomes, 344
Liquid crystals, 312, 313
Local environment, 86
Lumichrome, 93
Luminescence
　acid and conjugated bases, 72
　charge-recombination, 98

Magnetic field vector, 27
Membrane, 356, 377
Manganese
　in water oxidation, 387, 390

Marcus distribution, 47
Marcus equation, 47
Marcus theory, 36, 362, 75
Markovnikov, 88
Mechanism
　oxidative quenching, 268
　reductive quenching, 268
Medium polarity, 330
Membranes
　protein bound, 222
　Merrifield technique, 222, 233, 234
Mesoporphyrins
　as sensitizers, 232, 238, 252
Metal
　carbonyls, 141-142, 153-156
　colloidal, 278
　clusters, 184, 186, 189, 202
　complexes, 135-165, 232, 238, 243, 244, 245, 246, 252, 256
　ions, 3
　oxide powders, 278
Metals
　colloidal, see Colloidal metals
　supported, 277, 278, 285
Metal-carbon
　cleavage, 113
Metal complexes, 273, 274
Metal ions
　as electron donors, 277
　as electron relays, 272, 274, 299
Metal-metal bonds, 151-152
Metal-metal
　cleavage, 113, 151-152
Metal powders oxides, 277, 278, 285
Metalloporphyrins
　as sensitizers, 146
Metalloproteins
　cytochrome c, 51, 381
Methoxybenzenes, 64, 65
Methoxyquinoline, 86
　protonation rate, 87
Methylviologen, 119, 120, 152, 237, 256, 271, 273, 275, 279, 296, 299, 300, 342
　dyad, 295
　hydrogenation, 281, 283
　hydrogenation poisons, 284
Methylene blue, 224, 225, 226, 247
Micelles, 312, 343
Microemulsions, 315-320, 324, 334, 344
Microemulsion phase diagram, 336, 337
Microparticles, 348

INDEX

Mirror symmetry, 35
Mitomycin, 13, 107
MnO_2, 278
Model system
  classical, 279
  for hydrogen production, 269, 271, 297
  for oxygen production, 269, 270, 284, 297
  sacrificial, 268, 271
Molecular dipoles, 27
Molecular orbitals, quantized set, 20
Molecular states, 26
Molecular system
  homogeneous, 267, 277, 286, 299
  microheterogeneous, 265, 267, 274, 285, 299
  quasi-homogeneous, 267
Monte-Carlo simulation
  light absorption rate, 246
Montmorillonite, 225
Morse oscillator, 83
Motions
  twisting, 41
  stretching, 41
MV^{2+}, 274

Naphthalene, 35
Naphthol
  in color photography, 212
  deprotonation rate, 87
  in photocatalysis, 64
N-N bond breaking, 19
N-demethylation, 123
Negative (photographic) image, 184, 205, 212
Ni colloidal, 278
Nitrogenase, 277
NO insertion
  in C-C bond, 116
NO_2 group
  i.s.c. enhancer, 90
Non-adiabatic reactions, 32
Non-aqueous microemulsions, 333
Non-harmonic oscillator, 25
Norbornadiene
  valence isomerization, 145-146, 223, 236, 246, 248, 250, 251, 257
Normal modes
  vibration, 23
Nuclear motions, 20, 22

Offset silver process, 213
Offset printing, 213
Offset development, 214
Optical density, 18
Orbital interaction
  and energy transfer, 45
Organized media, 9, 295, 298, 311, 322, 323, 324
Oscillator strength, 29, 44
Osmium colloidal, 278
Oxazines tetrahydro
  synthesis, 111
Oxidation, photo-
  of alcohols, 250
  of alkenes, 117, 120, 145-146, 225, 234, 235
  of amines, 123, 247,
  of arenes, 119, 121, 231, 252, 257
  of histidine, 236
  of sulfides, 251
  of methionine, 226, 247
  of thiols, 257
  of tryptophan, 223
  of α, β-unsaturated ketones, 147-148
Oxidation of water, 386
Oxygen
  active species, 394, 243
  evolution, 386
  singlet, 117, 223, 224, 225, 229, 231, 232, 233, 234, 235, 239, 245, 246, 253, 254
Oxygen molecular
  quenching by, 42, 367, 254
  singulet, 367, 223, 224, 225, 229, 231, 232, 233, 234, 235, 239, 245, 246, 253, 254
  triplet, 42
Oxygen production
  model systems for, 269, 270, 284, 297
  requirements, 269
Ozonide, 122

Parabolic relationship
  between driving force, 379
  and rate, 47
Paraquat, see Viologen
Particles, see also Colloidal
Pauli exclusion principle, 26
Pauling relation
  bond order vs bond length, 77

Pb
  colloidal, 278
Peroxides, 119, 225, 228, 234
Perylene, 42
pH jump
  experiments, 70, 84
Phase diagram, 319
Phenanthrolinium, 274
Phonons, 38
Phenothiazine, 233
Phenylacetylenes, 88
  photohydration, 71
Phosphorescence, 32, 255, 257
Photo-
  activated catalysis, 327, 182
  catalysis
    definition, 1, 326, 59, 136
    heterogeneous, 9, 135–165
    IUPAC definition, 4
    kinetic aspects, 2
  chemical: 170
    conversion of sunlight, 356
    decomplexation, 6
  deoxygenation, 115
  emission measurements, 255, 257
  enhanced catalysis, 5
  generated catalysis, 6, 140
  hydrogenation, 153–156
  induced electron transfer, 7, 361, 152, 229, 232, 249, 252, 254, 255
  isomerization, 331, 141, 158–161, 223, 228, 229, 236, 245, 248, 249, 250, 251, 253, 257
  oxidation
    of organic compounds, 145–146, 223, 224, 225, 228, 231, 232, 233, 235, 236, 238, 246, 249, 257
  oxygenation, 117
  polymerization, 6, 141–143, 174
  reactions
    assisted, 59, 145
    cycloadditions, 11G, 100, 143, 157–162
    dimerization, 100, 144, 160
    of hydrocarbons, 162–165
    isomerization, 141, 158–161, 228, 231
    oxidation, 145–146, 147–148, 223, 224, 225, 228, 231, 232, 233, 235, 236, 238, 246, 249, 257
    sensitized, 367, 144–146

rearrangement, 100
sensitive materials, 176
sensitive receptors of human eye, 178
sensitization, 43, 144–146, 243
sensitizers, 103, 182
substitution, 106, 107
synthesis, 355
  bacterial, 355
  natural, 355
Photoamination, 106
Photoassisted, 59, 328, 145
Photochemical conversion of solar energy, 263
Photochemical systems, 263
Photocatalysis
  classification, 327, 140–148
  by semi-conductors, 4, 10
  definitions, 327
  generalized definition, 4, 140–148, 136
  Henning def., 5, 59
  heterogeneous, 10
  IUPAC def., 4
  Wubbel's definition, 2, 60
Photocyclizations, 107, 157–162
Photodamage, 391, 394
Photoelectrochemical
  cells, 3
Photogenerated catalysis, 6, 329, 140–143, 180
Photography, 169
  adiabatic, 88
  anti-Markovnikov, 90
  color, 207
  definition, 176, 205
  development, 186
  fixing, 188
  history, 171
  silver film, 176
Photoinduced
  chain reaction, 6, 141
  electron transfer, 99, 369, 150–153, 229, 233, 237, 249, 254, 256
  electron transfer, book, 128
  water splitting, 298
Photoinduced catalysis, 6, 59, 328, 140, 202
Photoketonization, 91
Photons, 176
  absorption, 18, 246, 251
  different roles, 6
  flux, 177
  impinging, 11

INDEX

Photopolymerization, 174
Photo-reductions, 105
Photosensitive materials, 176
Photosensitive receptors in retina, 178
Photo-Smiles reaction, 60
Photosensitization, 43
Photosensitizer, 144–146, 182, 267, 273
   solid adsorbed, 223, 224, 243, 244, 245
   types of cycles, 8
   vs catalyst, 58, 59
Photosynthesis, 341
   artificial, 263, 395
   natural, 286
Photosubstitutions, 106
Photosystem II, 286
Phototautomerization, 93
Photozymes, 230, 253
Phthalocyanines as photosensitizers, 273
   aggregation, 256
   sensitizers, 243, 244
Phycobiliproteins, 361
Pigments, 359
Pinacols
   photocleavage, 112
pK_a
   difference ground, 73
Platinum colloidal, 278
Planck equation, 16
   pOH jump experiments, 70, 84
Poisons, 284
   definition, 265
Polarizability, 28
Polarized light, 17
Polyad, 288, 290, 292, 293
Polyatomic molecules
   electronic states, 20
   vibrations in, 26
Polyethylene terephthalate
   sensitizer, 229
Polymers
   aqueous solutions, 347
   as sensitizers, 229, 236, 247, 248
   as supports, 229, 234, 235, 236, 237, 238, 248, 249, 253, 254
   multifunctional, 229, 230, 231, 235, 236, 248, 249
   degradation, 250, 251, 255
   photophysics, 248, 251, 256
Polymeric films, 229, 231, 251
Polymerized surfactant vesicles, 346
Polymethine
   laser dyes, 29

Polyoxometalates, 164–165
Porphyrins,
   adsorbed on vesicules, 46
   aggregation, 256
   metallation, 348
   as photosensitizers, 273
   as sensitizers, 232, 234, 237, 238, 252, 255, 256
Positive image (direct), 212
Potential difference, 202
Potential energy curve
   for a reaction, 8
Potential energy surface, 21
   adiabatic approximation, 26, 38
   crossing, 39, 41
   for several molecules, 42
   non-adiabatic, 32
Precomplexation, 143
Pre-exponential factor, 22
   and diabatic processes, 90
Previtamin, 237
Probability of transition, 25
   Einstein, 29
Procatalyst, 4
Protein, 358, 375, 388
Protein electron transfer, 51
Proton transfer, 64, 363, 372, 381
   adiabatic, 62
   adiabatic rate, 87
   intramolecular, 93
   vs diabatic, 92
Protonation rate, 61
   diffusion control, 69
   difference. GS excited state, 89, 91
Pt
   colloidal, 272, 279, 299
Pt O_2, 278
Pt salts, 277
Pulse radiolysis, 193
Pyrazolone, 211
Pyrene, 42
Pyrylium salts, 103, 229, 230, 233, 253

Quantum effect, 50
   on electron transfer, 50
Quantum efficiency, 225, 235, 236, 245, 251, 252, 253, 254
Quantum numbers
   angular momentum, 26
   magnetic, 26
   principal, 26
   spin, 26

Quantum size effect, 190
Quantum theory, 19
Quantum yield, 364, 180
  and catalytic effects, 59, 140-141
  and crossing between states, 60
  fluorescence, 33, 34
  in photocatalysis, 140-141, 225, 236
  IUPAC definition, 11, 35
  of oxygen production, 288
  in valence isomerization, 145, 245, 250
Quenching, 33, 41, 47, 98, 250, 251, 252, 253, 254, 255, 256
Quenching bimolecular, 62, 33, 41
Quenching process, 250, 251, 252, 253, 254, 255, 256
Quenching reactions, 33, 41
  static, 29
Quinine bisulfate, 35
Quinones, 362, 365, 371, 380, 391
  diimines, 210

Radiationless deactivation, 33, 38
Radiative deactivation, 32, 34
Radiationless processes, 38
Radicals
  cations, 117, 233
  metal-containing, 151
  ions, 229, 249
  pairs, 110, 118, 385
Radical anions
  aromatic, 107, 118
  photooxygenation, 120
  photosensitized dissociation reactions, 112
  synthetic applications amines, 110, 118
Radical cations
  from alkenes, 99
  from amines, 108
  nucleophilic substitution, 110
  intramolecular, 110, 112
  strained rings, 115
Radical chain reaction
  and catalysis, 4
  and photocatalysis, 4, 151
  photostimulated, 4
Radical ion pair, 98, 385
  contact, 99
  solvent separation, 99, 48, 249, 254, 255
Radiography, 193
Raman resonance, 98

Rate
  areal, 9
  specific, 9
  volumetric, 9
Rate constant, 41
Reactions
  catalyzed, 139
  electron transfer-induced chain, 105
  in organized systems, 326
  photoinduced catalysis, 140, 250
Reaction centers, 359, 370, 373
  PS1, 386
  PS2, 391
Reaction order
  and catalysis, 56
Reagent photoactivated
  and heterogeneous photocatalysis, 4
Recombination
  of electron/hole pairs, 181
Redox cofactors, 376
Redox energy levels, 365, 48, 152
  of metal clusters, 198
  of developers, 196, 201, 202
  size-dependent, 191
Redox potentials, 48
Reduction, 43
  of water, see Water, cleavage
Refractive index, 17
Rehm-Weller treatment, 47
  experiments, 49
Resonance effects, 79
Retina of human eye, 178
Rhodamine 101, 35
Rose bengal
  origin, 223
  immobilized, 223, 224, 225, 229, 231, 232, 234, 235, 236, 239, 245, 246, 251, 252, 253, 254, 255, 256
RPE, 98
$Ru(bpy)_3^{2+}$, 342, 144-145, 225, 231, 233, 245, 246, 252, 256, 279
  amphiphilic, 346
  as photosensitizer, 271, 272, 273, 279, 284, 293, 296, 298
RuO_2
  as catalyst, 277, 278, 284, 285, 298, 299

Sacrificial reagents, 144
Sacrificial systems, 268

INDEX 411

Salt effects, 99, 103, 123
  book, 129
Schrödinger equation
  time dependent, 23
Selection rules, 30, 40, 43
Selectivity
  in development, 196, 201
  regio-, 124
  stereo-, 124
  β-Selenine synthesis, 103
Semiconductors, 179
  metal halides, 178
  photocatalyst, 3
Semi quinone, 187
Sensitivity
  human eye, 178
  spectral sensitivity of AgBr, 179
Sensitization
  book, 13
  by energy transfer, 7, 58, 144–145, 223, 224, 225, 226, 228, 229, 230, 231, 232, 233, 234, 235, 236, 238, 239, 245, 248, 249, 250, 252, 253, 256
  by electron transfer, 7, 98, 59, 144–145, 233, 239
  redox, reviews, 243
Sensitizers, *see* Photosensitizers
  chromatic, 177
  sensitizer interaction, 144, 226
  sol-gel, 296
  supported, 8, 223, 224, 225, 226, 231, 232, 233, 234, 235, 236, 237, 238, 239, 240, 242, 243, 244, 245, 246, 247, 249, 250, 251, 252, 253, 254, 255, 256
Silica
  bound sensitizers, 223, 224, 239, 240, 242, 243, 244, 245, 256, 257
Silica gel
  as a support, 223, 224, 239, 240, 242, 243, 244, 245, 256, 257
  Rose Bengal, 223, 224, 239, 245, 256
Silver
  bromide, 176
  behanate, 188
  atoms, 181, 183, 194
  clusters, 184, 186, 189, 190, 191, 198, 202
  speck, 184
Singlet-singlet transition, 32, 38
Singlet, *see* SiO_2
Singlet state, 27, 226, 365

Singlet-triplet transition, 32, 42
Size-dependence
  cluster redox potential, 199
Solar energy
  conversion of, 265, 288, 356
  storage, 265, 288, 301, 395
Solvation
  clusters, 191
  shell, 71
  step, 68
Solvent
  optical and static dielectric constants, 48
  static and dynamic effects, 48
Solvent relaxation, 37
Solvent reorganization energy, 48
Solvent separated ion pair
  SSIP, 48, 99
Space of representation
  for a molecule, 21
Speck, 184
Spectral overlap
  and rate of energy transfer, 44, 45
Spin, 385
Spin correlation 26
Spin-orbit coupling, 32
States
  auto-ionizing, 20
  coupling of, 40
  correlation diagrams, 90, 92
  electronic, 20
  electronically bound, 20
  energy gap between, 40
  ground state, 22, 25, 27
  ionized, 20
  molecular, 26
  singlet, 27, 366
  triplet, 27, 366
  vibrational ground state, 123
  vibronic, 22
Stationary states, 23
Stereoselective
  supported sensitizer, 246
Steric hindrance
  in supported sensitizers, 246, 248
Stern-Volmer equation, 41
Stilbenes, 228, 233, 236, 248, 249, 251, 253
Stokes shift, 35
  dynamic, 37
  and pKa*, 73
Storage, of energy, 356

Structure defect in AgBr crystals, 179
Structural change, 384
Styrene
　photohydration, 88
Substitutions, in organic molecules, 65, 74, 91
Sulfides
　photooxidation of, 225, 232, 246, 251
Sulfonamides
　cleavage, 114
Sulphur compounds (sulfides)
　as electron donors, 277
Sum-frequency, generation, 37
Supermolecule
　definition, 264
Superoxide anion, 117
Supported
　metals, 277, 278, 285
　metal oxides, 277
Supported vs unsupported sensitizers, 228, 239, 246, 247, 254, 256, 257
Supports
Supramolecular chemistry
　definition, 264
Supramolecular systems
　for charge separation, 288
　hydrogen-generating, 294
Surfactants, 321
Symmetry, 35
　El Sayed's rules, 40
Systems
　photocatalytic, 8
　gain in complexity, 8
　tachysterol
　　photoconversion, 21, 237

Tabular crystals, 177
Taube's classification, 46
TEA, 277
Temperature, 383
TEOA, 277
Tetracyanoanthracene
　photosensitizer, 103
Thermal catalysis, 89, 137–139
　of a photogenerated intermediate, 61, 147
Ternary interactions, 105
Thiosulfate (sodium), 189
Time-dependent rate
　constant, 45, 46
　exciplex, 29

Time-resolved experiment, 33, 37
TiO_2, 278
Transition metal complexes, 273, 274
　excited states, 148–153
　photocatalysis by, 3, 6, 139–148, 153–165
Transition moment, 27
　on polymethines, 29
　vector, 28, 30
Transition moment vector, 30
Transition probability, 34
Transition state, 22
　energy lowering, 78
　and ISM, 78
　theory, 57
Transmittance, 18
Trap (electron, hole), 182, 183
Triad, 288, 289
　definition (see Polyad), 265
Triethanolamine
　as electron donor, 277
Triplet quenching
　by electron transfer, 42, 239
　by energy transfer, 43, 228, 247, 248, 249, 254, 367
Triplet-sensitization, 248
Triplet state, 27, 226, 246, 247, 248, 251, 257, 367, 385, 389
Triplet–triplet transition, 31, 247
Turnover
　number, 10
　rate, 10
Turnover number, 284, 299, 300
　definition, 283
Two phase model, 245
　sensitizers, 245
Tyrosine, 389

Upconversion technique, 37

Vesicles
　energy transfer on the surface of, 46
Vibrational
　energy spacing, 23, 40
　relaxation, 38
　quantum number, 23, 40
　overlap integral, 25
Vibrational spectroscopy, 384
Vibrations
　normal modes, 23

INDEX

Vibrons, 38
Vibronic states, 22
Viologens, 273, 288
  molecular structure 275, 276
  methylviologen, 119, 120, 152, 233, 237, 252, 256, 342
  redox potentials, 275, 276
  sulfonato-propyl viologen, 196

Water
  cleavage, 341, 345
  oxidation, 386
  see Cleavage splitting, 263, 265, 297, 301
  requirements for, 266
  splitting, *see* Water, cleavage, 386
Water soluble, 229, 231, 235, 239, 253, 256
  polymers, 347
Wavefunctions, 19
Wavepacket, 23

Wave equation, 19
Webster
  catalysis definition, 1
  catalyst definition, 1
  photon definition, 1
Weller equation, 98
W-heteropolycompounds, 164–165
WO_3, 278

Xanthene dyes
  as photosensitizers, 273
  X-ray crystallography, 375

Zeolites
  as constraining media, 296, 278
  and photosensitization, 233, 252, 253
  silver, 287
  as support, 285, 286
Zn tetraphenylporphyrin
  as a sensitizer, 232, 252